CIVIL ENGINEERING CONTRACTS AND ESTIMATES

FIFTH EDITION

B S PATIL

Universities Press

CIVIL ENGINEERING CONTRACTS AND ESTIMATES, FIFTH EDITION

UNIVERSITIES PRESS (INDIA) PRIVATE LIMITED

Registered Office
3-6-747/1/A & 3-6-754/1, Himayatnagar, Hyderabad 500 029, Telangana, India
e-mail: info@universitiespress.com

Distributed by
Orient Blackswan Private Limited

Registered Office
3-6-752 Himayatnagar, Hyderabad 500 029, Telangana, India

Other Offices
Bengaluru / Chennai / Guwahati / Hyderabad / Kolkata
Mumbai / New Delhi / Noida / Patna

© Universities Press (India) Private Limited 1972, 1988, 2006, 2015, 2026
First published 1972
Reprinted 1981, 1983, 1984, 1986
Second Edition 1988
Reprinted 1990, 1991, 1994
Third Edition 2006
Reprinted 2009, 2010, 2011, 2012, 2013
Fourth Edition 2015
Reprinted 2016, 2019
Fifth Edition 2026

ISBN: 978-93-49750-05-0

Cover and book design
© Universities Press (India) Private Limited 1972, 1988, 2006, 2015, 2026

Set in Minion Pro 11 on 13 *by*
Bookcraft Publishing Services (I) Pvt Ltd, Chennai

Printed at
B.B Press, Tronica City, Ghaziabad, U.P. 201 103

Published by
Universities Press (India) Private Limited
3-6-747/1/A & 3-6-754/1, Himayatnagar, Hyderabad 500 029, Telangana, India

502274

Disclaimer
Care has been taken to confirm the accuracy of the information presented in this book. The authors and the publisher, however, cannot accept any responsibility for errors or omissions or for consequences from application of the information in this book, and make no warranty, express or implied, with respect to its contents.

Dedicated to

the memory of my loving parents,
and to my wife and daughters

Foreword

It would not be incorrect to classify 'construction of civil engineering works' as one of the major industries. Many of the basic requirements needed to run an industry smoothly and efficiently are, therefore, applicable in the activity of this industry also. Most construction works, whether in the public sector or in the private sector, are carried out through the agency of a contractor. The terms and conditions according to which the contractor agrees to carry out the work on behalf of the owner need to be set out clearly and unambiguously. When the owner himself or his authorised representative is an engineer, as is the case with the Central or state PWD works, the contract agreement is signed by the contractor and the engineer. In other cases, the contract is executed between the owner and the contractor, but the owner takes the assistance of an architect and/or engineer to supervise the work to ensure that it is being carried out properly.

The contract agreement is obviously the most important document, because the smooth sailing and timely execution of any construction work depends on it. It is therefore essential for both the engineer and contractor to be thoroughly acquainted with the basic principles of contracts, so that the agreement entered into by the parties is as flawless as possible. A thorough study of the contract agreement, both by the engineer and the contractor, is also equally important if subsequent disputes and complications are to be avoided.

It is evident that the author has taken great pains to bring all the important features of the contract agreement within the compass of a moderate-sized book. He has also added two very useful chapters, one dealing with contract organisation and management, and the other on the PWD procedure of executing works. This extends the usefulness of the book beyond the student world to that of contractors and engineers in the field. The sets of questions appended to each chapter will undoubtedly enhance the utility of the book for the student.

The author deserves compliments for the trouble and care he has expended on the book, which bears the unmistakable stamp of both study and experience. I heartily commend his efforts in producing this very valuable and useful book for the guidance of students and engineers and contractors engaged in the building industry.

Pune

K V JOGLEKAR
Chief Engineer (Retired)
Buildings and Communications
Maharashtra

Preface to the **F**ifth **E**dition

The fourth edition of this book was published in 2015, ten years ago. The rates and costs needed to be updated. The AICTE issued new guidelines to the universities for revising the syllabus. Since these changes needed to be incorporated, a new edition became necessary. As in the past, the publishers responded positively. It is gratifying to note that some of the suggestions made by the AICTE had already been foreseen by the author many years ago, and to that extent, revising the contents has been easier. The author is grateful to the AICTE for including his book in the list of recommended titles.

As before, sincere thanks are due to friends and associates who provided material for updating this edition. At the top of the list, I acknowledge Shree Mahalaxmi Construction Co. Ltd. for agreeing to revise the rates and costs wherever needed, and for sparing the services of its Senior Engineer, Mr Vinayak Janawade, who also prepared the estimate of the tunnel for the guidance of readers. Thanks are also due to Mr Balaram Mahajan, government-approved valuer from Kolhapur, who, besides supplying the standard form of valuation report, suggested and included a questionnaire to be raised along with answers, which will be especially useful to students and beginners.

Acknowledgements would be incomplete without expressing sincere thanks to Mr Madhu Reddy and Mr Kallol Das, Universities Press (India) Private Limited. Sincere thanks are also due to Dr Gita S Dattatri, the editor, for her painstaking efforts, meticulous attention to detail and patience in editing the book. I am deeply grateful for her support in bringing this book to completion which gives me the confidence to place it in the hands of my readers.

Pune
September 2025

B S PATIL

Section I

Preface to the First Edition

The construction industry, one of the single largest private businesses in the world, represents a heterogeneous and vital enterprise in the economy of a country. A programme of development in an underdeveloped country like India necessarily entails a large volume of construction activity. Roughly 40 to 50 per cent of the Five Year Plan expenditure in this country falls within the purview of construction activity carried out mainly by private agencies. It, therefore, logically follows that if this amount is to be properly and economically utilised, the organisation of the private engineering and contracting industry, trade and profession should be a matter of great importance. With this background it is surprising that there has been, in India, very little literature on the subject of engineering contracts.

In the absence of comprehensive books on the subject one has to refer to scores of books (most of them written by foreign authors), each dealing exhaustively with only some of the aspects of contracts. Being disconnected, the information so collected fails to give a complete picture of the construction industry and profession. Students who have to prepare for many subjects in their examinations have, therefore, no alternative except to depend upon the meagre information given in the books on 'estimating and costing'. When they enter the profession, they almost always have to work in one capacity or other in the construction industry. Inadequate knowledge and information of the fundamental principles and rapid developments in the field undoubtedly mar the progress not only of the individual engineer or architect but ultimately of the nation as a whole.

In this book an attempt has been made to provide, in a compact and handy form, a comprehensive and lucid account of contract procedure which is of immediate use to civil engineers, architects and builders in their day-to-day work.

The discussion on the tendering procedure, contract documents and the different kinds of contract is in such detail that it will be truly informative and useful. The chapter on 'Conditions of Contract' points out the benefits of avoiding one-sided conditions of contract. It includes ideal conditions of contract, which, if adopted, would help to cut down cost of construction. The basic principles of specification-writing have been dealt with. These would help the reader prepare fair and practical specifications. Specimen specifications are thorough and complete.

The chapter on 'The PWD Procedure of Executing a Project' has been written with a view to being useful to students. But the treatment is elaborate enough to be helpful to engineers and builders in the field as well.

The author respectfully acknowledges the assistance gained from well-known standard works on the subject. At the end of the volume some of them are suggested for further reading and for reference.

The author is highly obliged to Shri K V Joglekar, Retired Chief Engineer (Buildings and Communications), Maharashtra, for kindly going through the book and writing a foreword.

In the preparation of this volume the author has been assisted and encouraged by suggestions from his friends and colleagues. In particular he is indebted to Principal G P Nagarkar, College of Engineering, Pune, and Principal B D Choudhary, College of Engineering, Aurangabad, for their kind encouragement and guidance; Prof. M V Shrotriya, Government Polytechnic, Sholapur, and Prof. A G Joshi, College of Engineering, Aurangabad, for useful and constructive criticism and advice during the months of detailed preparation before publication; his wife for her help in the physical preparation of the manuscript.

Above all he wishes here to acknowledge the debt this book and he owe to Prof. S C Marathe, Head of Civil Engineering Department (Retired), College of Engineering, Pune. The original conception, the final scope and contents owe much to the counsel, advise and help given by Prof. S C Marathe.

Aurangabad   **B S PATIL**
15 August 1972

Section II
Preface to the First Edition

The success of an engineer or a contractor depends, in the main, upon two things. The first is the accuracy of his estimate and the second is the efficiency of his site organisation and management. It is the purpose of this volume to deal with the first part only. The second part has been dealt with in the first volume.

It is the experience of the author, as a student, as a field engineer and as a teacher of this subject to degree and diploma students for the past many years, that a book containing several illustrative estimates has little utility as either a text or a reference book. This is obvious because no two works are identical. In the opinion of the author, what a beginner needs most is a thorough understanding of first principles, and an explanation and illustration of measurement in the simplest form. Once the beginner has a good grounding in first principles, just one typical i!lustrative estimate of each type of work may suffice to give him the knowledge, the understanding and the confidence to prepare an estimate of a similar work. By devoting a major portion of this book to this objective the author has attempted to achieve this aim.

The need of a book of this type has been felt far more than ever before since the Standard Method of Measurement of Building Works (Indian Standard 1200 of 1964) was published. This standard has specified the methods of measurement for various trades which in many cases are different from those existing in many Government departments and construction agencies. Interpretation of the various clauses and the direct use of the standard in the preparation of the estimates of building works are found difficult by students and also by field engineers who are used to the conventional methods of measurement followed by them.

No attempt is made to repeat in detail the clauses of the Indian Standard which, it is now recognised, must be in the hands of every student. The author has aimed at giving an interpretation of the clauses, and concentrating on the method and procedure of taking out with simple illustrative examples. A detailed estimate of a typical building is prepared in accordance with the *standard method of measurement* to which a beginner can refer after studying the isolated examples. Estimates of other civil engineering works are also included to increase the utility of the book.

In deciding the scope of this book, the interest of a civil engineering student and a young engineer beginning his professional career was uppermost in the mind of the author. The chapter on valuation of land and buildings is included to cover the syllabi specified by different universities and technical education boards, and it is also hoped to satisfy the needs of a young engineer who may be called upon to fulfil his duties as a valuer.

To increase the utility of the book, in the chapter on analysis of rates, valuable information is presented concisely in a tabular form. The use of the tables is illustrated by analysing prices of typical items of work. Labour hour is used as a unit rather than a labour day. The information in regard to prices of materials and labour rates is approximate. It is included to give a general idea of the prices prevailing in 1968 when the text was written.

The author is grateful to Prof. M V Shrotriya, Government Polytechnic, Sholapur, for kindly reading the manuscript and making a number of valuable suggestions. The author cannot but add a word of gratitude to Principal G P Nagarkar, College of Engineering, Pune; and Principal B D Choudhary, College of Engineering, Aurangabad, for their encouragement and advice. His grateful thanks are due to: Shri S N Gokhale, College of Engineering, Aurangabad, for his help in preparation of sketches and drawings; Shri Sham Khose for excellent typing of the manuscript; a number of engineers, his friends and colleagues from whom he received help and encouragement; his wife for her help in the physical preparation of the manuscript.

Aurangabad   **B S PATIL**
15 August 1972

Contents

Foreword v

Preface to the Fifth Edition vii

Section I: Preface to the First Edition ix

Section II: Preface to the First Edition xi

SECTION I   Civil Engineering Contracts

1. Introduction 3
2. The Law of Contract 7
3. The Tender 27
4. Contract Documents 52
5. Contract for Engineering and Architectural Services 60
6. Contract between Owner and Contractor 75
7. Conditions of Contract 95
8. Specifications 134
9. Contract Organisation and Management 152
10. The PWD Procedure of Executing a Project 181

Appendices 193

References 198

SECTION II   Civil Engineering Estimates

Introduction 201

1. Approximate Estimates 206
2. Valuation 222

3. General Principles of Taking out Quantities 252

4. Methods of Taking out Quantities 257

5. Excavation and Foundation 270

6. Concrete 282

7. Stone Masonry and Brickwork 300

8. Doors and Windows 309

9. Roofs 319

10. Floors and Pavings 328

11. Plastering and Pointing 333

12. Stairs 338

13. Plumbing, Drains and Sanitary Fittings 351

14. Electrical Work 357

15. Structural Steelwork 362

16. Measurement of Earthwork 367

17. Squaring the Dimensions 376

18. Abstracting and Bill of Quantities 382

19. Factors Affecting Cost of any Work 387

20. Analysis of Rates 392

Estimates   I: A Two-Storeyed Residential Building 435

Estimates II: A Two Span Arch Bridge 476

Estimates III: Earthwork and Roads 496

Estimates IV: A Canal 510

Estimates V: A Railway Track 520

Estimates VI: A Tunnel 525

Section I

Civil Engineering Contracts

[Along with an introduction to The Arbitration and Conciliation (Amendment) Act, 1996]

1 | Introduction

1.1 GENERAL

The field of civil engineering is tremendous in scope and encompasses the whole of man's endeavour to shape and improve his physical environment. The civil engineer may be called upon to design and carry out the construction of a large variety of works such as residential and public buildings, roads, bridges, railways, dams, docks, airport, power stations, water supply and drainage schemes, tunnels, sea defence, land reclamation projects and so on. The gigantic scope of civil engineering will be evident from the following facts.

The 12th Five Year Plan has fixed the total plan size of ₹47.7 lakh crores out of which the share of the construction industry would amount to nearly 45% to 50%—roughly ₹23 lakh crores or say ₹4.77 lakh crores per annum.

1.2 PUBLIC AND PRIVATE WORKS

Most construction works fall under one of the following two groups:

1. **Public Works**

   All works, which are executed in the interest of the people at large, fall under this heading. Typical examples of such works are highways, railways, dams and public buildings financed by the Government or public bodies.

   However, certain works which are capable of giving returns immediately, on completion and putting to use, but for which funds are not immediately available with the Government department or the public body are constructed by Public Private Participation (PPP), such as on Build, Operate and Transfer (BOT) by which the project is initially financed by private companies and after the capital is recovered by collecting toll, and so on, the works are transferred to the Government department or the public authority concerned. Examples of this type of works are highways, bridges and flyovers.

2. **Private Works**

   Works, which are conceived, designed and constructed for an individual, or a group of individuals or a corporation or body that is or proposes to be in private business, are called private works. The cost of these works is met from private funds.

1.3   OWNER/EMPLOYER, ENGINEER AND CONTRACTOR

Whether the work is public or private, the scheme must, in the first instance, be set in motion by some individual or somebody having the necessary capital to pay for it. The term 'owner' used hereafter in this book refers to the individual, firm or public body which initiates and finances the work.

With the advent of internationally used Standard Contract Forms such as the FIDIC, the owner is generally designated as 'Employer' which is appropriate in contract terminology because it is he who 'employs' the engineer, architect, contractor, and so on.

The design and supervision of civil engineering works are carried out by firms of consulting engineers or architects and/or by civil engineers in the employment of private and public bodies. The choice between an architect or engineer depends upon the type of work. For example, in the case of a residential or public building involving architectural features, an architect will be a proper choice; for an industrial structure requiring a design specialist, an engineer will be preferred. The term 'engineer' (which includes architect) used hereafter refers to the civil engineer, or a firm of consulting engineers/architects, employed by the owner in order to advise him, to negotiate for him, and to protect his interests.

The contract terms usually provide for appointments of the engineer's representative and assistants to help the engineer or his representative in discharging the duties assigned to an engineer. The firm of engineers retained for supervising the construction work is sometimes referred to as the Project Management Consultants (PMC) and the site representative in charge of the works is called Project Director (PD). Where the firm engaged is of architects, the representative is sometimes called 'Clerk of Works'. It has to be noted that the engineer or architect remains responsible for and retains control of the works though he may engage the services of assistants. The duties of the engineer/architect, besides preparing and supplying drawings and designs, supervision and quality control, include taking measurements of work done and certifying bills for payment.

The responsibility of organisation and execution of the work is shouldered in many cases by independent businessmen who enter into a suitable agreement with the owner. Individuals pursuing this business are known as 'contractors', and are so designated hereafter.

Sometimes a part of the work is allowed to be sublet to another person with the consent of the engineer/employer. Such an 'other person' is called the 'subcontractor'. Occasionally, the terms of the contract between the employer and the contractor include name of such a firm or person chosen by the employer who is then called the 'nominated subcontractor'.

1.4   CONTRACT

The relationship between the above-stated parties to a construction activity is governed by an agreement entered into by and between the parties. The agreement is generally between:

1. the employer and the engineer,
2. the employer and the contractor, and
3. the contractor and the subcontractor.

An agreement, the terms of which are enforceable by law is called a 'Contract'. It includes, besides expressly agreed terms, such further documents as may be expressly incorporated in the agreement.

1.5 DEVELOPMENT OF A PROJECT

It is obvious that no scheme of appreciable size can be planned in a day. Most schemes take months, or sometimes years, before they reach the construction stage. During this period of development, the project goes through the following stages:

1. If the owner is an individual or a body without its own engineering staff, the owner employs an engineer to investigate and plan the project for him.
2. The engineer carries out preliminary investigations, prepares an approximate estimate of the cost, and determines the financial feasibility of the project.
3. If the engineer recommends the project as technically and financially feasible, the owner may give his consent to go ahead with it. In case the owner is a government, the approximate estimates have to be approved by the concerned administrative authority under the process called 'administrative approval'.
4. The engineer then undertakes the work of detailed investigation and getting ready all other papers containing the data needed to enable the contractors to bid on the job. All such papers which include tender drawings, instructions to bidders, Bill of Quantities, specifications, Conditions of contract and so on are referred to as Notice Inviting Tenders (NIT) documents which, after the contract is signed, become 'contract documents'.
5. In case the owner is a government, the detailed plans and estimates have to be checked and approved by a competent authority. The process is known as 'technical sanction'.
6. When all is set for the construction of the project, contractors are invited to bid for the job. The contractors, after making all the business arrangements, submit their offers, which are known as 'tenders'. The engineer scrutinises the proposals and selects a particular tenderer, who then enters into an agreement (the 'contract') with the owner.
7. The contractor sets his organisation into motion to carry out the project in the most efficient and economic way. For the efficient and economic construction of a project, all parties to the contract must have the knowledge of the law of contracts and the conduct of business. Equally important is the knowledge of the latest developments in the methods of contract management and organisation; similarly, the knowledge of the procedure adopted by Government departments (which are the owners of some of the largest projects) in the execution of a project is also essential to those who desire to undertake the construction of such projects.

All these important aspects of civil engineering contracts are dealt with in this book.

The second chapter gives, in just sufficient detail, the provisions of the law of contract, useful for anyone concerned with construction activity. The third chapter deals with the manner in which the offers called tenders are invited, evaluated and the work awarded.

The fourth chapter deals with the contract documents. The fifth chapter covers introduction to contract for engineering and architectural services. Chapter 6 deals with types of contracts generally entered into by and between the employer and the contractor. Important conditions of contract between the employer and the contractor are discussed in chapter 7. Chapter 8 introduces rules of specification writing and specimen specifications for some selected items. Contract organisation and management is the subject matter of chapter 9. The last chapter 10 introduces the highlights of the PWD procedure of executing a project.

Section 2 deals with preparation of cost estimates and methods of valuation of properties/assets.

The Law of Contract

2.1 INTRODUCTION

The law of contracts pertains to that branch of the law, which determines the circumstances in which an agreement shall be legally binding on the persons entering into one. It is too vast a subject to be dealt with within the narrow scope of a single chapter. However, it is proposed in the present chapter—for the benefit of those readers who are not legally qualified—to give a brief account of the general principles of the law relating to the formation of contractual obligations. These principles are embodied in the Indian Contract Act, 1872.

2.2 DEFINITION OF CONTRACT

There are many accepted definitions of the term 'contract'. The Indian Contract Act, 1872, defines contract as 'an agreement enforceable by law'. The same term 'contract', as used in the engineering field, can be defined as:

An agreement entered into by two competent parties, under the terms of which one party agrees to perform a given job for which the other party agrees to pay. In order to make this agreement valid, there must be a definite offer and there must also be an equally definite and unconditional acceptance of this offer.

2.2.1 Offer

Section 2(a) of the Indian Contract Act defines, 'offer', as:

When one person signifies to another his willingness to do or to abstain from doing anything, with a view to obtaining the assent of that other to such act or abstinence, he is said to make a proposal.

An offer or proposal must be made 'with a view to obtaining 'the assent' of the person to whom it is made. In other words, an offer must be something which invites, and is intended by the offerer to invite, 'acceptance'. Thus, a simple advertisement or circular stating that certain goods are for sale by tender, amounts to a mere proclamation that the owner is willing to negotiate for the sale of goods. He is not bound to accept any offer received in response to the advertisement. As a matter of fact the person responding and submitting a proposal is said to have made an offer to the advertiser. A tender notice inviting tenders falls under the same category. All that the law requires is that the offer should be made known to the acceptor, and thus it can be oral or written or partly written and partly oral. A person can make a proposal valid for a limited time which will automatically terminate with the lapse of time. For example, a contractor can submit his tender with a validity of 90 days within which time it must be accepted.

2.2.2   Acceptance

Section 2(b) of the Indian Contract Act defines acceptance as:

'When the person to whom the proposal is made signifies his assent thereto, the proposal is said to be accepted. A proposal, when accepted, becomes a promise'.

To be binding and effective, the acceptance must be unconditional and absolute. A conditional acceptance amounts to a counter-offer. Like a proposal, an acceptance also may be made expressly in words (orally) or in writing, or implied by conduct. If the proposal prescribes the manner in which it is to be accepted, the proposer can insist that his proposal be accepted in the prescribed manner, and not otherwise. For example, if the person submitting a proposal insists that the acceptance must be in writing, he can refuse to rely on oral acceptance and can insist on written acceptance. But if he fails to do so, he accepts the acceptance. The performance of the conditions of the proposal also amounts to an acceptance of the proposal. For example, the proposal to buy goods is deemed to have been accepted if the seller supplies the goods and the buyer accepts and retains the goods so supplied. This is a case of acceptance by conduct.

2.3   WITHDRAWAL OF AN OFFER OR ACCEPTANCE

A proposal or its acceptance can be withdrawn by the party making it without committing any liability before the time specified under the law for valid revocation elapses. Sections 4 and 5 of the Indian Contract Act deal with the aspect of completion of communication of offer and acceptance, and acceptance and withdrawal of an offer or acceptance. The relevant provisions together with illustrative examples are given below.

Section 4   The communication of a proposal is complete when it comes to the knowledge of the person to whom it is made.

The communication of an acceptance is complete:

> as against the proposer, when it is put in a course of transmission to him, so as to be out of the power of the acceptor;
> as against the acceptor, when it comes to the knowledge of the proposer.

The communication of a revocation is complete:

> as against the person who makes it, when it is put into a course of transmission to the person to whom it is made so as to be out of the power of the person who makes it;
> as against the person to whom it is made, when it comes to his knowledge.

Illustrations

(a) A proposes, by letter, to sell a house to B at a certain price. The communication of the proposer is complete when B receives the letter.
(b) B accepts A's proposal through a letter sent by post.
The communication of the acceptance is complete, as against A, when the letter is posted; as against B, when the letter is received by A.
(c) A revokes his proposal by telegram.
The revocation is complete as against A when the telegram is dispatched. It is complete as against B when B receives it.
(d) B revokes his acceptance by telegram. B's revocation is complete as against B when the telegram is dispatched, and as against A when it reaches him.

Section 5 *Revocation of proposals and acceptances*: A proposal may be revoked at any time before the communication of its acceptance is complete as against the proposer, but not afterwards.

An acceptance may be revoked at any time before the communication of the acceptance is complete as against the acceptor, but not afterwards.

Illustrations

(a) A proposes, through a letter sent by post, to sell his house to B.
(b) B accepts the proposal also through a letter sent by post.
A may revoke his proposal at anytime before or at the moment when B posts his letter of acceptance, but not afterwards.
(c) B may revoke his acceptance at any time before or at the moment when the letter communicating it reaches A, but not afterwards.

The above illustrations make the provisions of the Sections self-explanatory. In short, a valid and binding contract comes into existence at the time when and the place where the letter of acceptance is posted. However, if the acceptance is revoked before it reaches the person to whom it is sent, no binding contract results.

2.4 ESSENTIALS OF A VALID CONTRACT

Section 10 of the Indian Contract Act lays down requirements of a binding contract. According to the provisions of the said Section, an agreement, in order to constitute a valid contract, must fulfil the following essential requirements:

1. parties to the contract should be competent,
2. the subject matter of the agreement must be legal and definite,
3. there must be a proper proposal and its acceptance,
4. parties to the agreement must give their free consent, and
5. there must be a lawful consideration.

The foregoing requirements are further discussed in the following sections.

2.5   THE PARTIES TO A CONTRACT

There must be at least two parties (although a greater number may be involved) to an agreement, each of whom must be legally capable of playing his intended part. The law cannot enforce the agreement on someone who does not have the legal capacity to enter into an agreement. Such incapacity may be due to several causes including the following ones:

1. *Infancy*: A person who is not a major according to the law to which he is subject cannot make a valid contract, unless it is to meet his essentials or 'necessaries' or for his benefit. A minor's contract (unless it relates to so-called 'necessaries') is according to the Indian law 'void', that is, not enforceable by law.

2. *Lunacy*: A person is said to be of 'unsound' mind for the purpose of determining the validity of a contract if, at the time he makes the contract, he is not capable of understanding it and of forming a rational judgement as to its effect upon his interests. The contract is voidable if the person was of unsound mind at the time it was made and the other person knew it. The contract becomes binding if it is made or confirmed by a person during the lucid interval (that is, when the lunatic is of sound mind) or relates to the 'necessaries' supplied to him or his wife.

3. *Drunkenness*: A sane man who is so inebriated that he cannot understand the terms of a contract, cannot legally enter into one whilst such drunkenness lasts. Any contract made by an intoxicated person is invalid, especially if the other party takes advantage of his intoxication. Of course, the degree of intoxication will decide the validity of such a contract. Again, as in the case of infants and lunatics, an intoxicated person will be liable for the value of 'necessaries' received by him.

4. *Contracts of a corporation 'ultra vires'*: The nature of a corporation and the terms of its incorporation impose some restrictions upon its contractual powers, If a corporation enters into a contract which is beyond the scope of its contractual powers, such a contract is *ultra vires* and is ordinarily not enforceable by law.

2.6   THE SUBJECT MATTER OF A CONTRACT

The subject matter with reference to which the parties desire to contract must be lawful and agreements must describe what is to be done or what materials are to be supplied with sufficient clarity and enough detail so that there is no doubt as to what is wanted. Agreement, the meaning of which are not certain or are not capable of being made certain, are not enforceable by the law and are called 'void'. For example, a contract that involves the erection of a building designed or located in violation of municipal regulations is not binding.

2.7   PROPOSAL AND ACCEPTANCE OR 'MEETING OF THE MINDS'

When one person signifies to another his willingness to contract (by submitting a tender which may be oral or in writing) he is said to make a proposal. Once the proposal has been communicated, it gives to the other party a continuing power to create by his acceptance, a legally enforceable contract; unless the proposal is withdrawn for some reason before it is considered or automatically terminates with the lapse of time. Tender notices generally specify the time limit during which the proposals would remain valid and open for acceptance. There is a meeting of the minds of the parties to a contract when each of them understands the 'same thing' in the 'same sense.' Same thing means the whole content of the agreement. No effective and binding contract comes into existence unless the parties are *ad adem*.

In order to convert a proposal into a contract, the acceptance must be absolute and unqualified and expressed in some normal and reasonable manner. In case of engineering contracts, both the proposal and acceptance are generally expressed and communicated in writing. The very nature of the agreement involves voluminous documents consisting of terms and conditions, bill of quantities, specifications drawings, and so on.

2.8 FREE CONSENT OF PARTIES

Section 13 of the Contract Act defines consent in these words, 'two or more persons are said to "consent" when they agree upon the same thing in the same sense'. For example, a successful bidder for three items was offered orders for two items which he refused. Thereafter, a third order was placed and he was told that each order is a subject matter of independent contract. The bidder can successfully decline to perform the contracts because the parties did not understand the same thing in the same sense. The bidder bid for a single transaction under one contract whereas the acceptor wanted three contracts for three items.

Consent is said to be free when it is not caused by force or undue influence or fraud or misrepresentation. A contract, wherein the consent of one or both parties has been given under circumstances such as to render it not a bonafide expression of intention, is void. However, a contract is voidable at the option of the innocent party if the misrepresentation or fraud has been induced by the improper conduct of the other party. In other words, the innocent party can refuse to follow the contract but such an option is not available to the other party.

2.9 CONSIDERATION

The term 'consideration' signifies something of value received by or given at the request of the promisor in reliance upon and in exchange for his promise. By way of illustration of the concept of consideration, assume that A offers to sell B a well-built house for a sum of ₹15,64,00,000 and when B asks for ten days' time in which to consider the offer, A promises not to sell it to anyone else during that period. However, A sells his house to C for a sum of ₹16,10,00,000 within three days of his promise to B. Under the circumstances, B cannot complain of breach of contract because A's promise to him was not supported by any consideration. If B had paid A a certain sum of money in order to keep the offer open for the ten-day period, A's act of selling the house would have been against the law. This payment by B to A is known as 'consideration'. According to the Indian Contract Act of 1872, the consideration, given in exchange for a promise, by the other party may be one of the following:

1. something that has been done or an abstinence from doing something,
2. a present act or abstinence, or
3. a promise to do something or abstain from doing it sometime in the future.

For a contract to be legally binding there must be a consideration and this consideration must be lawful, definite and certain. Every agreement, whose object or consideration is unlawful, is void, that is, not binding. To illustrate this point, assume that A, an engineer, promises to accept the tender of B, a contractor, against the best interests of the owner, and in return, B promises to pay ₹1,00,000 to A. The agreement is void, as the consideration for it is unlawful.

In civil engineering contracts, the consideration for the contractor's promise to carry out the work is usually the owner's promise to pay the agreed cost of construction either lump sum or worked out at the agreed rates for the quantities of different items as per actual measurements taken at the site of work.

2.10   EXPRESS AND IMPLIED CONTRACTS

Section 9 of the Indian Contract Act reads:

Section 9   *Promises, express and implied:* In so far as a proposal or acceptance of any promise is made in words, the promise is said to be express. In so far as such a proposal or acceptance is made otherwise than in words, the promise is said to be implied.

In other words, the law recognises that the contracts may be either express or implied. The difference is not one of legal effect but simply by the manner in which the consent of the parties is manifested. A contract is known as 'express' when its terms are stated in words by the parties. It is often said to be implied when its terms are not so stated as, for example, when a passenger is permitted to board a bus; from the conduct of the parties, the law implies a promise by the passenger to pay the fare, and a promise by the operator of the bus to carry him safely to his destination. An implied promise may result from a continuing course of conduct as well as from particular acts. Thus, an agreement between the parties to vary the terms of the partnership contract may be implied from a uniform course of dealing.

2.11   VALID, VOID, VOIDABLE AND UNENFORCEABLE CONTRACTS

Contracts may be valid, void, voidable or unenforceable. Section 2 of the Indian Contract Act includes the definitions of these terms.

2.11.1   Valid Contracts

Section 2(h) defines a valid contract as follows: 'An agreement enforceable by law is a contract'.

2.11.2   Void Contracts

Section 2(g) of the Indian Contract Act declares: 'An agreement not enforceable by law is said to be void'. An agreement is void under the following circumstances:

1. when there is unlawful consideration or object,
2. when it is entered into without consideration, unless it is covered by exceptions made under Section 25 of the Indian Contract Act; which Section makes some exceptions to the rule by declaring contracts in writing and registered for compensating for something done or a promise to pay a debt barred by limitation law as valid and binding,
3. in restraint of marriage of a person not a minor,
4. in restraint of trade,
5. in restraint of legal proceedings with an exception of reference to arbitration,
6. by way of wager,
7. when it is of uncertain meaning,
8. when it is entered into so as to perform an act which is impossible or becomes impossible,
9. when there is a mistake on the part of both the parties as to essential matters of fact, and
10. when there is a mistake on the part of both the parties as to foreign law.

A void contract produces no legal effect whatsoever. Neither party is able to sue the other on the contract.

2.11.3   Voidable Contract

Section 2(i) of the Indian Contract Act reads: 'An agreement which is enforceable by law at the option of one or more of the parties thereto, but not at the option of the other or others, is a voidable contract'.

A voidable contract, thus, is one wherein one or more parties thereto have the power, by a manifestation of election to do so, to avoid the legal relations created by the contract. In accordance with the Indian Contract Act, the contract may be voidable at the inception itself at the option of the party whose consent has been obtained by fraud, or coercion, or misrepresentation or by undue influence. A valid contract may become voidable by subsequent default, for instance, where the offer of performance is not accepted or when one party prevents the performance of the reciprocal promise or when a party fails to perform at the time fixed if time is of the essence for the contract.

2.11.4 Contingent Contracts

Agreements in which the parties agree that the performance of the contract would be dependent upon the happening of a future event are termed as 'contingent contracts'. Section 31 of the Indian Contract Act defines contingent contracts in these words:

A 'contingent contract' is a contract to do or not to do something, if some event, collateral to such contract, does or does not happen.

Illustration

'A contracts to pay B ₹10,000 if B's house is burnt'.

The contract of life insurance and similar other contracts are contingent contracts within the meaning of this Section.

In building and engineering contracts, the main contractor sometimes enters into a contract with the subcontractor, in the event of his tender being accepted and he securing the work. This contract is dependent upon the future event of the main contractor succeeding in securing the contract.

Sections 32 to 36, of the Indian Contract Act deal with enforcement of contingent contracts. The said Sections are self explanatory and provide:

Section 32   *Enforcement of contracts contingent on an event happening:* Contingent contracts to do or not to do anything if an uncertain future event happens, cannot be enforced by law unless and until that event has happened. If the event becomes impossible, such contracts become void.

Illustration

 (i) A makes a contract with B to buy B's horse if A survives C. This contract cannot be enforced by law unless and until C dies in A's lifetime.
(ii) . . . . (iii) . . . .

Section 33 provides for an eventuality of contracts contingent on an event not happening. Such contracts can be enforced when the happening of that event becomes impossible. The Section gives the illustration as follows: A agrees to pay B a sum of money if a certain ship does not return. The ship is sunk. The contract can be enforced when the ship sinks.

Section 34 provides for when a contract is contingent to be deemed impossible, if it depends upon the future conduct of a living person. The Section gives the illustration as follows: A agrees to pay B a sum of money if B marries C. C marries D. The marriage of B to C must now be considered impossible, although it possible that D may die and C may afterwards marry B.

Section 35 deals with contracts becoming void which are contingent on the happening of a specified event within a fixed time. Such contracts become void if, at the expiration of the time fixed, such an event has not happened, or if, before the time fixed, such an event becomes impossible.

Section 36 declares that an agreement contingent on an impossible event is void. The illustration (b) of the section reads:

A agrees to pay B ₹1000 if B will marry A's daughter C. C was dead at the time of the agreement. The agreement is void.

2.11.5 Government Contracts

Contracts by the Government need to be specially discussed because they raise some problems, which do not or cannot possibly arise in the case of contracts entered into by private persons. In order that public funds may not be depleted by clandestine contracts made by any and every public servant, there should be a definite procedure according to which contracts must be made. The law that governs the procedure to be followed is embodied in Art. 299 of the Constitution of India. Art. 299 reads as:

1. All contracts made in exercise of the executive power of the Union or of a State shall be expressed to be made by the President, or by the Governor of the State, as the case may be, and all such contracts and all assurances of property made in the exercise of that power shall be executed on behalf of the President or the Governor by such persons and in such manner as he may direct or authorise.
2. Neither the President nor the Governor shall be personally liable in respect of any contract or assurance made or executed for the purposes of this Constitution, or for the purpose of any enactment relating to the Government of India heretofore, in force, nor shall any person making or executing any such contract or assurance on behalf of any of them be personally liable in respect thereof.

The words 'expressed to be made' and the word 'executed' suggest that there must be a deed for a formal written contract. A contract by correspondence or an oral contract is, accordingly, not binding upon the Government. The use of the word 'shall' is intended to make the provision itself obligatory and not directory.

However, it is now well-established that if the contract can be spelt out from the tender and its acceptance, provided the tenders were invited for and on behalf of the President of India, in the case of Central Government works or the Governor of the State in the case of State works, and the tender is accepted by the person authorised to do so and he does so for and on behalf of the President or the Governor, as the case may be, the contract would be valid and binding on the government, even in the absence of a formal deed.

Effect of Non-Compliance with Requirements of Art. 299(1)

The provisions in Art. 299 above make it plain that non-compliance with the requirements of Art. 299(1) renders the contract void and as such the plea of implied contract or Estoppel or ratification or specific performance or damages for breach for contract is not permitted. The question for consideration then remains as to what relief, if any, a party to such a void contract will be entitled to if such a party has done work, provided services or supplied materials to the Government. In such an event a person may be entitled to the cost of goods supplied or work done under Section 70 of the Indian Contract Act.

Section 70 of the Indian Contract Act:
'Where a person lawfully does anything for another person or delivers anything to him, not intending to do so gratuitously and such other person enjoys the benefit thereof, the latter is bound to make compensation to the former in respect of, or to restore, the thing so done or delivered.'
   The essential conditions to be satisfied before a claim can be allowed are:

1.   the person should lawfully do something for another person,
2.   the other person for whom something is done must enjoy the benefit thereof, and
3.   the thing done must not be done fraudulently or dishonestly nor must it be done gratuitously.

The section will not be applicable if any one of the three requisite conditions is not satisfied.

2.11.6   Contract of Guarantee

In big construction projects large sums of money have to be invested in mobilisation of men and machinery. Although this is primarily a responsibility of the contractor undertaking such projects, provisions are made in contracts where under the owner advances specified sums to the contractor for mobilisation. Invariably such sums are advanced against the security of bank guarantees. The law applicable to such guarantees is embodied in Chapter VIII of the Indian Contract Act. The relevant Section reads:

Section 126   A 'contract of guarantee' is a contract to perform the promise, or discharge the liability of a third person in case of his default. The person who gives the guarantee is called the 'surety'; the person in respect of whose default the guarantee is given is called the 'principal debtor' and the person to whom the guarantee is given is the 'creditor'. A guarantee may be either oral or written.
   In the case of contract of guarantee, the surety need not be called upon to pay unless the 'principal debtor' has committed a default. In the construction contracts, the bank guarantees taken are generally unconditional that is payable on demand without demur.

2.11.7   Contract of Indemnity

The contractor is required under certain provisions incorporated in construction contracts to indemnify the owner against any loss caused by the contractor's men or machinery to the property of others. The relevant provisions of the Indian Contract Act read as:

Section 124   A contract by which one party promises to save the other from loss caused to him by the conduct of the promisor himself, or by the conduct of any other person, is called a 'contract of indemnity'.

Illustration

A contracts to indemnify B against the consequences of any proceedings which C may take against B in respect of a certain sum of 200 rupees. This is a contract of indemnity.

PERFORMANCE OF CONTRACT

2.12   BY WHOM THE PROMISE MUST BE PERFORMED

If it appears from the nature of the agreement that it was the intention of the parties to the agreement that the promise should be performed by the promisor himself, such a promise must be performed by the promisor himself. Otherwise, the promises bind the representatives of the promisor in case of death of the promisor before performance.

An example of the former type is the contract between an employer and an engineer. On the death of either party, the contract becomes void. A contract between an owner and a contractor is an example of the latter type. On the death or incapacity of the contractor, his representatives or executors will be obliged to ensure that the work is completed.

2.13   JOINT LIABILITIES

When two or more persons make a joint promise, then, unless a contrary intention appears from the agreement, all such persons, or their representatives, or survivors must fulfill the promise jointly.

When a contract has been made with one particular member of a firm, the other may be held responsible under the Partnership Act. However, the contract will not be binding on the firm if the partner has acted beyond the scope of his authority.

2.14   PERFORMANCE OF RECIPROCAL PROMISES

When a contract consists of reciprocal promises, to be performed in a specific order, they have to be performed in that order. Where the order is not expressly fixed, the reciprocal promises are to be performed in that order which the nature of the transaction requires. However, no promisor needs perform his promise unless the promisee is ready and willing to perform his reciprocal promise. The following illustrations will clarify the aforementioned points:

1. Suppose A (a builder) promises to construct a building for B in consideration of a sum of ₹2,00,50,000 payable on completion of the work. In this agreement, the order of reciprocal promises is specified and A must fulfil his promise first.

2. In another case, A and B contract that A shall build a house for B at a price to be paid by monthly instalments equal to 90 per cent of the value of the work actually performed during a month's period. A need not proceed with the work unless B is capable and willing to pay the first instalment on completion of the part of the work during the first month.

2.15 TIME OF PERFORMANCE

When, by a contract, a promisor has to fulfil his promise and no time for performance is specified, the engagement must be performed within a 'reasonable' time. The reasonable time in each particular case is a question of fact. The court will allow a reasonable time for the performance, keeping in view the interest and convenience of the parties to the contract, and also the nature, scope, and circumstances of that particular case. If a delay occurs for no fault of either party, where no time limit is mentioned, the loss must remain where it falls.

In case there is a time fixed, either expressly or implied, and if time is to be the essence of the contract, or where the nature of work implies it to be so and the promisor fails to perform the obligation, the contract becomes voidable at the option of the promisee.

Time will not be of essence to the contract where there is provision of extension of time under certain contingencies and a penalty for delay or bonus for the expedition mentioned in the agreement. In that case, the contract cannot be terminated by the promisee, if the promisor fails to fulfill his promise but can recover the penalty for the delay as stipulated in the agreement. From this, it seems that time is not regarded as being of essence in the case of a typical building contract. Many such contracts call for a certain amount per day as liquidated damages, which are to be deducted from the agreed total price if the work is not finished within the stipulated time limit. In such contracts, a reasonable extension of time for the completion of the work must be given and that extended time can be made the essence. On expiry of the extended time, the contract can be terminated if the work remains incomplete.

EXCUSES FOR DELAY OR NON-PERFORMANCE

2.16 NON-PERFORMANCE

A contract need not be performed if the parties to it agree to:

1. alter the original agreement,
2. rescind the original agreement, or
3. substitute a new agreement in the place of the original agreement.

The following provisions of the Indian Contract Act dealing with this aspect are self-explanatory.

Section 62   *Effect of novation, rescission and alteration of the contract:* If the parties to a contract agree to substitute a new contract for it, or to rescind or alter it, the original contract need not be performed.

Illustrations

(a) A owes money to B under a contract. It is agreed between A, B and C that B shall henceforth accept C as his debtor instead of A. The old debt of A to B is at an end, and a new debt from C to B has been contracted.

(b) A owes B 10,000 rupees. A enters into an agreement with B, and gives B a mortgage of his (A's) estate for 5,000 rupees in place of the debt of 10,000 rupees. This is a new contract and extinguishes the old.

(c) A owes B 1,000 rupees under a contract. B owes C 1,000 rupees. B orders A to credit C with 1,000 rupees in his books, but C does not assent to the arrangement. B still owes C 1,000 rupees, and no new contract has been entered into.

Section 63   *Promisee may dispense with or remit performance of promise:* Every promisee may dispense with or remit, wholly or in part, the performance of the promise made to him, or may extend the time for such performance, or may accept instead of it any satisfaction which he thinks fit.

Illustrations

(a) A promises to paint a picture for B. B afterwards forbids him to do so. A is no longer bound to perform the promise.

(b) A owes B 5,000 rupees. A pays to B and B accepts, in satisfaction of the whole debt, 2,000 rupees paid at the time and place at which the 5,000 rupees were payable. The whole debt is discharged.

(c) A owes B 5,000 rupees. C pays to B 1,000 rupees and B accepts it in satisfaction of claim on A. This payment is a discharge of the whole claim.

(d) A owes B, under a contract, a sum of money, the amount of which has not been ascertained. A without ascertaining the amount, gives to B, and B, in satisfaction thereof, accepts the sum of 2,000 rupees. This is a discharge of the whole debt, whatever may be its amount.

(e) A owes B 2,000 rupees, and is also indebted to other creditors. A makes an arrangement with his creditors, including B, to pay them a compensation of eight annas in the rupee upon their respective demands. Payment to B of 1,000 rupees is a discharge of B's demand.

Impossibility

A contract which is impossible to perform cannot and need not be performed. The law stipulates that 'an agreement to do an act impossible in itself is void'. Impossibility may be (1) subjective or (2) objective.

1. *Subjective impossibility*: When impossibility is due to the incapacity of the party who has undertaken the work, it is known as 'subjective impossibility'. Such incapacity may be due to death or serious illness. Subjective impossibility becomes an excuse for the non-performance of a contractual obligation in contracts for strictly personal services, for example, an engineer's contract with his employer. If the service involved is of such a nature that it can well be performed by someone other than the promisor, death or serious illness of the promisor will not invalidate the contract. Ordinary building contracts come under the latter category.

2. *Objective impossibility*: Destruction of subject matter is one example of objective impossibility. Where the contract is such that fulfillment thereof absolutely depends upon the continued existence of a particular item or object, the destruction or non-availability of that item or object relieves the promisor of his contractual obligations. For example, if a contractor undertakes to carry out the work of providing additions and alterations to an existing building, and this building is completely gutted by fire before the work can be executed by him, he is freed of his obligation, provided of course, that he is not to blame for the mishap.

Neglect of Promisee

Neglect of the promisee to provide the promisor reasonable facilities for the performance of his contractual obligation will be an excuse (which the law recognises) for the non-performance of the contract by the promisor. For example, the work of carrying repairs to the existing structure can only be executed if the premises are available for carrying out the repairs.

2.17   DELAYS IN PERFORMANCE

Unless a construction contract clearly provides otherwise, delay is not excused on the grounds of bad weather, natural calamities (storms or floods), strikes and so on. Such contingencies are supposed to be part of a contractor's calculated risks. However, most construction contracts provide for an extension of time for performance on these counts. If any provision is contained in the contract for extra works, and extra works are ordered for, a reasonable delay in performance cannot be penalised, unless the contractor has agreed to complete the work within the specified time, whatever the additional work that may have been ordered for.

2.18   BREACH OF CONTRACT

Breach of contract means the failure to perform it. However, not every failure to perform an obligation amounts to a true breach, as there are a number of valid excuses for non-performance, as mentioned in Section 2.16. When a contract has been broken, without sufficient reason or justification, the party who suffers due to such a breach is entitled to receive, from the party in default, compensation for any loss or damage caused by such a breach. In general, every contract provides for the precise consequences which shall follow upon breach. The provisions offer the injured party a choice, and once the other party has taken any action in reliance upon that manifestation, he will have no recourse to any alternative.

2.18.1 Anticipatory Breach

When one party to a contract renounces liabilities under the contract or by his own act makes it impossible that he could fulfill them, such a party is repudiating the contract by refusal to perform or by his inability to perform. This repudiation may take place not only during his course of performance but also while the contract is still wholly executory, that is, before either party is entitled to demand a performance of the promise by the other. The breach in such circumstances is usually called an 'anticipatory breach'. The law in this respect is laid down in Section 39 of the Indian Contract Act. The provision reads as follows:

'When a party to a contract has refused to perform, or disabled himself from performing his promise in its entirety, the promisee may put an end to the contract, unless he has signified, by words or conduct, his acquiescence in its continuance.'

Illustrations

(a) A, a singer, enters into a contract with B, the manager of a theatre, to sing at his theatre two nights every week for the next two months, and B agrees to pay her ₹100 for each night's performance. On the sixth night, A wilfully absents herself from the theatre. B is at liberty to put an end to the contract.

(b) A, a singer, enters into a contract with B, the manager of a theatre, to sing at his theatre two nights every week for the next two months, and B agrees to pay her at the rate of ₹100 for each night. On the sixth night A wilfully absents herself. However, with the assent of B, A sings on the seventh night. In this case, B has signified his acquiescence in the continuance of the contract, and cannot now put an end to it, but is entitled to compensation for the damage sustained by him as a result of A's failure to sing on the sixth night.

It can be inferred from the wording of Section 39 that the refusal by a party to enter into a contract can be express or it can be evidenced by conduct. Whether an intention is evidenced by conduct or can be ascertained by whether the party enunciating has acted in such a way as to lead a reasonable person to the conclusion that he does not intend to fulfill his part of the contract. When there is an anticipatory breach of contract committed by one party, the other party is entitled to take one of the following two courses:

1. he may treat the renunciation as discharging him from further performance and suing for damages forthwith, or
2. he may wait till the time for performance arrives and sue at that stage.

Thus, in short, the injured party is allowed to anticipate an inevitable event and is not obliged to wait until it occurs.

2.18.2 Fundamental Breach

A contract may contain several terms. The expression 'fundamental term' indicates a term of the contract which is, by reason of intention of the parties or as the result of an operation of a rule of law, considered to be of such importance that any breach of it entitles the innocent party to treat himself as discharged.

A breach of a fundamental term of the contract is called 'fundamental breach'. The expressions 'fundamental breach' and 'the breach of fundamental term' are relatively recent judicial concepts and, in many cases, may be, and have been used, interchangeably. These concepts constitute a substantive rule of law by which no party to a contract would exempt himself from the responsibility for a fundamental breach. In short, a fundamental term seems now to be merely an alternative way of referring to a condition, that is, a term which goes to the root of the contract so that any breach of it entitles an innocent party to treat the contract as repudiated, and fundamental breach merely a breach of sufficient gravity to entitle the innocent party to do likewise.

2.19 DAMAGES—LIQUIDATED AND UNLIQUIDATED

When one party breaches a contract, the other party is likely to incur loss or suffer damage. The law provides that the injured party is entitled to payment of compensation by the wrongdoer. The compensation is called damages. Damages consist of a sum of money which will, as far as it is practical, place the injured party in the same position in which it would have been if the contract had been performed.

In the case of civil engineering contracts, generally, the parties, when entering into an agreement, agree that a certain sum shall be payable if a breach occurs. This sum is known as 'liquidated damages', and estimates the loss which is likely to result from the breach of contract. Provision made in this respect in the Indian Contract Act entitles the injured party, whether or not actual damage or loss is proved to have been caused thereby, to receive from the defaulting party reasonable compensation not exceeding the amount stipulated in the agreement. The following examples will illustrate this point.

Illustrations

(a) A (a supplier) contracts to supply B (a builder) 50,000 bricks at a price of ₹2,50,000 to be paid on delivery. A breaks his promise. B is then compelled to purchase bricks of the same quality from C at a price of ₹3,00,000. Now B is entitled to recover, by way of compensation, the sum of ₹50,000, by which the contract price falls short of the price paid by B to C.

(b) A (an owner) abandons his project when B (a contractor) has incurred expenses in preparation for the performance of the contract. B is entitled to receive, by way of compensation, the amount of expense plus profit that would have been realised as a result of full performance.

(c) A (a builder) contracts to construct a house by a stipulated date, in order that B (the owner) may give it on a rental basis to C (a tenant) on that date. A is informed of the contract between B and C. A builds the house so badly that it collapses before the stipulated date and has to be rebuilt by B, who has to pay damages to C, for breach of contract. In this case, A must make compensation to B for:

   (i)   the cost of rebuilding the house,
  (ii)   the rent lost, and
 (iii)   the compensation paid to C.

2.20 TERMINATION OF AGREEMENTS

A contract may be brought to an end in a variety of ways including the following:

1. full and satisfactory performance by both parties,
2. breach of contract, when default by one party releases the other party from the contractual obligations,
3. mutual agreement of the parties to terminate it,
4. impossibility excuses the parties from their obligations and duties, and
5. by the operation of law. Termination of a voidable contract by the party entitled to do so is, in a sense, discharge through the 'operation of law'.

2.21 DISCHARGE BY AGREEMENT

A contract rests on the agreement of the parties, and, as such, by their agreement it may be discharged. This mode of discharge may occur in one of the following four ways:

1. by release under seal,
2. by accord and satisfaction,
3. by rescission of a contract, which is still executory, or
4. by the operation of some provision contained in the contract itself.

1. *Release under seal*: Normally, consideration is necessary for the formation of a contract as well as for its discharge. For example, when one party has fulfilled the contractual obligations but the other party has not discharged his side of the obligations, the release by one party, from the contractual obligations, of the other party who failed to perform his part would be without consideration and hence invalid. The release is therefore required to be under seal because the seal dispenses with the need for consideration. A release not under seal requires consideration. In such a case, the agreement is discharged by accord and satisfaction.

2. *Accord and satisfaction*: Discharge of a contract in return for a consideration which consists in some satisfaction other than the performance of the original obligation is called 'accord and satisfaction'. It is, in fact, the purchase of a release from an obligation, arising under contract by means of any valuable consideration which is not the actual performance of the contract itself. It is effective to discharge any contract, whether executory or executed and even under seal.

3. *Rescission of a contract which is still executory*: A contract which is executory on both the sides may be discharged by an agreement between the parties that it shall no longer bind them. Such an agreement is formed as a result of mutual promises, and the consideration for each promise of each party is the abandonment by the other of his rights under the contract.

4. *Operation of some provisions contained in the contract itself*: A contract may contain within itself the provisions for its own discharge, express or implied. A continuing contract may contain a provision making it determinable at the option of one or both the parties upon notice. For example, a personal services contract may be determinable by three months' notice in writing on either side. Similarly, a partnership for no fixed time is determinable by notice.

Generally, in all legal systems, provision is made for the discharge of a contract subsequent to its formation when there is a change of circumstance, which renders the contract legally or physically impossible to perform. In English law, such a contract is provided for by the doctrine of frustration. In the Indian Contract Act, Section 56 provides for this contingency. It reads as follows:

Section 56   *An agreement to do impossible act:*
'An agreement to do an act impossible in itself is void'.

A contract to do act, afterwards becoming impossible or unlawful: 'A contract to do an act which, after the contract is made, becomes impossible, or, by reason of some event which the promisor could not prevent, unlawful, becomes void when the act becomes impossible or unlawful'.
Compensation for loss through non-performance of act known to be impossible or unlawful: 'Where one person has promised to do something, which he knew, or with reasonable diligence, might have known, and which the promisee did not know to be impossible or unlawful, such promisor must make compensation to such promisee for any loss which such promisee sustains through the non-performance of the promise'.

Illustrations

(a) A agrees with B to discover treasure by magic. The agreement is void.
(b) A and B contract to marry each other. Before the date is fixed for marriage, A goes mad. The contract becomes void.

(c) A contracts to marry B, being already married to C, and being forbidden by the law to which he is subject to practice polygamy. A must make compensation to B for the loss caused to her by the non-performance of his promise.

(d) A contracts to take in cargo for B at a foreign port. A's government afterwards declares war against the country in which the port is situated. The contract becomes void when war is declared.

(e) A contracts to act at a theatre for six months in consideration of a sum paid in advance by B. On several occasions, A is too ill to act. The contract to act on such occasions becomes void.

Questions

1. Define a contract.   (**Ans.:** Section 2.2)
2. Define the following terms: (i) offer; (ii) acceptance; and (iii) counter-offer.
       (**Ans.:** Sections 2.2.1 and 2.2.2)
3. A submitted his tender in competition with others for a public construction project. On opening of the tenders, it was realised by him that he had commited an error. He wants to revoke his tender. Can he do so without any liability under each of the following circumstances?
   (a) Before tenders are scrutinised and a decision is taken by the department.
   (b) Tenders are scrutinised and the department has decided to accept his tender but the letter of acceptance has not been posted.
   (c) On receipt of the letter of acceptance of his tender. Give reasons for your answer.
       (**Ans.:** Section 2.3)
4. What are the essentials of a 'valid contract'?   (**Ans.:** Section 2.4)
5. How many parties are required to form a contract? Who are incompetent to form a contract?
       (**Ans.:** Section 2.5)
6. Explain the term 'meeting of the minds'.   (**Ans.:** Section 2.7)
7. Define 'free consent'.   (**Ans.:** Section 2.8)
8. What is meant by the term 'consideration' as used in the law of contract?   (**Ans.:** Section 2.9)
9. Under what circumstances may a contract be unenforceable?   (**Ans.:** Sections 2.4 to 2.9)
10. When is a contract voidable and when is it void?   (**Ans.:** Sections 2.4 to 2.9 and 2.11)
11. Distinguish between 'express' and 'implied' contracts.   (**Ans.:** Sections 2.10)
12. Which promises bind the successors or representatives of the promisor?   (**Ans.:** Section 2.12)
13. Write short notes on: (a) joint liabilities of parties to a contract;   (**Ans.:** Section 2.13)
    (b) reciprocal promises.   (**Ans.:** Section 2.14)
14. A particular contract stated that 'time is of the essence'. What is the legal significance of the quoted wording?   (**Ans.:** Section 2.15)
15. A consulting engineer dies when he has a number of unfinished contracts. His son, who is also an engineer, takes over the business of his father. Will he be bound by the contract of his father? Why?   (**Ans.:** Section 2.16)
16. 'Impossibility may be an excuse for the non-performance of a contractual obligation; inconvenience is not'. Explain.   (**Ans.:** Section 2.16)
17. Distinguish between subjective and objective impossibility of performance.

18. What are the circumstances under which the delay in the performance of a contract cannot be penalised?   **(Ans.:** Section 2.17)
19. Explain the following terms:
    (a) breach of contract   **(Ans.:** Section 2.18)
    (b) anticipatory breach   **(Ans.:** Section 2.18.1)
    (c) fundamental breach of contract   **(Ans.:** Section 2.18.2)
20. What are 'liquidated damages'? What is the underlying idea in awarding damages in a contract? Illustrate your answer with examples.   **(Ans.:** Section 2.19)
21. Name and explain several ways in which a contract can be terminated.   **(Ans.:** Section 2.20.1)

3 | The Tender

3.1 GENERAL

The first step in the formation of a contract is a definite offer by one party—usually written, although it may be oral or partly written and partly oral—called a tender. It is an offer from a contractor to undertake a work in return for a certain sum of money. The work to be undertaken may be:

1. construction work of a project as defined by drawings and specifications,
2. supply of requisite materials of specified quality, and/or of labour, required to complete a specified work,
3. supply of machinery and equipment with or without fuel, operator, and so on, and
4. transport of materials.

3.2 THE INVITATION TO TENDER

When all the preliminaries mentioned in the first chapter under the heading 'Development of a Project' have been completed and the owner has decided to proceed with the work, tenders are invited. Legally, this is an attempt to ascertain if an offer can be received from interested bidder(s) to execute the work within the estimated limits of time and finances. But the owner does not, by the mere invitation of tenders, incur any liability for any expenses to which a bidder may put himself to in preparing and submitting his tender.

Tenders are invited in any one of the following ways:

1. negotiated tender,
2. limited competition or selective tendering, and
3. open competition: local competitive bidding (LCB) or international competitive bidding (ICB).

3.3 NEGOTIATED TENDER

In this method, the price to be paid is negotiated with a single contractor. This method has, therefore, the drawback of not providing the owner with competitive prices. It is likely that the cost of a work will be higher when contracts are negotiated with a single bidder. An owner may prefer to run this risk if he expects some real advantages by employing a particular firm whose methods and policies are known and which has, in the past, proved capable of satisfactorily fulfilling his special requirements. The higher cost may be offset by better quality of work and early completion. This procedure can be adopted to the owner's advantage under the following circumstances:

1. The firm chosen is one in which the owner and his engineer have confidence, and whose integrity and reliability are well established.
2. The work to be carried out is within the special scope and experience of the firm.
3. In certain limited types of work, there may be only one firm which is capable of undertaking the job, and negotiations with it will be the only reasonable method of obtaining a tender.

The procedure is as follows:

1. The engineer, on behalf of the owner, invites a firm to tender. This initial invitation includes information regarding the proposed contract procedure, a brief description of the scope of the work, the approximate dates of commencement and completion, and other essential information.
2. The engineer or the contractor prepares the priced bill of quantities. In many cases, it is more satisfactory that the contractor do the original pricing, as he is in a better position to judge the correct price, which may depend upon the plant and the methods to be employed in execution.
3. The priced bills are then handed over to the other party for consideration.
4. The rates are examined by the other party. Rates in dispute are compared with current rates for similar work, obtained in competitive tendering after allowing for the special features or peculiarities of the situation.
5. When the engineer and the contractor reach an agreement, the agreed schedule of prices is sent to the owner for his approval and assent.

3.4   LIMITED COMPETITION

In this method, the owner invites tenders from a few contractors who are known to have specialised in a particular type of work and from whom he has, on previous occasions, obtained excellent results. This procedure is usually adopted for private jobs, where the owner has the right to negotiate directly and to enter into an agreement with whomsoever he chooses. This procedure can be highly recommended in connection with structures of monuments, industrial construction, and other works that require specialised knowledge and equipment.

3.5 OPEN COMPETITION

In this method, tenders are invited by public advertisement. The open competition method is usually adopted for public works, as the law generally requires that government contracts and other public contracts be advertised publicly, to obtain the most advantageous terms. Any contractor who is willing to undertake a piece of work and who has the requisite finance and equipment to complete it satisfactorily is allowed to submit the offer. This competition is classified into two groups: local competitive bidding (LCB) when submission of tenders is restricted to Indian companies only and international competitive bidding (ICB): when it is open to all contractors including international firms.

Advantages

This method has the following advantages:

1. It safeguards the contractors from being denied tendering opportunities for unjust reasons.
2. It protects the government or the local authorities concerned from possible insinuations of favouritism.
3. It provides an opportunity for having the work done at a minimum cost to the public, by securing the benefit of keen bidding competition.

Drawbacks

The main drawback of this method is a possibility that the firm submitting the lowest tender may be unsuitable for carrying out the work, thereby resulting in considerable waste of time and effort. The lowest tender can be by design or by mistake. This drawback can, however, be eliminated by following the procedure of pre-qualification of bidders.

3.6 PRE-QUALIFICATION OF BIDDERS

In India, the system of pre-qualification of bidders has gained importance especially with many massive projects being financed by the World Bank. It has now become common for any major or even medium works. Irrigation works costing above ₹50 lakhs, tunnels, underground structures and so on, are executed by pre-qualification of contractors. Under this procedure, instead of inviting tenders publicly, the intending tenderers can be invited to apply for selection and registration as a qualified tenderer. For this purpose, the employer or his engineer/architect publishes an advertisement in leading newspapers. The intending bidders are supplied with pre-qualification documents. To review the qualifications of the tenderers, the information provided by them in the prescribed format supplied to them for this purpose includes:

1. past experience,
2. major works on hand,
3. technical personnel engaged,
4. financial capacity,
5. equipment to be used, and
6. any other relevant information.

The blank tender forms are then issued only to those contractors who, in the judgement of the employer or his engineer/architect, have adequate experience, personnel, equipment and financial and administrative capacity. This procedure can be adopted even when a government department has lists of approved contractors in various categories. In such a case, the above information may be called for along with the tenders but in a separate cover. Sometimes when the cost of the work is large, there will be very few contractors who can qualify. To overcome this difficulty and ensure healthy competition, the option is given to contractors to strengthen their hands by forming joint ventures (JVs) of two or maximum three firms. The 'lead contractor', the largest shareholding among the partners, generally heads the joint venture. Joint venture partners are jointly and severally responsible for the work.

3.7   CRITERIA FOR PRE-QUALIFICATION

Applications received for pre-qualifications are scrutinised by allotting points or marks out of 100–135. To pre-qualify the applicant must score, generally, at least 60% on each of the first four categories listed below and 70% of the total points:

1. financial strength,
2. past five or six years experience,
3. technical personnel engaged,
4. construction, plant and equipment, and
5. other considerations such as quality control standards, ISO certifications, and so on.

The first four points carry nearly 90% and the last generally 10%.

After evaluation, those contractors found qualified would be allowed to submit tenders in response to the advertisement.

3.8   ADVERTISEMENT FOR TENDERS

The purpose of advertising for tenders is to create interest for the proposed work amongst a considerable number of builders so as to secure the benefits of keen competition. The advertisement should come to the notice of the greatest number of potential bidders. City newspapers with a large circulation and appropriate technical journals usually provide the best medium. However, advertisement of small projects, which are likely to interest only local bidders, may be confined to local newspapers.

The advertisement should be published sufficiently often—either successively for a few days or at regular intervals of times, such as weekly or biweekly, for a month or so. In order to attract the attention of the readers of a particular group, the advertisements are generally given under a heading such as 'Tender Notice'.

The advertisement should appear well in advance in order to afford prospective bidders a reasonable time to make necessary inquiries and to collect any background information that they may require. A period of less than two or three weeks is seldom sufficient. For big projects, the period allowed is generally four to six months.

3.8.1   Pre-tender Conference

It is suggested that for the purpose of obtaining necessary clarifications, a pre-tender conference needs be held, preferably at a place near the site of the proposed work. The intending tenderers should be invited to attend the conference at their own cost. The special features of the project as also unusual or peculiar conditions and specifications, if any, should be clarified. The tenderers may also raise their doubts or seek clarifications in this conference so that they can submit their tenders on a firm basis after fully understanding the work, its nature, scope and conditions under which it is to be executed. This procedure, if followed, will restrict the seeking of clarifications at the post-tender stage to the minimum.

3.9   INFORMATION TO BE GIVEN IN A TENDER NOTICE

The tender notice must be as short and concise as possible because advertisement charges are generally high. The notice should, however, contain adequate information on the nature and scope of the proposed work and all essential details. The text of a good advertisement should include at least the following information:

1. **Mode of submitting tender:** Bidders should be asked to submit tenders in sealed covers, in order to maintain the secrecy of the quotations.
2. **Form in which tender has to be submitted such as item rate, percentage rate or lump sum, and so on:** It is advisable to get all the offers in the same type of form, so as to ensure uniformity in the basis of the different bids. This would facilitate the scrutiny and comparison of the offers received.
3. **Name of the inviting authority:** This helps a bidder know the person(s) he will have to deal with and the amount of cooperation he may receive, if his tender is accepted. This may affect his bid price.
4. **Nature of the work and its location:** These two factors are of obvious importance. If the nature and location of a work are within his area of operation, the prospective contractor will want to learn more about the job; else, he need not waste his time reading further details.
5. **Estimated cost of the work:** This indicates very briefly the magnitude of the work and enables the contractor to determine whether it is too small to interest him or too large for him to handle.

6. **Time limit:** The time limit within which the work has to be completed should be mentioned. This will be of great help to the bidders in working out a realistic price for the work. There exists a definite relationship between the cost and time of performance. The time limit may influence the type and number of machines and workmen to be employed and thus vitally affect the contractor's bid price.

7. **The availability of data and forms:** The following information should be included, for obvious reasons:
   (i) where, from whom, at what cost, and up to which date blank tender forms and specifications may be obtained, and
   (ii) whether a refund will be made upon their return in a satisfactory condition.

8. **Earnest money required along with the tender:** Security in the form of earnest money is demanded along with the completed bid to make reasonably certain that the bid is made in good faith and that the bidder is willing to sign the contract if the work is assigned to him. The advertisement should state:
   (i) *The purpose for which the security is required:* This fact is so generally understood that many advertisements contain no reference to it.
   (ii) *The character or the form of the security:* That is, whether the amount will be accepted in the form of cash, bank guarantees, cheque or a treasury receipt or challan. The name in whose favour the cheque or the challan is to be drawn must be specified.
   (iii) *The amount of security:* In case the successful bidder refuses to enter into contract, the delay is likely to cost the owner many incidental losses, in addition to the expenses of advertising anew. The amount of security mentioned, about 1% of the estimated cost of the work, is intended to cover these expenses.
   (iv) *Return of deposits:* The number of days within which the amount will be refunded to the unsuccessful bidders should be stated.

9. **Security deposit:** The amount which the successful bidder will have to deposit as security for satisfactory performance, and the procedure for recovering the security deposit, should be stated.

10. **Information regarding plans, blueprints, and so on:** The advertisement should state explicitly where and when plans and detailed drawings can be examined. They are often not supplied to bidders as this would involve considerable expense and any secret or confidential aspects may be revealed.

11. **Information regarding permission to import equipment not available in India:** Such information and the restriction on disposal of the equipment after completion of the work, if any, should be made available whenever relevant.

12. **Whether successful foreign contractors would be permitted:** To engage any approved local subcontractor or to enter into a joint venture with Indian contractors or firms. Information in this regard should be stated in the case of international competitive bidding (ICB).

13. **The last date, the place, and the time of receipt of sealed bids:** All these facts have to be mentioned in the text.

14. **The date, time, place and procedure of opening of tenders:** All these facts should also be mentioned.
15. **Reservation of the right to reject bids:** The owner should make it clear that the right to accept a tender which is higher than the lowest one, or to reject any or all tenders without assigning any reason, is reserved.
16. **Award of contract:** The time required for the consideration and scrutiny of tenders, after which the contract is likely to be awarded, should be stated. Normally, this period should be about 90 days though sometimes it is 120 days, from the date of the opening of tenders.
17. **Whom to contact for further information:** This should be clearly stated.

The advertisement should carry a date and should be signed by the proper person whose name and/or designation should also appear therein. It may seem that so much information is superfluous in a mere advertisement. However, a fairly complete description at this early stage of the proceedings is helpful to all concerned because:

1. It minimises the number of supplementary inquiries that the engineer may have to answer.
2. In the absence of a complete advertisement, a prospective contractor may hastily conclude that he is not interested. This adversely affects the interest of the owner by reducing potential competition.
3. The number of unfruitful requests for plans and supplementary information is reduced.
4. A sufficiently comprehensive and accurate advertisement protects the engineer against a possible claim that he misrepresented the nature and condition of the proposed contract.

From this legal standpoint, it is advisable to include the entire advertisement in the contract document as part of the contract.

3.10   SPECIMEN TENDER NOTICE

The content of the following specimen advertisements, the first for conventional 'Engineering, Procurement and Construction' (EPC) contracts and the second for Public Private Partnership (PPP) contract can be compared with that suggested in the preceding discussion:

For EPC Contract

Tender Notice No . . . . . . of 20 . . . . . .

SEALED tenders in the prescribed form for the following works are invited from registered contractors (in the appropriate class with the ______________ government, etc.) to reach this office up to or before ______________ a.m/ p.m on ______________ (date). The tenders will be opened on the same day at ______________ p.m, if possible, in the presence of the intending tenderers.

Blank tender forms will be available on payment of ₹_______________ (here state the tender fee), which is not refundable, on any working day from _______________ (date) to _______________ (date) during office hours, that is, _______________ a.m _______________ p.m on weekdays, and up to _______________ p.m on Saturdays. Contractors desirous of obtaining tender forms by post should remit by money order in advance and well in time the amount of the regular tender fee plus

₹_______________ to cover postage. The earnest money by way of treasury-receipted challan should accompany the tender, without which the tender will not be considered. Earnest money will not be accepted in cash or by cheque under any circumstances.[1] The contract documents can be seen by prospective bidders during office hours from_______________ (date) to_______________ (date) at the office of the undersigned. Further particulars in connection with the work can be had from the same office. The authority competent to accept the tender reserves the right to reject the lowest or all tenders without assigning any reason.

Name of the work	Estimated cost	Earnest money	Security deposit	Time limit	Tender form	Tender fee
1. Construction of a school building in Jawaharnagar Dist.	₹200 lakhs	₹50,000	₹10 lakhs	2 years	Item rate	₹1000

Date _______________   (_______ Name _______)
                                         _______ Designation _____
                                         Office address

PPP Contract (see Chapter 6)

(Name of the State)…. State Infrastructure Development (Name of the Authority) (Board/Authority)

Invitation for Proposals (Request For Qualification – RFQ)

National Competitive Bidding (or) International Competitive Bidding

Subject-(Name of the Project) in the State of — through Public–Private Partnership on Design, Build, Finance, Operate and Transfer (DBFOT) Basis

The … (Name of the Authority) … Authority is engaged in development of infrastructure and as part of this endeavour, the Authority has decided to undertake development of the below project through Public–Private Partnership on Design, Build, Finance, Operate and Transfer (Toll) basis, and has decided to carry out the bidding process for selection of the bidder to whom the project may be awarded. Brief particulars of the project are as follows:

---

1. Nowadays the earnest money deposit is also accepted in the form of bank guarantee rather than treasury challan.

Name of the project (example, Road)	Length in kilometres	Indicated project cost in Rupees
Four-laning/Two-laning		
with paved shoulders of NH (Location)....	*80*	*₹300.00 crores*

The Authority intends to prequalify and shortlist suitable applicants (the bidders) who will be eligible for participation in the bid stage, for awarding the project through an open competition in accordance with the procedure set out in the detailed tender notice.

Request for qualification proposal (RFQ) documents consisting of detailed scope of work, eligibility criteria, etc. can be obtained between ....hours and .....hours on all working days from ....(date)... to ...(date)... [2] at the address for communication mentioned below on payment of a non-refundable fee of ₹30,000 in the form of a demand draft favouring..... (Name of) ....Authority and payable at .....(Place)...... The detailed NIT and documents can also be downloaded from the website: (For example, *www.tenderwizard/NHAI*).

Sealed proposals should reach the Authority at the address for communication not later than ..(Time)…hours on .....(date)[3].

Address for communication

Name of the Person and his Designation
Name of the Authority
Telephone number
Email

3.11 EXTENSION OF TIME LIMIT FOR SUBMISSION OF COMPLETED TENDERS

The last date and the time of receiving the tender forms, complete in all respects, must necessarily be extended under the following circumstances:

1. If the period which the contractors get to prepare the tenders is inadequate—a fact brought to the notice of the inviting authority by the intending tenderers.
2. If a sufficient number of blank tender forms are not issued to the intending contractors.
3. If important changes in the tender are considered necessary since the issuing of the tender notice.

The extended date and time of issuing and accepting the tenders must be brought to the notice of all concerned. An advertisement in the form suggested below if inserted in the newspapers or trade journals in which the original advertisement appeared, will serve this purpose.

---

2. Allow about one month time between the two dates.
3. Allow about one month time from the last date of supply of RFQ documents.

CORRIGENDUM TO TENDER NOTICE

In partial modification of the tender notice issued under this office No. ______________ of /______________/ 20______________ it is hereby notified that the dates of issue and receipt of tenders for the work of '______________ (*insert name of the work*) ______________' have been extended up to (*insert date*) ______________ 20 ______________ and (*insert date*) 20 ______________ respectively.

All other conditions remain unchanged.

Place: Date: Signature and name of the authority

3.12 METHODS OF PREPARING A TENDER

The knowledge of the procedure adopted by a contractor in arriving at the figures he has quoted in his proposal is essential to all engineers, as it would enable them to attain better results while preparing contract documents. The methods followed by a contractor in the preparation of tender estimates are invariably complicated. Nevertheless, an attempt is made here to outline the procedure adopted for a typical large contract.

1. *Careful study of the contract documents:* The contract documents issued to the contractor or supplied for his inspection include items such as general conditions, drawings, specifications and bills of quantities. The contractor makes a careful study of these documents and notes down any unusual conditions, specifications, or any feature of the work demanding special attention during pricing.
2. *Subcontractor's work:* The contractor next identifies the subcontractor's work and also the materials necessary for its execution. For this purpose, he would need to know the relevant prices and so passes on the bills of quantities to an assistant for carrying out the inquiries in coordination with subcontractors and material suppliers.
3. *Visit to the site of work:* Having very carefully examined the drawings and specifications, the contractor next goes to the site of the proposed works. The visit may be arranged jointly with the owner's representative in order to obtain on-the-site clarifications of the contractor's queries, which would be later confirmed in writing and would form the site document, in case his tender is accepted. After inspecting the site, the contractor ascertains the conditions under which the work has to be carried out and their effect on certain unit prices, and also on the overall tender figure. There are a number of factors which could affect the tender figure; some of these are as follows:
   (a) difficulty in reaching the site of work,
   (b) limited space on the site for the movement of vehicles,
   (c) the type of soil and the depth of the water table therein,
   (d) availability of space for storing materials on the site in a convenient position, that is, close to the place of erection, thereby avoiding the cost of double handling and wastages,

(e) availability of materials, their sources and the prevailing market rates,

(f) local availability of skilled and unskilled labour, the extent of necessity of recruiting personnel from other localities and the prevailing wages of workmen,

(g) source and cost of water needed for construction work, and

(h) power and lighting (if available) and the cost of erecting, maintaining and eventually dismantling power connections on the site.

4. *Time for completion:* The contractor estimates the probable length of time required for completing the work and also the number and category of permanent staff indicated by the nature of the work. This helps him to calculate 'establishment charges' or 'overheads'.

5. *Unremunerative work:* The contractor determines the amount of unremunerative works that need to be done before the commencement of the construction job and also the amount of work done in clearing away items such as temporary offices, store sheds, access roads, water supply connections and so on after completion of the work. He also calculates the costs involved in wear and tear of the plant, transport, insurance, taxes and so on.

6. *Estimation of on-costs:* On-costs or overheads are those costs which the contractor has to incur apart from the cost of labour, machinery and equipment and materials for the work. Costs of these items are called direct costs. These can be estimated and included in the item rate of each item. However, there are certain costs which cannot be specifically attributed to an item of work such as salaries of staff, financing costs, and so on. There are one hundred odd items which contribute to overheads/on costs. These are worked out from the books of accounts of the tendering company. The balance sheets also show the total turnover of a company in a year and the overheads incurred during the same year. The percentage the overheads bear to the turnover is used to increase the direct cost and work out the tendered rates/costs. For example, if the turnover of a company in the previous year was ₹100 crores and the overheads ₹10 crores, the percentage will be 10%. If the firm expects 10% profit the combined percentage will be 20% of the tendered cost. It works out to 25% of the direct cost of materials, labour, plant and machinery. Thus, the direct costs of each rate will be increased by 25% to work out the rate to be tendered. In short, if the direct cost in the example of an item of work is ₹80 per unit, 25% of that cost works out to ₹20. The rate to be tendered will be ₹100 per unit of the work under that item. The rate tendered includes the margin of ₹20 for overheads and profit, which is 20% of the tendered rate/cost.

7. *Final estimate for tendering:* The procedures adopted by a contractor in the preparation of the final estimate for tendering would differ depending on the type of contract. Three different types of contracts commonly employed are lump sum contract, unit price contract and BOT contracts. The procedure in respect of these types is explained as follows.

(i) Preparation of lump sum tender

(a) The contractor first estimates the quantities of materials involved, and computes their probable cost.

(b) Next, he estimates the cost to be incurred for the entire labour force and the plant that would be required to perform the work.

(c) He estimates on-costs as explained above.

(d) He then decides the amount of profit which, in his opinion, can be obtained without losing the job to the competitors. Adding the costs of all inputs above he works out the lump sum to be quoted.

(ii) Preparation of unit price tender or measurement tender

The procedure followed is substantially the same as for the lump sum tender except for the following:

(a) The contractor may not have to be so careful in estimating the quantities of materials and labour involved.

(b) The unit price for each item is worked out separately by distributing the on costs and profit in suitable proportions over the various items. While doing so, the contractor will try to make the bid unbalanced to his advantages, that is, by increasing the rates of those items whose quantities are underestimated and decreasing the rates of those items whose quantities are overestimated. He may also put down enhanced rates for those items of work that would be performed early in the course of contract, so that partial payments made in the first few months would yield him generous return. This would help him generate working capital. This is called front loading of a tender.

(iii) Preparation of build-operate-and-transfer (BOT) tender

In this case also the cost of construction of the project is ascertained in the same manner as for the lump sum or item rate contract. The tenderer then works out the cost of financing the project and its recovery from the collection of toll or a similar method till the cost of construction together with the cost of finance is fully recovered. Each year's income will reduce the cost till the cost becomes zero or negative. The cost of maintenance of the project during the period in question has also to be added to the basic cost. The number of years required to recover the total cost is generally called 'concession period'. The concession period worked out by each tenderer will depend upon the cost of construction worked out by each tenderer and also the rate of interest considered for financing the project and expected profit. The tenderer who quotes the minimum period of concession is generally awarded the contract.

Sometimes, the authority fixes the period of concession and seeks offers on the basis of grants expected from the employer. The tenderer expecting the least grant is generally awarded the contract. In addition, income tax holidays for predetermined years to be selected by the tenders are offered to reduce the grant amount.

3.13   SUBMISSION OF TENDERS IN THREE ENVELOPES

The tenderer, while submitting the tender, is directed to submit the documents in three sealed envelopes as specified below.

Envelope No. 1

The first envelope clearly marked 'Envelope No. 1' shall contain the following documents:

1. Registration details of registered contractors/certificate as a registered contractor.
2. Information regarding pre-qualification requirements, pre-qualification documents with prescribed forms and so on.
3. Details of joint venture, if any, together with information called for in this respect by the terms of the invitation.
4. A cheque/DD or government treasury challan or 'deposit at call receipt' or bank guarantee of any scheduled bank for the amount of earnest money (as directed in the notice inviting tenders) or certificate of exemption for the payment of earnest money, if applicable.
5. In the case of public works, an income tax clearance certificate in original from the income tax office falling under the concerned circle, dated not more than three months prior to the last date fixed for the receipt of the tender, unless specifically exempted in this respect by the government.
6. The list of machinery and plant immediately available with the tenderer for use on this work and the list of machinery proposed to be utilised on this work, but not immediately available, and the manner in which such machinery is proposed to be procured.
7. Details of technical personnel on the rolls of the tenderer.
8. Details of works of similar type and magnitude that have been carried out by the tenderer.
9. Details of other works tendered for and those in hand with the value of work unfinished on the last date of the submission of the tender. The certificates from the heads of the offices under whom the works are in progress should be enclosed.

Envelope No. 2

The second envelope clearly marked 'Envelope No. 2' shall contain the following:

1. Covering letter for the offer.
2. Special features that the contractor feels are worth mentioning about his offer.
3. Assumptions, conditions, deviations, and so on put forth by the contractor, if any, as distinct from the provisions in the tender papers issued to him.
4. Tenderer's own design where so allowed.

This envelope shall not contain the amount or rates offered by the tenderer.

Envelope No. 3

The third envelope clearly marked 'Envelope No. 3' shall contain only the main tender document mentioning the amount/rates offered by the tenderer. Before presentation, the tender is rechecked in order to avoid mistakes and then signed and sealed.

The three sealed envelopes mentioned above should again be put together in one common cover and sealed. On the cover, the following information should be provided:

1.   the title of the project,
2.   the last date of submission of the tender,
3.   full name and address of the tenderer, and
4.   name of the authorised agent delivering the tender.

If the tender is submitted by post, the sealed cover containing the three envelopes mentioned above, and also marked as specified above, should be sent by registered post with acknowledgement due, so as to reach the competent authority before the expiry of the time limit.

In a two-cover system, the contents of the second envelope can be called for in the first envelope with the bid in the second envelope.

3.14   ON LINE TENDERING

With the advent and advancement of the internet technology, tenders are sometimes invited and submitted on line. The concerned authority engages the services of a specialised firm to develop and customise the software for facilitating the process of e-tendering. The authority thereafter makes available an e-tendering portal which can be accessed by logging on to the website such as www.onlintenders.co.in, http://www.tenderwizard/NHAI, and so on. It is mandatory for all applicants to have a digital signature certificate issued by any licensed certifying agency, in the name of the person who will sign the application. For the list of licensed certifying agencies one may log on to www.cca.gov.in. The next step is to register with the website concerned by logging onto the website by taking the following steps:

1.   Click on 'Sign Up'.
2.   Provide all information including valid user email-id, PAN Card number, and so on.
3.   Enter the password as per direction which can be obtained by paying annual registration charges to the service provider.
4.   Attach necessary documents such as the Registration Copy in case of a partnership firm or certificate of incorporation in the case of a limited company.
5.   Click on the 'Submit' button.

Thereafter, the registration approval mail would be received by the tenderer from the website. The tenderer can login to this website for purchase of the tender form.

Purchase of tender form: Only a registered tenderer can purchase the tender form by using his email id, password and digital signature, if applicable, by logging on to the website and clicking on **View Tender List**. It will display the current published tenders list. A tenderer can view the tender list and also view the tender document by clicking on **Tender Set** if the tender invitation authority allows viewing tender set before purchase of tender. To download the tender, enter the correct characters and click on the **Buy Form** button, it shows the **Terms and conditions** page. If the tenderer accepts the terms and conditions, he can purchase the tender by paying the cost of the form by deposit or credit card. The cost generally varies from ₹20,000 to ₹30,000 plus service tax. After purchasing tender, the tenderer gets the receipt, *sammatipatra* and tender set.

Tender submission: Submission of a tender is generally done in two stages.

1. Request for Qualification (RFQ)
    Tender invitation initially calls for applications requesting for qualification (RFQ). Documents such as Detailed Scope of Work, Eligibility Criteria, and so on, can be downloaded from the website of the authority inviting RFQ on payment of the cost indicated by the notice for RFQ. The application needs to be furnished on or before the last date for receipt notified. It is done by uploading the documents in electronic form and originals, bound in hardcover, to be submitted in physical form. The place and date where the pre-qualification conference would be held is also notified. This process is similar to pre-qualification of bidders as already mentioned in Art. 3.8.1 for the normal tendering process.

2. Request for Proposal
    This is a normal invitation of proposals from previously pre-qualified bidders on any one of the forms used for PPP contracts discussed in Chapter 6. A tenderer has to upload the technical documents. If those documents are be used every time during technical submissions, he has to use the **My Document** link. He can upload the documents in **My Document** by clicking on **Add document**. Then it will display the document description and document link under the file name heading. By clicking on the document link, the tenderer can view the uploaded document. He can edit and delete that document before the use of that document in any technical submission. He can upload the technical documents which are already in his account by clicking on this button to display the list of documents and by selecting the documents that are necessary for technical submission and clicking on the **Submit** button. He must wait for confirmation after clicking on the **Submit** button. If the uploaded document does not show up after confirmation, it is necessary to refresh the browser submission process.

    It is clarified that the process suggested above may vary from Authority to Authority and the reader must follow the procedure suggested by each Authority. For example, the National Highways Authority of India (NHAI) has published detailed instructions for use by applicants regarding procedure under e-tendering.

Submission of commercial details: To submit commercial details, the digital signature of the tenderer is compulsory. By clicking on this button, the tenderer can quote the rate and fill the details. He can quote the rate in 'above/below' format or attach the price list document. He may have to enter characters as per image and then click on the **Submit** button. If characters entered are wrong, the submission will not proceed further.

Opening and scrutiny of applications: The authority shall open the documents of applicants received in electronic form on the date and time appointed for the purpose in the presence of applicants who choose to attend. The documents are thereafter examined, evaluated and listed in accordance with the provisions of the RFQ.

Tender status: After opening of the tenders, a tenderer can view the tender status which he has purchased by clicking on the **Tender Status** button, which shows if the tenders have been opened or not opened and if opened it also shows commercial reports which include lowest (L1, L2, L3) tenderer.

3.15 OPENING OF THE TENDERS

In the case of public bodies, the tenders are usually opened and read aloud at the stated hour on the day notified in the tender notice. Usually, a representative of each bidder (or the bidder himself) will be present to safeguard his principal's interest. The opening and the public reading of the sealed tenders reveal the lowest bidder and by what margin his bid is lower than the others.

3.16 OPENING OF TENDERS SUBMITTED IN THREE COVERS

The tenders will be opened on the specified date—if possible in the presence of the tenderers or their authorised agents who may choose to be present—by adopting the following procedure.

Envelope marked 'WITHDRAWAL' are generally opened first. After a tender is submitted and before it is accepted, if any tenderer has withdrawn his offer and if the notice of withdrawal is found in order, the envelopes of such a tenderer will be removed and not opened. Of the valid tenders, the envelopes will be opened as follows:

Envelope No. 1

First, the Envelope No. 1 of all the tenderers will be opened to verify its contents as per requirements. If the various documents contained in this envelope do not meet the requirements of the tender invitation, a note will be recorded accordingly by the tender-opening authority and the said tenderer's Envelope Nos. 2 and 3 will not be considered for further action although they will be recovered. This system is followed when the bidders are not pre-qualified.

Envelope No. 2

This envelope shall be opened immediately after the opening of Envelope No. 1, only if the contents of the latter are found to be acceptable to the tender-inviting authority. The forwarding letter of the tenderer, his assumptions, conditions, deviations and so on, shall then be read out.

Subsequently, the tender-inviting authority will scrutinize the various assumptions, conditions, deviations and additional stipulations incorporated by the competing tenderers and a common set of deviations from the original conditions will be incorporated in the tender. The additional stipulations, which are considered as acceptable to the owner, will be formulated and all competing tenderers will be informed accordingly. The effect of this change in terms of money shall be submitted by the tenderers, in terms of a percentage above or below their originally tendered basic total amount. The tenderers while re-evaluating their bid price are expected to take cognizance of:

- the additional facilities offered to them in the common set and then reduce their bids, or,
- their conditions that have been rejected and then increase their bids in the form of a percentage above or below their original bids in Envelope No. 3.

The fresh proposal is to be enclosed in another cover properly addressed and marked Envelope No. 3A. This cover should not contain any fresh conditions. It should be submitted in the same manner as the original bid was submitted, before the expiry of the time limit.

When the contractor's own design is allowed for by the owner, broad details thereof shall be enclosed by the tenderer in Envelope No. 2. This shall be subjected to scrutiny by the engineer concerned to ascertain the technical feasibility and acceptability of the design. Doubts, if any, of the tenderers should be clarified by the owner or his engineer/architect. In case the engineer finds a design unfeasible, the quotations and offers in Envelope 3A and 3 shall not be considered by him or the owner.

Envelope Nos. 3 and 3A

Both these envelopes will be opened simultaneously on the specified date, in the presence of the tenderers whose offers are under consideration, so as to arrive at the final figures with respect to the offers of the various tenderers without any individual conditions except those included in the common set of conditions issued by the owner, which shall form a part of the contract documents.

3.17 CONSIDERATION AND SCRUTINY OF TENDERS BEFORE ACCEPTANCE

After the tenders have been opened, a comparative summary statement is prepared as tabulated:

S. No.	Name of firm	Amount of tender	Remarks/Position
1	A	₹10,40,99,990	III
2	B	₹10,30,77,770	II
3	C	₹ 9,99,99,990	Lowest
4	D	₹10,60,66,660	IV

The consideration and scrutiny of tenders will vary in respect of—lump sum contract and measurement contract.

Lump sum contract: In the case of a lump sum contract, not likely to involve any extras, it is of little or no importance to the owner to know how the different bidders have worked out the price as tendered. In such a case, the contract may be awarded to the lowest bidder, provided, of course, that this bidder has satisfied all other requirements and has not brought to the notice of the owner or his representative any error or omission committed by him while working out the cost. A serious mistake may sometimes be committed by a bidder, and, if he is awarded the contract, it becomes obligatory on his part to carry out the work, often in spite of great financial loss. If he refuses to carry out the work, not only will he forfeit his earnest money but he will also make himself liable to suitable legal action for breach of agreement. It is, however, in the owner's interest to grant the bidder an opportunity to revoke the offer or to correct and re-submit the same, if the latter communicates any such mistake before the tenders are opened and then wishes to revoke the tender or is ready to execute the work if an opportunity to correct the tender is given. The action to be taken would depend upon the owner or his architect/engineer. In such a case, if the owner is satisfied that the mistake is genuine and has been committed unintentionally, he may award the contract to the next lowest bidder and give an opportunity to the lowest bidder to save himself from embarrassment. He may, however, reject all the bids, recover any loss that he may suffer from the lowest bidder and recall the tenders.

Measurement contract: In the case of a measurement contract, the total cost is based on the estimated quantities mentioned in the schedule. As these quantities are approximate and likely to vary, the total amount tendered is not to be taken as the criterion for the award of the contract (as is the case with the lump sum contract). However, the rates quoted by the different bidders have to be meticulously scrutinised with a view to identifying any rates which seem extraordinary. The lowest tender may ultimately prove far more costly than the others if the tendered sum is worked out on the basis of an 'unbalanced bid'. An example will clarify this point.

Example of an unbalanced bid: See following table (only a few of the items in the contract are listed. The rates are for illustrative purpose only.)

Item No.	Description	Quantity	Tendered prices in rupees per unit Contractor		
			A	B	C
1.	Earth excavation	8000 cu. m	45	90	65
2.	Rock excavation	2500 cu. m	275	115	245
3.	Cement concrete M-15	1200 cu. m	2,000	2,000	2,150
15.	Reinforcement	100 quintals	2,200	2,150	2,390
29.	Cement paint	8000 sq. m	25	16	12
Total for the items considered			38,67,500	37,50,500	40,47,500
Relative position			Second	Lowest	Third

Notice that contractor B has quoted a relatively high rate for item 1, and a low rate for item 2. He evidently expects the quantity of earth excavation to exceed the estimate, and that of rock excavation to fall below the estimate. Now, if on the basis of the lowest tender, the contract is awarded to B and if the actual quantities of earth and rock excavated are as follows:

1. Excavation in earth 10,000 m^3
2. Excavation in rock 500 m^3

The total amount payable to B will be

$$₹10,000 × 90 = ₹9,00,000$$

$$₹500 × 115 = ₹57,500$$

$$\text{Total} = ₹9,57,500$$

Had the tender of A, which was initially higher than that of B, been accepted, the total amount payable to A would be:

$$₹(10,000 × 45) = ₹4,50,000$$

$$₹(500 × 275) = ₹1,37,500$$

$$\text{Total} = ₹5,87,500$$

The total amount payable to A would have been ₹3,70,000 less than the sum paid to B. The owner thus lost ₹3,70,000 as a result of the simple trick of submitting an unbalanced bid.

Further, notice that contractor C has quoted comparatively higher rates for items 1, 2 and 3 than normal rates, covering work that will be done early in the construction operation, whereas for item 29, which will be done towards the end, bid is low. This is obviously an unbalancing that is designed to obtain substantial partial payments for the work to be done in the early stages of construction. This is an example of 'front loading' of tender.

Let us now discuss a real-life example. In the early 1980s, a contractor in his late twenties came to consult the author with a problem. He intended to submit a tender for canal excavation work costing about ₹10 crores. He wanted to ascertain whether or not under the terms of the contract, a dispute regarding strata classification could be referred to arbitration. After carefully studying the terms, he was informed that the decision of the engineer-in-charge would be final and out of the scope of arbitration. He thanked the author and went away leaving the latter guessing as to his intentions. A few months later, he confided to the author that he had been awarded the contract, a copy of which had been sent for perusal. There were three major items:

1. excavation of the canal in all varieties of soil, including sand, soft and hard *murrum*: 2,372,700 cu. m,
2. excavation of the canal in soft rock: 8,50,200 cu. m, and
3. excavation of canal in hard rock: 1,396,500 cu. m.

Against each of the above three items, the rate tendered by the young technocrat was ₹15.50 per cu. m. He clarified that he had ascertained the actual possible quantities under each category and worked out the actual cost and in the end worked out the average rate. He had avoided a sure source of future dispute, that is, about strata classification, by unbalancing his tender.

From the foregoing example, it is clear that the tenders for a measurement contract need to be carefully scrutinised. Any tender that is unbalanced so as to cause considerable disadvantage to the owner has to be rejected.

The authority inviting tenders sometimes ask the tenderer, generally the lowest, for clarification about his tender including breakdown of the unit rates tendered. No change in the quoted rate is permissible. There is a practice coming into vogue to obtain from the contractor who has bid low on some items, additional security by way of a bank guarantee, which would be forfeited if he leaves the work of the said items incomplete. This would not help, if the contractor obtains extra benefit under the items bid high, because he would not leave the items bid low incomplete to retain the overall benefit expected by him.

Correction of Errors

Tenders considered as substantially responsive will be checked for any arithmetic errors following the rules framed in that respect such as that if amounts in figures and words, are varied the amount in words will govern, multiplication errors will be corrected by rate in words considered correct, and so on. The amount finally tendered will be adjusted in accordance with the procedure for correction. If the tenderer does not accept the corrected figure, his tender is generally rejected with the risk of forfeiture of earnest money.

3.18   SPECIAL CONDITIONS OFFERED BY THE CONTRACTOR IN HIS TENDER

The tender submitted by a contractor is said to be unqualified when he accepts, unreservedly, all the conditions of the contract. He may, however, submit his tender in cases where the three-cover system is not adopted, subject to certain conditions which he specifies in lieu of, or in addition to, the conditions set out by the invitation to tender. In such a case, the owner, or his architect/engineer, should consider very carefully the alternative conditions offered, in order to assess whether they are of ultimate advantage to the owner.

It should be noted that if the contractor's conditions are accepted either tacitly or explicitly, they take precedence over the original conditions. Although most tender invitations expressly stipulate that tenders containing a deviation from the contractual terms and conditions will not be considered, the contractors, while quoting for large projects, make their tenders conditional in respect to the following important aspects:

1.   the payment of mobilisation advance by the employer with or without interest,
2.   the payment of security deposit by way of bank guarantee,
3.   the provision of a suitable escalation clause to take care of variations in cost of materials, labour and fuel, oil and lubricant components,
4.   a reasonable limit on ordering of variation by the employer, and
5.   an arbitration clause for adjudication of disputes.

Nowadays, the tenders for large projects do incorporate the provisions in respect of the above aspects. Mobilisation advance as also machinery advance is given against bank guarantee to the tune of 10% to 15% of the estimated cost. Security deposit is also accepted in the form of a bank guarantee. Escalation payment is assured if the project completion time is more than one year. Variations are limited to plus minus 25% of the quantities in the Bill of Quantities. Dispute resolution provisions incorporate reference to Dispute Resolution Expert (DRE) or Dispute Resolution Board (DRB). There is one DRE whereas the DRB consists of 3 members, one to be nominated by each party and the third by the two members. If any party is not satisfied with the decision of DRE or DRB, the matter can be referred to arbitration of a sole arbitrator or three or any odd number of arbitrators.

In this connection, it is safe to remember that, legally, a one-sided attempt to modify the original conditions is null and void. Acceptance of the modified conditions by the other party must be in writing; mere silence or inaction does not amount to acceptance. The contractor, to avoid possible trouble on this ground, should insist upon the inclusion of the letter forwarding the tender incorporating the conditions, if any, proposed by him and accepted by the employer, in the list of contract documents.

Preference for Domestic Tenderers

When tenders for public works are invited in international competition, a condition is inserted giving preference to domestic tenderers. Domestic tenderers are those registered in India, have majority ownership by Indian nationals and who agree not to sublet more than 50% of the works in terms of value to foreign contractors. The preference is given by adding a predetermined percentage say, 7.5% of the evaluated tender price to all tenders other than domestic tenders and then determining the order of merit. If a joint venture with a foreign company is allowed to be classified as a domestic tenderer, in addition to performance security, 'domestic preference security' is to be obtained to ensure that the profit and loss distribution and work-sharing arrangements are not changed subsequent to acceptance of tender.

3.19   ACCEPTANCE OF THE TENDER

Before accepting a tender, the engineer makes sure that the successful contractor's tender is in order and also ensures that his resources, equipment, experience and sureties are genuine and satisfactory. Thereafter, the engineer notifies the successful contractor of the acceptance of his proposal and that a formal contract between him and the owner has to be signed at a specified date and place. This notice to the contractor is written and delivered to him through a messenger, by registered post, or by telegram. Sometimes the letter of acceptance itself mentions the order to commence work. In other cases, the work order is issued after the agreement is signed. In the former case, the letter of intent (LOI) mentions that till the signing of a formal agreement, the tender and its acceptance would constitute a binding contract.

3.20   REJECTING THE LOWEST TENDER

A tender will be liable to be rejected outright if, while submitting it, the following contingencies occur:

1. The tenderer proposes, in the tender, any alterations in the work specified or in the time allowed for carrying out the work or any other conditions that are not acceptable to the owner.
2. Any of the pages of the tender are removed/replaced.
3. The tenderer or, in the case of a partnership firm, each partner thereof does not sign or the signature/s is/are not attested by a witness in the space provided for the purpose. In case of a tender issued by a company, it should be signed in the legal name of the company, by the person authorised to bind it in the matter.
4. The tenderer in the tender makes any erasures.
5. If all corrections and additions or pasted slips are not signed by the tenderer.

The fact that the owner wants to get his job done as economically as possible, does not necessarily mean that he should award the contract to the lowest tenderer. We now discuss some of the considerations on account of which the lowest tender may be rejected, in cases where the pre-qualification procedure was not adopted.

1. *Improper offer:* If a proposal is not prepared as per specifications, there is a just and proper cause for rejection. An offer to execute a work at a cost of 'so many rupees below the lowest offer' is not a proper offer at all and should be rejected outright.
2. *Inadequate finance:* The lowest tenderer may not have sufficient finance to execute the job. It may, therefore, prove more economical in the long run to award the contract to the second lowest bidder than to the lowest bidder who may not be able to carry out the work to completion.
3. *Lack of experience in a particular type of work:* The lowest tenderer may not have sufficient experience to execute the kind of work specified. If his tender is accepted, inadequate experience on his part may not only hamper the progress of the work but also affect its quality.
4. *Unsatisfactory reputation:* In the opinion of the engineer, the lowest tenderer may have an unsatisfactory reputation; for example, the contractor may in the past have been found careless, un-cooperative, and unscrupulous. In some public works contracts, the tenderer is asked questions such as the number of arbitral proceedings to which the tenderer had been a party. In such cases, it is suggested that the information should also be asked in how many cases the tenderer had succeeded. Many times arbitral proceedings are necessitated due to failure to take a decision or an erroneous decision by the engineer giving no option to the contractor except to invoke arbitral proceedings. It will be wrong to declare a person litigant, and to ignore his tender, in spite of his having succeeded in litigations.
5. *Inadequate staff and equipment:* If the lowest bidder lacks the special equipment and the trained staff required for the satisfactory completion of the work, accepting his tender would result in trouble for all concerned.

While rejecting the lowest tender, the engineer has to be very careful and ensure that his action is correct and can be defended, if necessary, particularly if he is acting on behalf of a public or corporate body.

3.21   REJECTING ALL TENDERS

All the tenders received may be rejected if:

1. there exists a well-grounded suspicion of collusion between tenderers or some other form of fraud has been detected,
2. radical changes in design are found necessary during the interval preceding the opening of the tenders,
3. the lowest tenderer has quoted a figure, which is higher than the funds available for the execution of the work,

4. sickness, financial reverse or an unforeseen incident has occurred, which makes it inadvisable for the owner to proceed in accordance with his original intentions, and
5. the minimum number of tenders to assure adequate competition has not been received.

3.22 RE-ADVERTISEMENT

When all tenders have been rejected, re-advertisement becomes essential, unless it is proposed to abandon or postpone the project. In case of such re-advertisements, the entire design or some major features of the work need to be substantially changed so that the new tenders can be submitted on an independent basis. Without such a change in the work, the tenders that will be received would be influenced by the knowledge obtained from the former proposals and no useful purpose will be served by the re-advertisement.

Questions

1. What is a tender? **(Ans.:** Section 3.1)
2. Does the issue of an advertisement for tenders mean that the owner has to accept one of the tenders received? Why? **(Ans.:** Section 3.2)
3. In what ways other than advertising may an engineer request tenders?

    (Ans.: Sections 3.3 and 3.4)
4. Why should an engineer advertise for tenders on a public construction project?

    (Ans.: Section 3.5)
5. What is the main drawback with respect to invitation of tenders by the open competition method? How can it be eliminated?

    (Ans.: Section 3.5)
6. Explain briefly the procedure of pre-qualification of contractors. State its advantages and drawbacks. List the main points to be considered while pre-qualifying the contractors.

    (Ans.: Section 3.6)
7. Write a brief note on:
    (a) advertisement for tenders **(Ans.:** Section 3.8)
    (b) pre-tender conference **(Ans.:** Section 3.8.1)
8. List the information which should customarily appear in an advertisement for tenders.

    (Ans.: Section 3.9)
9. What influence may the following factors have on whether prospective contractors will be interested in submitting the tenders?
    (a) the location of a project
    (b) the character and magnitude of the work
    (c) the name of the owner **(Ans.:** Section 3.9)
10. Explain the purpose of 'earnest money' deposit. **(Ans.:** Section 3.9)
11. Draft a tender notice for the work of construction of a hospital building costing ₹1520 lakhs.

    (Ans.: Section 3.10)
12. Under what circumstances is the time to submit tenders extended? How is this notified to the contractors? **(Ans.:** Section 3.11)

13. Explain, step by step, the procedure followed by a contractor in the preparation of a tender.
(**Ans.:** Section 3.12)

14. Distinguish between a contractor's estimate of cost and his proposal on the work.
(**Ans.:** Section 3.12)

15. Explain the proper procedure for the submission and opening of tenders for construction contracts. (**Ans.:** Sections 3.13 and 3.14)

16. Explain fully the three-cover system of submission of tenders. Give your suggestions in case the two-cover system is to be adopted. (**Ans.:** Section 3.13)

17. Briefly outline the procedure to be followed for submission of tenders on line.
(**Ans.:** Section 3.14)

18. Briefly outline the procedure to be followed while opening tenders that have been submitted in three covers.
(**Ans.:** Sections 3.15 and 3.16)

19. On what basis is the selection of a contractor determined in the case of
    (a) a lump sum contract?
    (b) an item rate contract? (**Ans.:** Section 3.17)

20. Under what circumstances may the withdrawal of a tender be permitted? (**Ans.:** Section 3.17)

21. What is an 'unbalanced lender'? What is the purpose of making such a tender?
(**Ans.:** Section 3.17)

22. What is meant by 'unqualified tender'? What action should the engineer take when a contractor offers conditions in his tender? (**Ans.:** Section 3.18)

23. Your company is required to submit a tender for an irrigation project estimated to cost ₹120 crores and is to be completed in 5 years. Suggest draft conditions to be submitted in the second cover along with your tender for consideration to your company's chief executive (marketing).
(**Ans.:** Section 3.18)

24. Explain and distinguish between: letter of intent (LOI) and work order. (**Ans.:** 3.19)

25. Under what conditions may the lowest tender be rejected? (**Ans.:** Section 3.20)

26. Why may it be desirable to reject all tenders? (**Ans.:** Section 3.21)

27. Should a contract be re-advertised 'as is it' in case the tenders previously received are too high? If not, what procedure should be followed? (**Ans.:** Section 3.22)

Contract Documents

4.1 GENERAL

A contract may consist of a number of documents which contain collectively all the essentials of a contract and which are usually linked together by cross-references. Engineering contract documents usually contain the following:

1. tender notice,
2. general rules and directions for the use of contractors,
3. the letter of submission of tender,
4. the letter of acceptance of tender,
5. the addenda, if any,
6. form of contract,
7. conditions of contract,
8. schedule A—showing the details of materials, if any, to be supplied by the owner to the contractor,
9. schedule B—bill of quantities,
10. specifications:
    (a) general and (b) detailed or particular,
11. contract drawings, and
12. Public Private Partnership (PPP) contracts discussed in Chapter 6 contain, in addition to the above, additional documents mentioned in Section 4.11 below.

Consideration will now be given to the nature and purpose of each of these documents.

4.2 TENDER NOTICE

The tender notice, inviting bids from different builders, is included and forms a part of the contract document. It includes information such as the name of the work, estimated cost, amount of earnest money and security deposit, and time of completion, which is referred to in the conditions of the contract. The main purpose of reprinting the entire advertisement in the contract document is to protect the engineer against the possibility of a claim that he misrepresented the nature and conditions of the proposed work while inviting the tenders.

4.3   GENERAL INSTRUCTIONS

At the beginning of the contract documents, it is necessary to insert the rules and directions instructing the intending contractor as to how the tender should be completed and emphasising particulars, which may vitally affect his interest. The purpose of including the general instructions in the contract documents is to clearly define the requirements of the contract.

4.4   LETTERS EXCHANGED AND ADDENDA

After a tender is opened and before it is accepted, the parties sometimes indulge in negotiations on the price/rates to be agreed and/or conditions to be added or modified. The final contract is obviously subject to the additions or alterations agreed to by the parties and as such it is necessary that the correspondence exchanged and conditions modified are also made part of the contract documents.

4.5   FORM OF CONTRACT

Printed forms are normally used for civil engineering contracts; the language of such contracts being, in general, uniform. The Public Works Department and other public bodies have specific, pre-agreed forms of civil engineering contracts, and these are issued in printed forms. For example, the B-1 form is used for percentage rate contract; the B-2 is used for item rate contract, and so on. For major works, internationally accepted forms and conditions are adopted, for example, the FIDIC (French acronym for the International Federation of Consulting Engineers) form. These forms can be varied by agreement between the parties. After the tender of a contractor has been accepted, the owner and the contractor enter into a formal agreement. After it has been properly signed, witnessed and sealed, this formal agreement is binding on the both the owner and the contractor.

4.6   CONDITIONS OF CONTRACT

A set of well-prepared conditions of contract is invariably included in the agreement. These conditions define in detail the rights, duties and liabilities of:

1.   the contractor who carries out the work,
2.   the owner for whom the construction is to be executed, and
3.   the architects and engineers who design and supervise the work.

The conditions of contract are drafted by the engineer for the owner, and, quite naturally, he has to keep in mind the interest of his principal. The engineer has to ensure that the owner and he himself are protected against any contractor who may intentionally or unwittingly do something that is not fair to the owner. At the same time, it must be remembered that the proper execution of the contract cannot be achieved without the full cooperation of the contractor and, as such, no clauses, which are unfair to the contractor, should be included in the set of conditions. In India this subject has been debated in many seminars. Suggestions and forms were evolved including one by the Planning Commission. However, for local bidding of small projects, the age-old forms with a few modifications are in use. For major projects and especially those funded by the World Bank a different set of conditions, which are fair to all parties such as, the FIDIC are in use. There is no justification for this dual policy and it is recommended that the same form in use for large projects be used for all works—small or big.

It is expected that the contractor, after a careful study of the above-mentioned conditions, will know exactly what is expected of him, and this understanding will go a long way in the smooth execution of the contract. Should it become necessary, the engineer can resort to the provisions made in the contract to force a defaulting contractor to abide by its requirements. However, it must be mentioned here that the engineer should use the various provisions made in the contract only as a protective 'shield' and not as 'sword' to harass the contractor.

Great care needs to be exercised while drafting the conditions of a contract. All the conditions essential to a contract should be included. The wording of each clause should be properly chosen. If necessary, the help of a lawyer may be taken for this purpose since it is very important that each clause be capable of precise legal interpretation. To avoid the task of preparing anew the conditions of contract every time a new contract comes up, the engineer may use a standard set of conditions, which is printed along with the form of agreement. The standard conditions are so drafted that with a little alteration and addition, the same can be adopted for all types of work. For example, in the FIDIC form the variations are incorporated under the heading, 'Conditions of Particular Applications' (COPA). Care needs to be exercised while making alterations, since many of the clauses of a well-drafted form are mutually dependent, and on the insertion of new matter at one place, consequential alterations at other places are likely to be overlooked. Contractors, who are quite familiar with the standard form of conditions, may increase their prices to allow for the additional risks that they feel may arise when an unusual form of contract is adopted. From this viewpoint also, it is preferable to adopt the standard forms.

The conditions of contract are discussed fully in Chapter 6.

4.7   SCHEDULE A—MATERIALS TO BE SUPPLIED BY EMPLOYER

Schedule A is a list of materials, if any, agreed to be supplied to the contractor by the owner. The particulars commonly mentioned in it are listed as follows:

1. *Kind of material:* Cement and steel used to be two examples of materials agreed to be supplied to the contractor by the owner, in cases where the owner was a government department. With the decontrol of these materials and the possibility of their availability in the open market, the practice is stopped in recent contracts. The employer can agree to supply to the contractor any other material which he desires to be incorporated in

the works. The alternative used in some cases is to nominate the supplier in the contract from whom alone the contractor is required to obtain the materials.

2. *Approximate quantity:* Approximate quantities of materials under different categories, likely to be required for the execution of the proposed work, are determined and mentioned against the respective categories.

3. *Place of delivery:* The place at which the contractor will have to take delivery of the materials is normally mentioned. For example, in case of government works, the materials are issued to the contractor at the central stores of the department, from where the materials are required to be conveyed by the contractor to the site of work at his cost.

4. *Rate of issue and mode of payment:* Against each category of materials are written the rates per unit quantity at which the contractor will have to pay for the materials received. These rates are inclusive of the transport and storage charges in addition to the price paid by the owner. The cost of materials is recovered from the amount that becomes due to the contractor under work done bill, every month in the proportion of the approximate quantity of materials actually consumed on work.

5. *Ownership of materials:* The materials supplied by the owner to the contractor are supposed to be the property of the owner and the contractor cannot remove the materials from the site or use them elsewhere. The contractor who is responsible for the proper storage of the materials is also required to keep an account of the materials received and consumed in day-to-day work.

6. *Return of excess materials:* Materials which are supplied in excess of the actual quantities consumed during the execution of the work are to be returned to the owner in a satisfactory condition and the cost thereof, if already paid by him, is refunded to the contractor. In case of cement supplied in bags, it has to be clearly mentioned as to whether or not the rate of supply of cement is inclusive of the cost of the bags. In the latter case, the empty bags are to be returned by the contractor to the owner in a satisfactory condition.

4.8   SCHEDULE B—BILL OF QUANTITIES

The Bill of Quantities (BOQ) is a list of constructional items necessary for completing the work. The engineer has to be very careful in his choice of items for inclusion in the bill of quantities. The items to be included must be definite, readily measurable, of customary type, and of sufficient scope to cover the greatest practicable portion of the entire work. Each item is fully described in the description column and the approximate quantity of each is mentioned in the quantities column. The tenderer has to insert the following in the schedule:

1. tender rate per unit quantity (in a percentage rate contract the BOQ supplied to tenderers includes the rates),
2. cost of each item, that is, approximate quantity multiplied by the bid rate, and
3. the total of all items.

After signing the contract, these statements cannot be changed.

4.9 SPECIFICATIONS

Specifications are written instructions providing a clear and concise description of the materials, labour and workmanship to be employed in or around a construction, and the methods, precautions and so on, required thereof. It is not practicable to include all this information in the limited space of description in the bill of quantities. Similarly, not all the features and requirements can be depicted in the drawings. Every civil engineering contract should therefore contain detailed specifications as a part of the agreement.

The specifications should be as exhaustive as required, but they should not contain irrelevant material nor should they repeat what is shown on the drawings. The specifications are related to the bill of quantities and drawings. The following types of specifications are normally included:

1. general specifications, and
2. particular specifications.

1. *General specifications:* These describe, in general, the nature of the proposed work, the qualities of the materials and the workmanship for the work as a whole. They form a part of the conditions of contract.
2. *Particular specifications:* For every item mentioned in the schedule of quantities, detailed specifications are given, specifying the materials to be used, the workmanship, and the results to be produced, the mode of measurement and payment, and so on. The specifications for the various items are written in the same serial order in which the items are listed in the BOQ. This facilitates cross-reference.

In the absence of complete specifications, the contractor's obligation is limited to the performance of only what is actually called for in such incomplete specifications. As such, great care has to be taken in preparing the specifications. It is not uncommon to give cross reference to printed standard specifications published in the form of books. Because of the importance of specifications in contract documents they form the subject matter for a separate chapter and are fully discussed in Chapter 7.

4.10 CONTRACT DRAWINGS

Contract drawings include three types: (i) tender drawings, (ii) working drawings and (iii) record (as built) drawings.

1. Tender Drawings

A complete set of drawings which give full particulars of the proposed work are prepared prior to invitation of the tender (unless the work is to be let out right from the stage of investigations on a turnkey basis). These drawings, called 'tender drawings', are used to give the contractor the necessary information, which can be transmitted more effectively diagrammatically than by written description. They enable the contractor to visualise the nature of the work and anticipate probable constructional and other difficulties. It is made known that the tender drawings are for the purpose of preparation and submission of tender and 'NOT FOR CONSTRUCTION'. Sometimes these drawings are stamped 'NOT FOR CONSTRUCTION'.

The conditions of contract usually specify that the contractor has to carry out the work as shown in the drawings. The drawings on the basis of which work is to be executed are called 'WORKING DRAWINGS'.

2. Working Drawings

One of the conditions of contract specifies that the drawings, which form an integral part of the contract, are subject to change and the contractor has to carry out the work as per the revised drawings called 'working drawings'. This is likely to result in disputes in future. To avoid such disputes, it is desirable for the owner to preserve at least one set of the original contract drawings for record on the basis of which tenders had been called for. These original drawings are helpful in working out the adjustment in payment to be made to the contractor in case the drawings are revised, for they facilitate the comparison between what the contractor had bid on and what was actually constructed. This condition works both ways, that is, the original set can be used to prove that the contractor's claim for extras is unwarranted.

The working drawings are generally supplied after the work order is issued or along with it. The working drawings can be supplied in a phased manner to match the programme of the work agreed to by the parties so as not to hinder the progress for want of drawings. To that end, the working drawings are expected to be supplied generally 30 days in advance of the commencement of the activity with which it deals. This imposes a great deal of responsibility on the person who prepares the working drawings.

Anything that is shown in the drawings has to be provided for and performed by the contractor, and what is not shown in the drawings is, by implication, left out from the contract. In case of dispute, what is actually shown in the drawings matters more than what should have been shown or what the engineer intended to show.

If the working drawings drastically differ from the tender drawings, extra items may crop up. For example, if the open foundation is shown for a bridge structure in tender drawing and the working drawing proposes pile foundation, the claim for extra item would be justified, subject to other conditions of the contract. However, if only the depth of the open foundation changes, the work may not be extra but additional. Extra work may call for new rate to be agreed where as additional work within the agreed variation limit may not justify new rate.

Unless otherwise specified, the drawings are the property of the owner although the engineer prepares them. Previously, the drawings were made on tracing cloth. Form such drawings, blueprints or ammonia prints could be easily made for use at the site of construction. Nowadays these can be prepared and printed using computers.

In addition to the above mentioned two types of drawings, there is one more set of drawings involved in a typical construction contract.

3. Record (as Built) Drawings

The conditions of the contract generally stipulate that the contractor should submit to the employer a set of 'as built drawings' on completion of the work. These drawings are called record drawings. These drawings can be referred to in the future if any repairs or additions and alterations are to be carried out and in general for maintenance of the work.

The newly evolved PPP contract form includes some documents which are not required in the conventional old forms of contracts. These are mentioned in Section 4.11. The reader is advised to read the contents of Chapter 6, Section 6.11 which would facilitate proper understanding of the contents below.

4.11   PPP CONTRACT DOCUMENTS

In addition to the usual contract documents listed above in this chapter, cash flow projections, giving details of how the contractor intends to generate finance and use it till the recovery of his capital, investment cost and profit during the concession period, forms an integral part of the BOT/PPP contract. The main agreement contains several schedules describing the project site, project facilities, site delivery schedule, design requirements, construction requirements, operation and maintenance requirement, cash flow projections, annuity payment schedule, performance security, state support agreement, substitution agreement and hand-back requirements, and so on.

State support agreement: The implementation of a BOT agreement requires extensive and continued support and grant of certain rights and authorities by the Government. For this purpose the state support agreement is signed and included in the contract documents.

Substitution agreement: The contractor generally obtains the large amount of finance required for the project from financial institutions such as commercial banks called 'lenders'. The lender proposes a person called 'selectee' who is approved by the employer and who is empowered to substitute the contractor for the balance period, in the event of the contractor's failure to cure the default leading to suspension of the concession in the contractor's favour. An agreement signed by the employer, the contractor and the financial institutions to this effect also forms part of the contract agreement.

Hand-back Requirements

This lists the minimum requirements and item-wise inventory of all assets to be handed to the employer by the contractor at the end of the concession period. This and other terms are explained in Chapter 6, Section 6.11.

Questions

1. Prepare a list of contract documents.   **(Ans.:** Section 4.1)
2. Why is it necessary to include the tender notice in the contract documents?

    (Ans.: Section 4.2)
3. What is the advantage of using a standard set of 'conditions of contract'?   **(Ans.:** Section 4.5)
4. What is Schedule A? What information does it contain?   **(Ans.:** Section 4.6)
5. Why should the owner or engineer preserve one set of original drawings?

    (Ans.: Section 4.10)
6. Write short notes on:
    (a) bill of quantities
    (b) specifications
    (c) contract drawings
    (d) working drawings
    (e) record drawings   **(Ans.:** Sections 4.8 to 4.10)
7. Explain briefly why and in what respect PPP contract documents differ from other contracts.

    (Ans.: Section 4.11)

Contract for Engineering and Architectural Services

5.1 GENERAL

When the owner of a proposed project does not have an engineering or architectural staff of his own, he has to make a choice between the following three alternatives:

1. If the proposed project is a small one and, if the owner is an individual, he may select a particular consultant (engineering or architectural) firm and negotiate a contract with it upon mutually acceptable terms.
2. On the other hand, if the project is an extensive one, the owner may find it desirable to engage, on a full-time basis, either an architect or an engineer to handle the project for him.
3. The employer, for a very big project, may advertise and select project management consultants (PMC) for the management of the contract.

In either case, the engineer or architect commissioned to carry out the project is in contractual relationship with his employer.

5.2 OWNER'S CHOICE BETWEEN ARCHITECT AND ENGINEER

Generally, the owner may make a contract with whomsoever he selects. His choice between an architect and an engineer will be governed by the nature of the proposed work. He may select an architect as the overall in charge of the project, if planning in terms of architectural treatment is more important. Examples are, individual houses, apartments, public buildings, monuments, parks and landscaping.

The owner would prefer to place an engineer in overall charge of a project where the engineering features are predominant. Examples are, hydraulic structures, water supply and drainage projects, industrial buildings, bridges and highways.

In practice, it is sometimes found that an individual or an agency is competent, by training and experience, in both these fields of professional work. The owner can then entrust the project in its entirety to such an individual or agency.

5.3   CONTRACT TO BE IN WRITING

The nature of the contract between the engineer and the owner is such that there is not much of a possibility of disputes arising, except on the question of fees. For this reason, formal contracts between the engineer and the owner are not common. However, it is desirable that the contract be in writing. The engineer should insist on a written contract under the seal of a corporation, whenever he has to deal with a limited company. In addition, he has to be careful to ensure that the instructions given to him by such a corporation are within the latter's scope of authority. A specimen agreement form is given in Section 5.9.

5.4   FEES TO BE PAID FOR ENGINEERING SERVICES

To avoid disputes in the future and to safeguard his own interests, the engineer should enter into an agreement with regard to fees at the outset. He should not agree to the payment of his fees being contingent upon the happening of an event in the future (such as starting of the work), for until such an event takes place—and it may never take place, for example, if the owner abandons the work—the engineer will not be entitled to his fees.

There are various methods of working out the amount of fees payable to the engineer for his services. The method commonly employed is to express the amount of fees as a percentage of the cost of the work, the percentage depending upon the nature and type of the work and the amount of labour and skill involved in it. One serious drawback of this method is that there is no incentive for the engineer to attempt to get the work done at the least cost. On the contrary, should an engineer do this work painstakingly and design the project to effect economy in the cost of construction, his reward for the hard work (ironically enough) is a reduced fee.

A better method for all concerned parties would be to engage the engineer on the basis of the cost of preparation of contract documents and execution of the work plus a percentage fee to cover overheads and profit.

5.5   NATURE OF CONTRACT BETWEEN OWNER AND ENGINEER

A contract between the owner and the engineer is a 'personal' contract. The engineer agrees to carry out his contractual obligations himself. In other words, it is a contract of 'agency'. However, when an architect or engineer is first employed on any project, his services will be mainly of an advisory nature. Only after he has prepared the drawings and has been instructed to proceed with the work by his employer, will he have any authority or even opportunity to act as an agent for the project owner. All the provisions made in the Indian Contract Act regarding the responsibilities and liabilities of agents are applicable in the case of a contract between the owner and also the engineer/architect. Some of them are listed in Section 5.6.

5.5.1   Authority of the Architect or Engineer

The authority of the architect or engineer will be generally set out expressly in the contract between the owner of the project and the contractor selected for the construction of the project. In the absence of such express authority, it remains important to determine the extent to which the architect, when carrying out his many duties on behalf of the owner, can commit or bind the employer vis-a-vis the contractor or to third parties. Even where the authority is expressly given in the contract between the owner and the contractor, the architect's decisions on many points will not be binding upon the owner or contractor in view of the arbitration clauses included in such contracts which invariably declare the intention to permit a review of the opinion, decision or certificate of the architect.

The consequences of the limitation on the architect's authority are many. To cite a few by way of example:

1. Acceptance of the work by the architect and/or the work having been carried out in his presence do not prevent the owner from claiming damages for defective work during the period of limitation.
2. Decisions of the architect regarding extra items or rates do not bind the owner or contractor, who can refer the matter to adjudication by arbitration.
3. The granting of extension of time by the architect in no way binds the employer. It must be remembered that many provisions in the contract are included by the owner so that he can ensure additional protection for himself. For example, the FIDIC conditions of contract expressly provide that the engineer shall have no authority to amend the contract and that when a decision of the engineer involves financial liability to the employer, the engineer would take prior approval of the employer.

5.6   LIABILITIES OF ENGINEERS/ARCHITECTS

1. One engineer cannot lawfully employ another engineer (without the consent of the owner) to perform his duties and it follows that, on his death, the contract is automatically terminated. The engineer may, nevertheless, make use of the skill and labour of assistants, but he must always retain complete control of, and assume responsibility for, their acts. The conditions of contract generally define the term 'engineer's representative' and in the field with the project management consultant being designated as the engineer, the representative includes the 'team leader'.

2. So long as the engineer acts within the scope of his authority, the owner is responsible for all his acts.

3. When the engineer does more than what he is authorised to do, and when a part of what he has done is within his authority and can be separated from the whole of his act, the owner is responsible and liable only for that part which he was authorised to do. When the two parts cannot be separarted, the owner is not responsible for the acts of the engineer. It is therefore necessary to ensure that the engineer never makes himself liable by exceeding the authority given to him.

4. The engineer is liable for the following:
    (i) any fraudulent misrepresentation of his authority,
    (ii) making secret profit,
    (iii) receiving a reward or a bribe. (The owner can sue the engineer for recovery of the benefits the latter may have received from illegal transactions with third parties.)

5. The engineer is bound to perform his professional duties with reasonable diligence and skill. The owner is entitled to recover any loss he suffers due to lack of skill or carelessness on the part of the engineer. However, the engineer is not supposed to have abnormal or exceptional skill, that is, if he makes an unfortunate mistake, in spite of due diligence and skill, he is not liable for his mistake, unless the owner proves that he failed to employ reasonable skill and care which are expected of engineers of average ability.

5.7 IMPLIED DUTIES AND LIABILITIES OF ARCHITECTS AND ENGINEERS

The duty owed by an architect or engineer to his employer arises in a contract. The contract of employment may be informal or may be in writing without the duties being expressly defined. The duties which will normally be implied from the fact of employment in the absence of any express terms to the contrary include the following, and the architect is expected to secure the interests of his employer by:

1. producing a design which is skilful and effective in achieving the employer's purpose within any financial limitations he may impose or make known and comprehensive, in the sense that no necessary and foreseeable work is omitted,

2. obtaining a competitive price for the work from a competent contractor, and accordingly placing the contract on terms which afford reasonable protection to the employer's interests in regard to both price and quality of the work,

3. providing efficient supervision to ensure that the work, as carried out, conforms to the design, drawings and specifications, and

4. administering the contract efficiently to achieve speedy economical completion of the project by carrying out duties or exercising authority, specified in or implied by the contract between the employer and the contractor.

If the architect fails to fulfil his main obligations and, if the failure is clear, he may be held responsible to his employer for any damage, which the latter may have suffered. However, the test to be applied would be whether other persons possessing the same professional knowledge and having the same amount of experience and skill would or would not have come to the same conclusion as the architect in this case. In short, he is expected to show reasonable diligence and skill. Some of the important specific duties that an architect is expected to fulfil apart from the above general duties are now discussed.

5.7.1 Preparation of Drawings and Designs

The architect will normally be given a free hand in preparing plans, drawings, designs and specifications, but the owners may press their own ideas upon him regarding materials to be used, plans to be followed, and so on. In situations where the employer's suggestions are likely to lead to an unsatisfactory result, the architect must give sufficient warning and, in most cases, his duty will be discharged.

Where a project involves new techniques and the architect is specifically instructed to use a new method or technique of construction, he is duty-bound to obtain the best advice available upon the use of such a new method or technique and to advise his employer of any potential risks. It must be remembered that, in normal circumstances, an architect or engineer will not be automatically relieved from the liability ensuing from his plans or designs by merely obtaining his employer's approval of them, if the design defect is one relating to construction or is technical in nature.

The measure of damage for breach of the design obligation will differ widely depending on the nature of the breach—it may be nominal if the error can be easily rectified at an early stage; it may be for loss of value or loss of commercial profitability in the case of non-standard suitability breach; and will usually comprise the cost of repair in the case of structural breaches. Since the duty owed by the architect arises in contract, for purposes of limitation, time begins to run from the date of the breach.

The employer is duty-bound to mitigate the damages and, as such, should normally give the architect or engineer an opportunity to correct the design, if an error is discovered at an early stage.

5.7.2 Supply of Drawings, Designs and Instructions

The general practice followed by architects or engineers is to supply relatively generalised drawings for tendering purposes and to supplement them with working drawings after the contract. This practice contains the seeds of possible disputes and could lead to the disruption of the contractor's programme. Normally, the contractor is entitled to receive a full set of working drawings at the commencement of the work or soon thereafter. Specifications for road and bridge works published by the Indian Roads Congress stipulate that drawings for any particular activity shall be issued to the contractor at least thirty days in advance of the scheduled date of the start of the activity. This may be considered as reasonable time recognised by the engineers in the field for supplying the drawings and designs in advance.

The architect may be entitled to a reasonable notice period for supplying drawings, designs or instructions. What constitutes a 'reasonable' period of time is a debatable question and varies from case to case. When the contract specifically stipulates that working drawings, designs and instructions shall be supplied by the architect from time to time and on demand by the contractor, it may be safely presumed that the architect would be entitled to a reasonable notice period, which is dependent upon the time limit within which the contractor is expected to complete the work. In such a case, the contractor should make a request for drawings, designs and details, giving the architect about two weeks' time to supply the requisite material. However, this should not be taken as a hard-and-fast rule.

Architect/engineers are well advised to prepare a schedule for supplying the drawings, designs and instructions commensurate with the total completion time limit and the agreed programme for completion of the work. It may then be easy to maintain a supply of drawings and information without disrupting the contractor's programme and without providing him any ground to put up claims for compensation on this count.

5.7.3 Knowledge of Building Regulations, Bylaws and Rights of Adjoining Owners

The architect or engineer is expected to have a reasonable working knowledge of the law relating to building regulations, bylaws and rights of adjoining owners and also of the requirements as to the serving of notice on legal or any other authorities. If an architect designs or constructs projects without proper reference to the rights of adjoining owners or without regard to any relevant public legislations or bylaws, he may render himself liable for prosecution due to negligence. Again, it should be remembered that the obligations of an architect as far as legal knowledge is concerned will not be of the same standard as those of a legal adviser and will be correspondingly lightened or eliminated if the employer decides to take legal advice in relation to the proposed project.

5.7.4 Excess of Cost Over Estimated Cost

At the time of retaining the services of an architect or engineer, the owner usually indicates or imposes limitations on the cost of the proposed project. Even in the absence of express stipulation, it is commonly believed that an architect is duty-bound to design works capable of being carried out at a reasonable cost, keeping in mind the scope and function of the proposed works. An architect or engineer will be held liable for negligence if, in fact the excess of cost is sufficient to show want of care or skill on his part. However, in a case where an architect has obtained tenders, which are substantially in excess of his estimated costs, it is suggested that, he should normally, be given an opportunity of obtaining further tenders or he should be allowed to propose reasonable modifications or omissions to bring down the price. Although the damage likely to be suffered by the owner at this stage may not be excessive, the architect must realise the importance of correctly estimating the cost.

5.7.5 Selection of Builder/Contractor

Normally, an architect does not guarantee the solvency or capacity of the contractor, but it becomes his duty to make reasonable enquiries regarding the solvency and capabilities of the contractor if the former is responsible either directly on indirectly for the selection of the contractor to carry out the work.

5.7.6 Recommending the form of Contract

An architect must, before it is submitted, familiarise himself with the basic provisions of any contract form that he recommends. Further, he should possess reasonable interest and competence to be able to understand the consequences and effect of the standard form recommended by his own professional body.

5.7.7 Supervision

It is the duty of an architect or engineer to supervise the works and inspect them with sufficient frequency in order to ensure that the materials and the workmanship conform to the contractual requirements. The duty to provide reasonable supervision himself means such supervision as to enable him to certify that the work has been executed according to the contract. The normal practice of an architect is to visit the site about once a week, although this period may vary considerably according to factors such as the nature of the contract, the distance of the site from the architect's office and so on. Depending upon the magnitude of the work, the architect may appoint a resident engineer or an assistant architect for day-to-day supervision at the site of the work. In large projects, with project management consultants handling the work, a number of engineers headed by the team leader are appointed to do the needful. However, from the legal point of view, it is submitted that the architect or engineer will be fully responsible to the owner for mistakes or errors committed by the architect's assistant. This liability may not extend to the errors committed by the subordinates appointed and paid by the employer.

Most conditions of standard contract forms including the FIDIC provide that any failure or omission on the part of the engineer's representative to disapprove any work, material or plant shall not prejudice the authority of the engineer to disapprove such work, material or plant and give instructions for rectification thereof.

If a contractor considers that any decision or direction or communication of the engineer's representative is questionable, he is advised to bring it to the notice of the engineer who is empowered to confirm, reverse or modify the contents of such communication.

5.7.8   Administration of Contract

The terms of the construction contract require an engineer/architect to take a number of actions. In all these actions, with the possible exception of his strictly certifying functions, he will owe a duty of care to the owner. It is suggested that an engineer/architect owes a duty to the owner to obtain his prior approval of the variations and their probable costs in all but trivial or emergency matters.

It is important to note that any instructions given by the engineer which are likely to result in cost implications must be issued in writing. In an emergency, the instructions can be given orally but they need to be confirmed in writing subsequently. If the engineer fails to confirm in writing, the contractor should put the said instructions in writing and submit them to the engineer and if he does not contradict the same, the instructions would be deemed to be given in writing by the engineer.

5.7.9   Issuing Certificates

Under the terms of a typical building contract, an architect/engineer acts in a dual capacity, namely, as an agent and as a quasi-arbitrator. In his capacity as an agent he is liable for negligence, but in his capacity as a certifier, in so far as he is determining a dispute formulated between the parties, he owes no duty of skill or care to his employer. This is because a potential liability of this kind would interfere with the free and unfettered exercise of his judgement while performing his function as a certifier. However, since such certificates or decisions are invariably open to scrutiny by an arbitrator, it is worthwhile to remember that the architect is duty-bound to apply the provisions of the contract exactly and strictly to the situation upon which he is required to give his decisions and his failure to do so will render his certificate or decision invalid.

One of the main duties of an engineer is to issue interim payment certificates to the contractor. Generally, the contractor is expected to submit his bill each month or fortnightly for scrutiny and issual of a certificate by the engineer. The engineer on receipt of such a bill is to deliver to the employer within the agreed time schedule, generally not more than four weeks, an interim payment certificate stating the amount due for payment to the contractor subject to the retention money and other bona fide deductions contractually due.

Before issuing the certificate, the engineer is entitled to demand from the contractor reasonable proof that all the payments due to nominated subcontractors or suppliers have been paid. The engineer is empowered to direct the payment to be released to the nominated subcontractor or supplier by the employer and deduct the same from the money due to the contractor.

The engineer is generally empowered by the terms of the contract to modify or correct any interim payment certificate previously issued in any subsequent interim payment certificate.

On completion of the work, the engineer has to check the final bill submitted by the contractor and issue a final payment certificate after giving credit to the employer for all amounts previously paid and all other payments that the employer may be entitled to.

The contract subsists till the issue of the defects liability certificate by the engineer to the employer. The engineer is to issue such a certificate generally within a month of the elapse of the defects liability period mentioned in the agreement.

5.8   DETERMINATION OF AUTHORITY OF ARCHITECT OR ENGINEER

The authority of the engineer is determined by the following factors:

1. the completed performance of his contract,
2. the expiration of the period for which it was given,
3. the death of the owner,
4. the owner becoming of unsound mind,
5. the owner being adjudicated an insolvent under any act in force, and
6. winding up of a registered company, which is the owner.

Much depends on the special terms of the contract concerned. Unless contrary provision is made in the contract, the contract of the engineer or the architect with the owner may be determined at any moment by one of the parties giving notice to the other without sufficient cause. However, the termination of the contract by either party operates without prejudice to any claim for damages that either party may have, one against the other, for breach of contract of employment. For example, if the owner terminates the services of the engineer and appoints another in his place, the engineer has no legal power to restrain the owner from doing so, his only remedy will be to claim damages from the owner for breach of contract of employment.

5.9   SPECIMEN AGREEMENT FORM

(The specimen agreement form given below may be used as a guide and after suitable modifications as and where necessary.)

THIS AGREEMENT made at ______________ (place) this ________________(day) of ______ (month) in the year two thousand . . . by . . . . and between . . . . (Owner) of . . . . (address) hereinafter called 'the Owner' of the one part and . . . . (name and address of the Architect) hereinafter called 'the Architects' of the other part.

WITNESSETH that whereas the Employer intends to construct/develop . . . . (name and place of the work) hereinafter called 'the work'.

NOW, THEREFORE, the Owner and the Architect, for considerations hereinafter named agree as follows:

A. The Architect agrees to perform, for the above named work, professional services as hereinafter set forth.

B. The Owner agrees to pay the Architect for such services a fee:
   - for building work . . . . %
   - for development work, that is, (roads, water supply and drainage) . . . . %
   - for layout . . . . %

   (Score out inapplicable part)

of the total construction cost of the work with other payments and reimbursement, as hereinafter provided, the said percentage being hereinafter called the 'basic rate'.

The parties hereto further agree to the following conditions:

1. *The Architect's (Engineer's) Services*
   The Architect's normal professional services consist of those described in the Schedule(s) A, B and C, appended herewith. Additional and special services not covered by this Agreement and the basis fee are also described in the said schedules).

2. *Reimbursements*
   The Owner is to reimburse the Architect the cost of transportation and living incurred by him and his assistants while travelling in discharge of duties connected with the work subject to the condition that no costs on this account shall be payable if the work is situated within a radius of . . . km from the office of the Engineer/Architect.

3. *Extra Services*
   If the Architect is caused extra drafting or other expense on account of additional and special services mentioned in Article 1 or due to changes ordered by the Employer during the actual execution of the work, he shall be equitably paid for such extra expense and the services involved. If any work, designed or specified under this Agreement, by the Architect is abandoned, or deferred in whole or in part, the Architect is to be paid for services rendered on account of it.

   If the whole or any part of the work which has been postponed, cancelled or abandoned is again proceeded with and additional services by the Architect are necessary in connection with the resumption of such work or any part thereof, the Architect shall be entitled to charge for such additional work on a time basis.

   If the Employer executes the work by purchasing materials, employing labour gangs, hiring plant and so on, directly and without letting out the work on a contract basis, the Architect shall be entitled to additional fee on a time basis for giving instructions or providing service which he is not normally required to provide when the work is executed by an independent and experienced contractor.

4. *Payments*
   Payments to the Architect on account of his basic fee shall be made in stages and at the rates of percentage shown in Schedule(s) A, B and C, subject to the provisions of Articles 3 and 5. Payments to the Architect, other than those on his basic fee, fall due from time to time as his work is done or as costs are incurred.

   No deductions shall be made from the Architect's compensation on account of liquidated damages, or other sums withheld from payments to Contractors. The schedules of charges apply to work let under a single contract. For any portion of the work let under a separate contract, the Architect is entitled to increase his charges by 50 per cent in respect of the work so let. If substantially all the work is so let out, the higher rate shall apply to the entire work.

5. *Construction Cost of the Work*

   The total construction cost of the work as herein referred to will be based on cost, as certified by the Architect, of all works including built-in furniture, installation and equipment, drainage, water supply, electricity works, painting works, compounds, and so on, executed under his direction subject to the provisions of Article 3 and the following conditions:

   5.1 The total construction cost shall not include Architects or Engineer's or Special Consultant's fees or reimbursements or the salary and allowances paid to a clerk of works.

   5.2 When any material, labour or carriage, is furnished by the Owner below its market costs, the cost of the work shall be computed upon such market costs.

   When work executed by the employer shall be other than item rate contract on lump sum basis, the total expenditure shall include all costs at the site which may have been incurred by a competent contractor on purchase of materials, labour, machinery and equipment and overheads.

   5.3 Where appropriate, the cost of old materials used in the work shall be computed as if they were new.

   5.4 Until the total cost of constructon is finally worked out in accordance with this Agreement, the cost estimated by the Architect shall be the basis for periodical payments to the Architect as mentioned in Article 4. The Architect does not guarantee that the total construction cost of the work shall not vary from such estimated cost.

   5.5 When the professional services of the Architect during the construction stage are not required by the the Owner, the Architect's own estimate based upon current market rates shall be taken as the total cost of construction.

6. *Information Furnished by Employer*

   The Owner shall, at his own cost, so far as the work under his agreement may require, and consistent with the provisions of Articles 1 and 3 furnish the Architect with the following information:

   A complete and accurate survey of building site(s), giving the grades, and lines of streets, pavements and adjoining properties; the right restrictions, easements, boundaries and contours of the building site and full information as to sewer, water and electric services. The Owner is to pay for borings or test pits or other tests when required.

   Additional services performed by the Architect and other consultants in assisting the Owner to provide such data and information and not covered by the Architect's normal services mentioned in Article 1 shall be the subject of separate additional charge on a time basis.

7. *General Administration*

   The Architect will endeavour by general administration of the construction contracts to guard the Owner against major defects and deficiencies, during execution of work.

The general administration of the Architect is to be distinguished from continuous on-site supervision of a clerk of works. For continuous on-site supervision the Owner shall employ at his own cost a clerk of works acceptable to both Owner and Architect at a salary agreed to by the Owner and paid by the Owner, upon presenting of the Architect's monthly statements.

8. *Copyright and Ownership of Documents*

   All drawings and contract documents and copies thereof as instruments of service are the property of the Architect whether the work for which they are made is carried out or not. Copyright in all drawings and contract documents as well as the works executed in accordance with them shall remain the property of the Architect and shall not be used on other work except by agreement with the Architect.

9. *Termination of Contract*

   This Agreement may be terminated by either party at any time by giving one month's notice to the other party. Upon expiry of the notice period, the Agreement shall stand terminated and the Architect shall be entitled to remuneration in accordance with the provisions of Article 4. No party shall be entitled to claim damages from the other on account of such termination of the Agreement.

IN WITNESS WHEREOF the parties hereto have hereunto set and subscribed their respective hands (and seals) the day and year first herein above written.

Signed, sealed and delivered by the within-named   **OWNER**
Owner in the presence of

[1]

[2]

Signed, sealed and delivered by the within-named   **ENGINEER/ARCHITECT**
Architect in the presence of

[1]

[2]

Schedule A: For Building Construction Works

Architect's Normal Professional Services, Special Services are not included in this Agreement and Terms of Payment of Fee. (See Articles 1, 3, 4, 5, 6 and 9.)

A. *Architect's Normal Professional Services*
   1. The Architect will take instructions from the Owner for preparing rough sketches and will prepare final sketch designs for the Owner's approval.
   2. The Architect will submit drawings, notice forms, and so on, for approval from local authorities.
   3. The Architect will obtain approval from local authorities for plans.

4. The Architect will take instructions from the Owner and prepare an approximate estimate of the cost of proposed work after preparing the working drawings. The Architect will call for tenders, if necessary, on the Owner's behalf and at the Owner's cost and will advise him on the selection of a contractor.
5. The Architect will assist in drawing up the contract agreement between the Owner and contractor (if necessary).
6. The Architect will give such periodical supervision and inspection as may be necessary to ensure that the works are in accordance with the contract drawings. For constant supervision, the Owner shall emplopy at his own cost a clerk of works nominated or approved by the Architect.
7. The Architect will advise the Owner on payment to be made to the contractor from time to time.
8. The Architect will obtain the completion certificate from the local authorities for the work.

B. *Special Services Not Included in this Agreement*
   The following are to be special services not included in this Agreement:
   1. preparation of perspectives for publication, and so on,
   2. preparation of scale models,
   3. copies of drawings in addition to the two sets normally required,
   4. services of specialists, if required, with approval of the Owner,
   5. major changes in design and revisions,
   6. collection of data, information, and documents required,
   7. services rendered after lapse of 'execution period,
   8. services mentioned under Schedules B and C, and
   9. obtaining Non-Agricultural (NA) Permission.

C. *Agreed Scale of Professional Charges*
   The estimate cost of construction is ₹______________ The said fees shall be paid in stages as follows:
   Stage

	Portion of total fee
1. A retainer to be paid in advance on taking instructions from the Employer	1/12th
2. On preparation of final preliminary drawings	1/6th
3. On submission of plans and notices to local authorities	1/12th
4. On approval of plans, and so on, from local authorities	1/12th
5. On preparation of working drawings and on calling for tenders	1/12th
The balance to be paid from time to time as the works proceed on bills of the contractor provisionally at half the total percentage agreed.	1/2

On submission of the final bills of all contractors or completion of the work when executed by the Owner, without employing the contractor, account will be made and final payment, if any, adjusted so that the total fee does not exceed the agreed percentage of the 'total construction cost' of works due and payable in full before obtaining the completion certificate from the local authorities.

(Provided that, if the work is executed by the Employer himself without employing the contractor, the balance fee is payable at the mutually agreed intervals during the progress of the work.).

Schedule B: For Developmental Works

(Architect's Normal Professional Services and Terms of Payment of Fee. (See Articles 1,3,4, 5, 6 and 9 of the Agreement.)

A. *Architect's Normal Professional Services*
   1. The Architect will take instructions from the Employer and prepare final drawings for roads/water drainage according to the layout sanctioned by the local authorities.
   2. The Architect will submit drawings, detailed specifications, bill of quantities and approximate estimates for the approval of local authorities.
   3. The Architect will obtain approval of local authorities for plans and estimates with respective departments.
   4. The Architect will call for tenders, if necessary, on the Employer's behalf and at the Employer's cost and will advise him on selection of a contractor.
   5. The Architect will assist in drawing up a contract (agreement) between the Owner/ Society and the contractor.
   6. The Architect will give such periodical supervision and inspection as may be necessary to ensure that the works are in accordance with the contract. If constant supervision is desired, the Owner shall employ at his own cost a clerk of works nominated or approved by the Architect.
   7. The Architect will advise the Owner/Society on payment to be made to the contractor from time to time in accordance with the terms of the contract.
   8. The Architect will obtain the completion certificate from local authorities for the work with the respective departments.

B. *Special Services Not Included in this Agreement*
   The following shall be special services not included in the Agreement:
   1. collection of data, information and documents required,
   2. major changes in the design and revisions and renewal after one year,
   3. copies of drawings and contract documents in addition to the two sets normally required,
   4. services of specialists, if required, with the approval of the Owner,
   5. services rendered after lapse of the 'execution period', and
   6. services mentioned under Schedules A and C.

C. Agreed scale of professional charges
    This may be similar to the scale and stages stated in Schedule A.

Schedule C. For Layouts and Subdivisions

(*Full draft is not given and left as an exercise. Hints are indicated below*)

The normal services are limited to taking instructions from the Employer, preparation and submission to the local authorities the final scheme for approval and getting the scheme approved.

The normal services do not include services such as demarcation, preparation of models, and so on.

The fee agreed is payable fifty per cent in stages till approval of the layout and the balance on the approval of the layout by the local authority.

Questions

1. What steps should the owner take if he does not have his own engineering staff to execute a project?   (**Ans.:** Sections 5.1 and 5.2)
2. What precautions will you take while planning to contract for rendering your personal services as an engineer?   (**Ans.:** Sections 5.3 and 5.4)
3. A consulting engineering firm is to be engaged to investigate, design and prepare plans and estimates for the construction of a large bridge. What should be the basis of making the payments for these services? Why?   (**Ans.:** Section 5.4)
4. Write a brief note on: (a) the nature of contract between the owner and engineer; (b) the authority of the architect/engineer.   (**Ans.:** Sections 5.5 and 5.5.1)
5. What are the duties and liabilities of architects and engineers?   (**Ans.:** Section 5.6)
6. Briefly explain the following:
    (a) Responsibility of the owner for the acts of architect/engineer
    (b) Liability of an engineer for his mistake in designing and for administration of a construction project for an employer
    (c) Liability of an architect in respect to preparation and supply of drawings and designs
    (d) The architect's liability to produce the design within the owner's estimate and budget
    (e) Duty of an architect in respect of:
         (i) Selection of contractor
         (ii) Supervision
         (iii) Administration of contract
         (iv) Issuing certificates   (**Ans.:** Section 5.7)
7. State the different ways in which the contract between the engineer and the owner may be determined.   (**Ans.:** Section 5.8)
8. Prepare a list of 'architect's normal and special professional services' for building construction works.   (**Ans.:** Section 5.9; Schedule A)

Contract between Owner and Contractor

6.1 GENERAL

It would be worthwhile, before proceeding further, to consider the definitions of, and the relationship between, the two parties involved in an engineering contract.

1. *Owner/Employer*
    The owner or employer in a building or engineering contract is the individual, firm or public body which finances a work. The owner is the most important party in the whole of a contract system because it is he who finances the work. He is also designated as 'employer' because it is he who employs all concerned for converting the blueprints of a project into reality.
2. *Contractor/Builder*
    The contractor or builder is the person or firm who, in pursuit of an independent business, undertakes to supply certain materials, or to perform any work or service at an agreed price or rate, for others. He does not submit himself to the control of others but does the work in a manner he thinks best.

6.2 RELATIONSHIP BETWEEN EMPLOYER AND CONTRACTOR

Though the owner employs the contractor, for execution of his project, the relationship between the two is not that which exists between a master and his servant. The master is liable to third parties for the misconduct of his servants and also possesses the authority to direct in respect of what is to be done, and how, where and when it is to be done. The contractor, on the other hand, is one who does not submit himself to the orders of the owner in respect of means, methods and manner by which the result is to be obtained except to the extent stipulated in the contract but only as to the result of the work. Also, the owner, who employs the contractor, is not liable for the latter's misconduct with regard to the work he has let out on contract. In short, the relationship between the two is that of two equal parties bound by their contract and as such can best be described as 'contractual'. It can be compared to a contract for sale of a product except that in this case the product say, a building, bridge, tunnel, highway or dam, is sold before it is built. The contractor undertakes to build the product and pass it on to the owner but at the cost predetermined by the agreement. For this reason the contract is complex in nature and consists of several documents which collectively define the duties and liabilities of each party to the contract. There are, therefore, many forms or types of contract.

6.3   TYPES OF CONTRACT

Contracts for the execution of civil engineering works may be broadly classified as under:

1.   all in contract/entire contract,
2.   lump sum contract,
3.   item rate or unit price contract,
4.   percentage rate contract,
5.   cost plus percentage rate contract,
6.   cost plus fixed fee contract,
7.   cost plus fluctuating fee contract,
8.   target contract,
9.   labour contract, and
10.   BOT/PPP contract.

Each of the above types is discussed in the subsequent sections.

6.4   ALL IN CONTRACT/ENTIRE CONTRACT

This is a rare form of a contract but is considered first because in this, the owner ceases to be the promoter and delegates a large firm or consortium to perform both design and construction. In this form of contract, the owner specifies his requirements and also the broad and general outline of the proposed work and the contractor has to submit full particulars of detailed investigations, designs and construction cost including maintaining the work for a limited period. The two parties agree to the terms and conditions of executing the project through all the above phases. This form of contract is suitable to some exceptional types of works and is seldom adopted for normal works. The works most suited to this form are industrial facilities where firms are in market with say patented processing plant.

6.5   LUMP SUM CONTRACT

A lump sum contract is the oldest form of contract and is still popular in USA. Its essential features include:

1. *Nature of Agreement*

    In a lump sum contract, the contractor agrees to carry out the entire work as shown in the drawings and described by specifications, supplying labour and materials, all for a specified lump sum. Sometimes, the agreement makes a provision for adjusting the 'fixed sum' allowing for the cost of extras, variations, omissions, and so on.

    The main feature of the agreement is that the contractor agrees to fulfill all his contractual obligations for the stipulated payment, no matter what trouble or expense he encounters or incurs in doing so.

2. *Contract Documents*

    All the documents, as outlined in Chapter 4, except the 'bill of quantities', form the basis of the agreement. A schedule of rates is sometimes included to work out the cost of extras or omissions.

3. *Mode of Payment to Contractor*

    A lump sum contract is usually an 'entire' contract and, as such, no payment can be recovered by the contractor until the whole of the work is completed. In case the contractor abandons the work before completion, he is not entitled to payment for the portion of work already done, however substantial it may be.

    But, invariably, the agreements include a special provision for a series of partial payments to the contractor in stages as the work progresses, rather than final settlement after the acceptance of the work by the owner. This helps to reduce the cost of the project inasmuch as the contractor would not be required to consider the interest on his investment and load it in his lump sum. The engineer's interim certificates form the basis of payment to the contractor.

4. *Advantages*

    From the owner's standpoint, such a contract has the following advantages:

    (i)   If no extras are contemplated, the tenders inform the owner exactly what the project will cost him. This provides a sound footing on which he can take a decision either to start the project or abandon the same.

    (ii) He need not employ a staff to maintain periodical accounts of the contractor's materials, labour and output. All that the engineer has to do on the owner's behalf is to see that the work is being executed exactly according to the terms of the contract and issue interim certificates to the contractor.

    (iii) From the contractor's standpoint, the greatest advantage of this form of contract is that whatever benefits the contractor can gain by excellent planning and efficient management on the site of work are his own.

5. *Drawbacks*

    This form has, however, the following drawbacks:

(i)   Before a contract is let out, the project has got to be thoroughly investigated and all the contract documents kept ready in every respect. This entails costly and time-consuming work, which is often difficult to accomplish.

(ii) If the nature, extent and details of the work are not properly defined by the contract document, many additional features may have to be determined and provided for as the work progresses. These features, not being part of the original agreement, give an opportunity to the contractor to claim payment at abnormal rates.

To overcome this drawback, a schedule of rates covering items not included, but likely to be provided for in the contract, is agreed to both by the owner and the contractor. This schedule is made part of the contract documents. Any extra work carried out as per the written orders of the engineer is measured and paid for over and above the fixed sum at the rates mentioned in this agreed schedule of rates. Similarly, the amount to be deducted for the work omitted is also worked out. In this form the contract is known as: Lump sum with Bill of Quantities.

(iii) The contractors may deliberately submit high tenders to protect themselves from the uncertainties of the work.

6. *Suitability of the Lump Sum Form of Contract*

A lump sum contract can be used successfully for the construction of such works wherein what is to be constructed is known exactly at the time of tendering and no change is likely to occur. It is also suitable to works with respect to which the contractors have had considerable experience and whose cost they can predict with reasonable accuracy. Examples of such works are public and private buildings, warehouses and workshops.

This type of contract is not suitable for works whose extent and nature cannot be predicted in advance. It would then be unfair to make the contractor assume all risks and uncertainties. Examples of works unsuited to a lump sum contract are:

(i)   Works involving difficult foundations, that is, where it is not possible to know in advance the characteristics and quantity of excavation, de-watering of foundations, shoring, and so on.

(ii) Emergency projects that have to be rushed through without adequate time to prepare a complete set of contract documents.

(iii) Works, such as additions and alterations to existing structures, which involve the maintenance of operations while the work is being performed.

(iv) Works subjected to unusual and unpredictable hazards, such as floods, which are beyond the control of contractors.

6.6   ITEM RATE OR UNIT PRICE CONTRACT

This contract falls within the group of measurement contracts and is the most common form in use. Its essential features are:

1. *Nature of Agreement*

    An item rate contract is one in which the contractor agrees to carry out the work as per the drawings, bills of quantities and specifications in consideration of a payment to be made entirely on measurements taken as the work proceeds and at the unit prices tendered by the contractor in the bill of quantities.

    A bill of quantities is prepared, giving, as accurately as possible, the quantities of each item of work to be executed and the contractor enters the unit rate against each item of work. The basis of agreement is thus the unit rate of each item, certain reasonable variations in the quantities being accepted by both parties. For example, the bill of quantities includes an item of cement concrete and the estimated quantity is 11,000 m^3. The contractor's tendered rate represents the amount that he is to be paid for each cubic metre of concrete in the finished work, whether the actual quantity is 10,000 m^3 or 12,000 m^3. The variation limit is plus/minus 25% of the estimated quantities of an item of work mentioned in the Bill of Quantities, for which the rate will remain unchanged.

2. *Contract Documents*

    All the documents, as mentioned in Chapter 4, are invariably included in the agreement.

3. *Mode of Payment to Contractor*

    The contractor has to take measurements of the work carried out, prepare the bill and submit it periodically (generally once every month). The engineer either accompanies the contractor or his representative while the measurements are being taken or checks the measurements and quantities afterwards. If no discrepancy is found, he recommends the bill for payment. The amount recommended by the engineer is the cost of the work less the retention money and the cost of materials supplied to the contractor by the owner and actually used on the work during the period under consideration. If any item which is not included in the bill of quantities, occurs during execution, it is usually possible to construct that item by a combination of other items already in the bill, otherwise it is paid for as an 'extra item' at a specially analysed rate.

4. *Advantages*

    An item rate contract has the following advantages:

    (i) The owner can avoid the delay that becomes necessary in making a large number of contract drawings to show in detail everything that would be needed as in the case of a lump sum contract. The detailed drawings can be prepared after the contract is awarded, but sufficiently in advance to enable the contractor to secure all information in time.

    (ii) The bill of quantities, which forms a part of the contract documents, greatly assists in keeping the tendered sum as low as possible. This is obvious because in the absence of the bill of quantities (for example, lump sum contract), every contractor submitting the tender has to assess the amount of work involved, usually in a very short period, in addition to the normal work. Under these circumstances the contractor, unless he is in great need of work, is almost bound to price high in order to allow himself a sufficient margin of cover for any items which he may have missed.

(iii) This type of contract allows, within limits, variations in the quantities of different items. Thus, what the owner pays to the contractor is the actual cost of the work at the agreed rates. This arrangement is fair to both the parties.

5. *Disadvantages*

   (i) The owner cannot be absolutely sure of the total cost that he would have to incur until the work is completed. In case the quantities mentioned in the bill are found to be inaccurate, the cost of the work will vary considerably from the estimated cost. If the actual cost increases, the owner may be put into financial difficulty even leading to suspension of the work.

   (ii) Both the owner and the contractor have to do considerable computation and book-keeping during the progress of the work. Both the parties are required to appoint staff to record the measurement of the work done.

   (iii) As the quantities are likely to vary, there is a possibility of the contractor submitting an unbalanced tender on the basis of shrewd anticipation or perhaps outside information. If the contractor's anticipation proves to be correct, the owner would stand to lose heavily. A contract of this nature therefore requires very careful scrutiny and consideration by a skilful and experienced engineer or architect before it is entered into.

   (iv) The 'extra items' are often a source of trouble. The contractor invariably presses for higher rates than he would have tendered in the beginning, and, in most of the cases, the owner has to meet his demand to some extent that increases the cost.

6. *Suitability of Item Rate Contract*

   Item rate contracts are widely used in the execution of large works financed by public bodies or the government. This form of contract is suitable for works which can be split into separate items and the quantity under each item can be estimated with reasonable accuracy such as buildings, roads, bridges, tunnels, and so on. By carrying out a careful scrutiny to avoid acceptance of an unbalanced tender, this type of contract can be conveniently used for works which are unsuited to lump sum contracts.

6.7   PERCENTAGE RATE CONTRACT

This form of contract is similar to the item rate contract in almost all respects except the method of tendering the unit rates. The bills of quantities supplied to all the intending bidders include, in addition to the approximate quantities and detailed description, the unit rates as estimated by the engineer. While tendering, the contractors do not have to write the rate or estimated cost of each item, but a percentage figure by which the estimate unit rates are to be increased or decreased, the same percentage figure being applicable to all the items.

Contract documents are the same as those included in an item rate contract. At the time of making the payments to the contractor, the engineer works out the cost of the work done in a manner similar to an item rate contract by multiplying the quantities by the rate in the bill of quantities, but increases or decreases the final amount by the percentage figure above or below quoted by the contractor, respectively.

Advantages

The only additional benefit claimed by this form over the unit price contract is that there is no scope for a contractor to submit an 'unbalanced tender'. The work of scrutinising the tenders before acceptance is much simplified. For the above-mentioned reason, this form is used in preference to the item rate contract for works where the estimated quantities are uncertain and likely to change considerably.

Discussion in respect to all other points of the item rate contract applies to this type also.

6.8   COST PLUS TYPE CONTRACTS

It is seen while discussing the above common forms of contracts that neither party is sure of the final cost of the project as will be built, till the completion of the project. In these days of spiralling price rise and absence of a satisfactory solution to ascertain and pay escalation in the cost on mutually agreed basis, contracting can be reduced to gambling. If the contractor is likely to lose heavily, he abandons the work midway to the detriment of the employer. If the employer has to pay in excess of the agreed contract sum he runs the risk of financial difficulty. An alternative form of contract, which does not throw much risk on either party, is the cost plus type contract.

1. *Nature of Agreement*

    The cost plus type contracts, as the name suggests, are based on the payment of actual cost plus something additional to cover the contractor's overheads and profit. In other words, the owner agrees to pay to the contractor, the actual cost of the work plus an agreed percentage of the actual or allowable cost to cover overheads, profit, and so on. Those items, expenses under which will constitute the actual cost, are to be carefully defined in the agreement.

    The total cost of a construction project when broken down can be grouped under the five heads of 'M' as follows:

    (i)   materials,
    (ii)   men, that is, cost of labour,
    (iii)   machinery and equipment deployment,
    (iv)   management: called 'overheads', and
    (v)   money: that is, profit to the contractor.

The first three heads are called direct costs, which can be ascertained with reasonable degree of certainty for each item of work. In a cost plus type contract, the cost should be defined to mean the direct cost alone. The last two heads should be agreed on the basis 'plus' amount payable. This plus can be agreed as a fixed fee, or percentage of the direct cost or fluctuating fee giving rise to different forms of contract under the cost plus type.

Ordinarily, the plus to be paid to the contractor should cover the costs of the following:

(i)   salaries of the contractor's supervisory staff,

(ii) contractor's general office expenses such as stationery, postage and telephone accounts,

(iii) salaries or portion of salaries or travelling expenses of the officials of the firm who may visit the work, and

(iv) charges for the use of any equipment that the contractor would not normally use for the performance of the work.

The reason for not including the above (i) to (iv) heads in the cost is to induce the contractor to effect efficiency in managing the project and keeping the overheads down. If he quotes high against the 'plus' he may not get his tender accepted. Once he quotes a 'plus', he will try to maximise his profit by keeping the costs of plus items down.

2. *Contract Documents*

Of the various contract documents mentioned in Chapter 4, the bill of quantities, drawings and specifications are seldom ready when the contract is entered into. The contractor agrees to execute the work in accordance with the drawings, specifications and other necessary information that will be supplied to him from time to time during the progress of the work.

3. *Mode of Payment to Contractor*

The contractor generally keeps all records of costs, and presents them periodically to the engineer (who also maintains similar accounts) for checking and approval. The owner, on the advice of the engineer, makes the payments to the contractor as the work proceeds. Obviously, the payment consists of the actual cost incurred plus the agreed amount for overheads and profit.

4. *Advantages*

The cost plus contracts, in general, have the following advantages:

(i) *Early completion of the work:* The work can be started even before the drawings, designs, estimates and specifications are ready, these items being prepared as the work progresses. Also, decisions can be taken quickly. These factors lead to the rapid execution of the work.

(ii) *Quality of work:* The contractor is assured of a reasonable amount of profit even though the prices of materials and the labour charges are subjected to fluctuations.

Similarly, the use of another type of material, which is inferior to that specified and the hurried completion, resulting in poor workmanship, does not increase the contractor's profit. The contractor will be induced to perform the work in the best interests of the owner. The final result would, therefore, produce a better quality work.

(iii) *Disputes arising due to extra work are eliminated:* No work, or a portion thereof, that the contractor is directed to perform can be called an 'extra work'. This flexibility allows the adoption of alternative ways and methods of construction in order to choose the most economic one for the work as a whole. This may reduce the cost of the work.

5. *Drawbacks*

Although the cost plus percentage rate contract appears to be an attractive contract from the foregoing discussion, it suffers from many severe drawbacks, some of which are mentioned below:

(i) *Lack of incentive:* There is no incentive for the contractor either to complete the work speedily or to effect economy in the construction cost by proper planning and efficient management. On the contrary, any increase in the construction cost due to delay, wastage of materials, changes in the drawings and designs results in increase of his profit, if he has quoted plus in the form of percentage of the cost.

(ii) *The final cost of the work:* Like other contracts (except the lump sum type contract where no extra or changes are ordered), here also, the final cost of the project to the owner cannot be foretold. This may lead the owner to financial difficulties.

(iii) *Account keeping:* Both the parties have to do a lot of account keeping regarding the materials purchased and used, the labour and plant employed and other miscellaneous expenses incurred during the course of the work.

(iv) *Illegal with respect to public works:* Where the owner happens to be a public body or a government department, this form of contract is not being adopted except during emergencies.

6. *Suitability of Cost Plus Type Contract*

In spite of the serious drawbacks pointed out above, this form of contract can be used suitably for:

(i) Emergency works that require construction in a hurry without time to develop plans for it. In the absence of plans and estimates, the other forms of contract as discussed earlier cannot be used.

(ii) Works (in situations which are not emergencies) wherein no one can foretell with certainty just what troubles would be encountered and wherein correct decisions cannot be taken except during the progress of the work.

(iii) Construction of expensive structures, for example, palaces or similar structures where the cost of the work is of no consequence, but the materials to be used and the methods to be adopted are to suit the choice and taste of the owner.

7. *Selection of Agency*

The owner (through his engineer/architect) should exercise great care in selecting a contractor to carry out the work for him. A person (or firm) of known integrity, who (which) will not exploit the weakness of this type of contract to his (its) advantage, will be the best choice.

Types of Cost Plus Contract

As already stated the cost plus contracts are divisible into different types depending upon the mode of agreement of deciding the plus to be paid to the contractor.

6.8.1   Cost Plus Percentage Contract

In a cost plus percentage contract, the employer agrees to pay to the contractor the actual cost incurred by him for materials, labour and machinery equipment and so on, plus an agreed percentage of the said cost. The percentage may vary from 10% to 30% depending upon the magnitude of the work. The larger the project, the lesser will be the agreed percentage.

The main drawback of this form as already stated is the lack of incentive to the contractor to keep the cost down and execute the work economically by reducing the wastage.

6.8.2   Cost Plus Fixed Fee Contract

The cost plus fixed fee contract differs from the cost plus percentage type contract in respect to the determination of the fee to be paid to the contractor to cover overheads and profit. The agreement specifies the fixed lump sum to be paid to the contractor by the owner over and above the actual cost of the work. In other words, the fee does not fluctuate with the actual cost of the work. This factor overcomes the possible weakness of the cost plus percentage type contract.

In this case, however, there is no incentive for the contractor to do the work efficiently and to effect economy in the cost of construction of the work. If the fixed fee is to include the salaries of the contractor's staff, the contractor will try to complete the work as speedily as possible so as to make maximum profit. Even if the salaries of his staff are to be classified as a part of the general costs of the work, he will try to expedite the project so that his men can be available for another contract to make additional profit therefrom. In respect of all other points, this form of contract is similar to the cost plus percentage type contract.

6.8.3   Cost Plus Fluctuating Fee Contract

By introducing an element of incentive for the contractor to carry out the work in the most economical way, an attempt is made in this form of contract to overcome the main drawbacks of the previous two types of cost plus contracts. This is achieved by suitably changing the nature of the agreement in respect of the fee payable to the contractor. The amount of fee is determined by reference to some form of a sliding scale. Thus, the higher the actual cost, the lower is the value of the fee that the contractor receives and vice versa. From the owner's point of view this is one of the best of the 'cost plus' type contracts. Its developed form is a target contract.

6.8.4   Target Contract

This form of contract is comparatively of recent origin. It combines the best features of the cost plus percentage and the cost plus fluctuating fee type contracts. The contractor is paid on a cost plus percentage basis for work performed under the contract plus or minus a certain amount, which is an agreed percentage of savings or excess effected against the target value. The target value is arrived at by measuring the work on completion and valuing it at the rates agreed to earlier.

Illustration

A—a contractor—agrees to carry out the work for B—the employer. B agrees to pay the actual cost of the work plus 15% of the actual cost as the basic fee. In addition, B agrees to pay 25% of the savings effected against the target cost that will be arrived at by previously agreed rates. In case the actual cost is more than the target value, A will permit deduction at 25% of the excess from the basic fee payable to him.

A carries out the work, whose target value works out to ₹60 million. The actual cost of the work is found to be only ₹50 million. According to the agreement, B is to pay A 15% of ₹50 million, that is, ₹7.5 million plus 25% of ₹10 million (being the difference between the target cost and the actual cost), that is, ₹2.5 million. The total fee payable will be ₹10 million.

Had the contractor not used his skill and experience and had he not completed the work at a cost of ₹50 million as done but incurred ₹60 million, he would have received 15% of the cost of ₹60 million amounting to ₹9 million.

However, in the above example, if due to carelessness, wastage of materials and inefficiency, the contractor had completed the work at a cost of ₹70 million, he would have received: 15% of ₹70 million, that is, ₹10.5 million less 25% of ₹10 million totaling to ₹8 million.

The above illustration makes it clear that this type of contract encourages the contractor to use his skill and experience in keeping the cost as low as possible.

6.9   LABOUR CONTRACT

The special feature of this type of contract is that the owner agrees to supply all the requisite materials to the contractor and the latter agrees to supply all the labour and workmanship with tools and tackle, necessary to complete the work according to the drawings and specifications. This form of contract is suitable for those cases where the owner is in a position to buy large quantities of materials at favourable rates and also deliver them to the site of work most economically; for example, spreading and compacting metal and laying and fixing sleepers on a railway track.

6.10   SUBCONTRACT

Definition

A subcontract may be defined as an agreement between two contractors under the terms of which the one—subcontractor—undertakes to carry out for the other—general (main) contractor—a particular portion of the work which the latter has undertaken from the owner.

In the case of construction projects, usually, the successful contractor enters into agreements with specialist subcontractors to carry out that portion of the work in which each subcontractor has specialised himself. For example, in the case of a building work, waterproofing, air-conditioning, providing and fixing lifts, and so on, are sublet to specialised agencies.

The main contract between the general contractor and the employer, however, seldom permits the general contractor to merely become the supervisor of the subcontractors. In fact, he has to perform a substantial portion of the work with his own organisation. To ensure this condition, the contractor is not allowed to sublet any portion of the work without the permission of the owner or his authorised representative. Some contracts limit the extent of work that can be sublet to a maximum of 50%, that too with the permission of the employer/engineer. In some cases, the employer nominates the agencies from whom the materials should be procured or work must be got executed by the main contractor.

Where a part of the work (for example, air-conditioning in a building work) can be carried out in a more professional manner by a specialist in that line than the general contractor, it is to the advantage of the owner to permit subletting of such portion of the work.

The contractor, if permitted, may shop for bids and enter into an agreement with a firm approved by the owner or his engineer. The practice of a general contractor entering into an agreement with subcontractors before submitting his tender is not uncommon. In these cases, the contractor builds up the price on the basis of the terms of subcontracts. The subcontractor is bound by his offer if the employer accepts the main contractor's tender. In addition, the following are the special features of this type of contract:

1. The general contractor is responsible for the contract, the product, and payment of his subcontractors.
2. The owner's claims for damages are to be made against the general contractor, who, in turn, may recover them from the subcontractor if the latter is to blame.
3. No claims for damages can be brought against the owner by the subcontractor. They are to be made against the general contractor, who, in turn, may bring the corresponding claims against the owner.
4. Any special conditions that may be included in the general contract in respect of subcontractors will form the basis for settling disputes between the general contractor and the subcontractor.

To avoid errors due to omission and addition, the subcontract should generally provide for back-to-back application of the relevant provisions of the main contract, such as description of the work, specifications, and so on.

6.11   BOT CONTRACT/PUBLIC PRIVATE PARTNERSHIP (PPP) CONTRACT

Infrastructure development projects in the past were thought of as a way of providing infrastructure at no direct extra cost to the public. Public utility projects earned little or no revenue. For this reason no private party ventured into the field of construction of such projects, and therefore the projects used to be generally investigated, designed and financed by the Public Works Department of the State or Central Government. The projects needed several years for realisation from the stage of conception to become operational. After the need for a project would be felt, project report based on preliminary survey and investigations used to be prepared and based on the approximate estimates, its feasibility determined and administrative approval accorded by the concerned department. After the administrative approval, provision would be made in the annual budget for sanctioning the funds for preparation of detailed investigation, designs and drawings and estimates for approval or technical sanction. The project would then wait for requisite finance to be provided in the annual budget. By the time budget provision was made, it would be several years rendering the estimates outdated, thus requiring further revision and further sanctions. Even the ongoing projects were required to be suspended or slowed down for want of funds. This resulted in claims and litigations and further delays and escalation in cost.

When it was realised in recent times that the users of the facilities created by such projects would not mind paying for the facility, the traditional scenario changed dramatically. Over a period of time, four models emerged:

1.   Build, Operate and Transfer (BOT) Toll
2.   Build, Operate and Transfer (BOT) Annuity
3.   Build, Operate and Transfer (BOT) (Hybrid)
4.   Design, Build, Finance, Operate and Transfer (DBFOT) Hybrid

Before dealing with each of the above forms of the contract, it is necessary to understand the meaning of certain terms commonly used in the PPP Contracts.

6.11.1   Definitions and Explanations of Terms commonly used in PPP Contracts

1.   **PPP Contract**
      PPP stands for Public Private Partnership. The Indian Partnership Act defines 'Partnership' thus:

'Partnership' is the relation between persons who have agreed to share the profits of a business carried on by all or any of them acting for all.

Public–Private Partnership (PPP) thus involves a contractual relationship between a public sector authority and a private party, in which the private party provides a public service or project and assumes substantial financial, technical and operational risk in the project.

2. **SPV/SPE: Special Purpose Vehicle/Entity**

   SPV/SPE also known as Bankruptcy-remote Entity is usually a subsidiary company registered by an existing company with a limited object of acquisition and financing of specific assets. With its assets/liability structure and legal status, its obligations are secured even if the parent company goes bankrupt. It is invariably used by existing firms to finance large projects without putting the entire firm at risk.

3. **Project Finance**

   In simple words, it is a loan structure that relies primarily on the project's cash flow for repayment; with the project's assets, rights and interests held as secondary security or collateral. It is attractive to private sector because large projects can be financed off the balance sheets.

4. **The Concession**

   The concession means and includes exclusive right, license and authority to (design), fiancé (fully or partially), construct, operate, maintain the project for a given period and regulate use thereof by third parties on payment of appropriate fee (toll) for use of the facility or part thereof and refuse use of the facility if the fee due is not paid.

5. **Concessionaire**

   It is a party in whose favour the concession is granted. It is generally a company incorporated under the Companies Act, 1956 generally called SPV/SPE. The contract defines the term to include its successors and permitted assigns and substitutes.

6. **RFQ (Request for Qualification)**

   The concerned Authority issues notice to bidders requesting for qualification (RFQ) and short listing, if qualified. Where a bidder consists of two or more independent entities collectively called 'Consortium', it has to name one of its members as the 'Lead Member'. After the receipt of responses from the intending bidders, those who qualify are included in the short-list.

7. **RFP (Request for Proposal)**

   The public Authority, after finalisation of technical and commercial terms and conditions, invites bids from the shortlisted bidders pursuant to RFQ for undertaking the project (RFP). The selected bidder is issued a letter of award.

8. **Letter of Award (LOA)**

   If the Authority inviting bids, after scrutiny of bids received, decides to accept the bid of the selected bidder, issues to the bidder its letter of award (LOA). The letter, inter alia, calls for execution of the Concession Agreement within the time allowed from the date of issue of LOA.

9. **Concession Agreement**
    The Agreement entered into by and between the Authority and the SPV/SPE promoted and incorporated by the selected bidder for undertaking and performing obligations and exercising the rights conferred on the concessionaire is called 'concession agreement'.

6.12 ESSENTIAL FEATURES OF CONCESSION AGREEMENT

Nature of Agreement

In PPP type of contract, the only major difference from the other types of contract is in respect of the mode of payment of cost of the work to the contractor. The contractor undertakes to design, finance, construct, operate and maintain the works for a concession period in consideration of the exercise and/or to enjoy the rights, powers, benefits, privileges, authorities and entitlements including the amount receivable from the collection of charges levied on the beneficiaries who use the work and in some cases annuity payment each year. For example, in the case of a road or bridge, the contractor constructs the structure and is entitled to collect toll from the road or bridge users for the concession period.

The cost of construction of the project is ascertained in the same manner as for the lump sum or item rate contract. The tenderer then works out the cost of financing the project and its recovery from the collection of toll or similar method till the cost of construction together with the cost of finance is fully recovered. Each year's income will reduce the cost till the cost becomes zero or negative. The cost of maintenance of the project during the period in question has also to be added to the basic cost. The number of years required to recover the total cost is generally called the 'concession period'. The concession period worked out by each tenderer will depend upon the cost of construction worked out by each tenderer and also the rate of interest considered for financing the project and expected profit. The tenderer who submits the 'cash flow projections' seeking the minimum period of concession is generally awarded the contract.

Sometimes, the authority fixes the period of concession and seeks offers on the basis of grant in the form of 'annuity' expected from the employer. The tenderer expecting the least grant is generally awarded the contract. In addition, income tax holidays for pre-determined years to be selected by the tenders are offered to reduce the grant amount. The parties agree for release of the grant in a phased manner as required by the contractor. Income tax holidays for a limited period are also agreed for the years demanded by the contractor.

Models of BOT Contract

1. *Build, Operate and Transfer (BOT): Toll*
    In this category, the concessionaire agrees to build, finance, operate and transfer the facility created by the project (as per the design, drawings and specifications prepared and supplied by the Authority) in consideration of exclusive right, license and authority

to demand, collect and appropriate toll/fee collected from users of the facility during the agreed concession period.

2. *Build, Operate and Transfer (BOT): Annuity*

This is a special form developed in India by the National Highways Authority to cater to the projects where revenues from tolling are uncertain or will be insufficient to attract BOT operators. Such projects have to be got executed by the Government traditionally by employing conventional 'Engineering, Procurement and Construction' (EPC) contracts which entail little to no risk on the part of the private sector.

Annuity Concessions are a variant of the BOT model in which the private operator is remunerated via a fixed, periodical payment (annuity) from the Authority rather than through toll proceeds. Under these contracts, the private operator is responsible both for constructing the road, as well as for operating and maintaining it for a fixed period of time (typically ten years). This form of PPP transfers both responsibilities for bridge financing and performance risk to the private sector. In addition, because the annuity payments are not indexed, the private sector retains any risks associated with higher than anticipated operations and maintenance (O&M) costs. The Authority retains the right to set and collect tolls. Thus the revenue risk is retained by the Authority.

The Authority procures Annuity Concessions using a competitive bidding process (including prequalification based on the bidders' experience and financial capability) and awards the contract to the bidder offering to perform the work for the lowest annuity payments from the Authority. Although selection is primarily based on the financial bid, technical proposals must comply with predefined design, construction, Operation and Maintenance (O&M) and hand-back requirements.

The form has an advantage inasmuch as the Authority, unlike typical construction contracts, is not required to make any advance payment to the private operator. Instead, payment of the annuity starts after the road is constructed in accordance with the quality standards set out in the contract. This form provides for rewards for early completion and provides the private operator with a built-in incentive to ensure that the road is constructed in a way that minimises long term O&M costs while meeting quality standards. This focus on performance reduces the amount (and cost) of monitoring and supervision required of Government during the construction period. It has also resulted in construction costs that are on an average 12% to 35% lower than normal costs as per the NHAI's estimates.

The form of PPP has proven popular with private investors in India. and considering the advantages, this form has the potential to attract private finance in other sectors—such as water and solid waste—which often do not generate sufficient revenues to support BOT or concession type models namely PPCP (Public–Private Community Partnership) aimed at social welfare schemes eliminating the prime focus on profit.

3. *Build, Operate and Transfer: BOT-Hybrid*

To avoid the liability of paying annuity, the search for an alternative form resulted in a hybrid model which combines the features of the BOT (Toll) and BOT (Annuity) models.

Where a project is not capable of generating revenue by toll alone, under this scheme, the gap between the revenue required and generated is covered by the fund provided by the Authority called VGF (Viability Gap Funding) generally limited to 40% and by payment of annuity if the gap exceeds 40%. The concessionaire's right to collect and keep the toll is not affected.

4. *Design, Build, Finance, Operate and Transfer (DBFOT) Hybrid*
The concessionaire agrees to design, build, finance, operate and transfer (in the DBFOT form) the facility created by the project in consideration of exclusive right, license and authority to demand, collect and appropriate toll/fee collected from users of the facility during the agreed concession period.

To establish the fact that the Authority is the owner of the project and facility and the concessionaire is a licensee, the agreement provides for payment to the Authority of concession fee of Re. 1 per annum. Depending upon the financial forecast of the project, the concessionaire may offer to pay additional fee to the Authority for each year of the concession period as an agreed sum as premium out of the gross revenue as a share of the Authority. On the other hand, in some cases the Authority agrees to provide the concessionaire cash support by an outright grant (generally by way of equity support).

In all the above forms, the project is financed by the concessionaire in the first instance and the investment made is recovered with profit over a period of concession by—toll collection revenue or by payment of annuity by the Authority or partly by toll collection and partly by annuity payment by the authority respectively.

In the first three categories, the project is surveyed, designs and drawings, specifications and so on, are prepared by the Authority and the construction work assigned to the concessionaire. In the fourth category, all activities including the design stage and thereafter are to be carried out by the concessionaire; similar to 'all-in type contract' with the addition that the project is financed in the first instance by the concessionaire and the investment recovered by toll revenue, annuity or by both.

6.13   PARTS OF CONCESSION AGREEMENT

A Concession Agreement necessarily involves several heads and is therefore generally divided into several parts.

1. *Preliminary*: It includes besides Recitals, definitions and interpretation.
2. *The concession*: This part defines the scope of the project, grant of concession, conditions precedent, obligations of the parties—the concessionaire and the Authority, Representations and Warranties and Disclaimer.
3. *Development and operations*: This part deals with the right of way, obstructions at the site such as utilities, trees, encroachments and so on, construction of the project and

its monitoring till completion, entry into commercial services, O&M, safety, traffic, its regulation, census and sampling. It also includes the provision for appointment of an independent Engineer, his duties and functions, remuneration, termination, dispute resolution, and so on.

4. *Financial covenants*: This part incorporates provisions including financial close, grant, premium, concession fee, user fee, revenue shortfall loan, variation in traffic growth, construction of additional toll ways, escrow accounts, insurance, accounts and audit, and so on.

5. *Force majeure and termination*: Besides Force Majeure and Termination this part also deals with breach of contract, compensation for breach, suspension of concessionaire's rights and defects liability after termination.

6. *Miscellaneous provisions*: These include assignment and changes including changes in law, liability and indemnity, rights and titles over the site, dispute resolution, redressal of public grievances, and so on.

Contract Documents

In addition to the usual contract documents listed in the earlier chapter, cash flow projections, giving details of how the contractor intends to generate finance and use it till the recovery of his capital, investment cost and profit during the concession period, forms an intergal part of the contract. The main agreement contains several schedules describing the project site, project facilities, site delivery schedule, design requirements, construction requirements, operation and maintenance requirement, cash flow projections, annuity payment schedule, performance security, state support agreement, substitution agreement and hand-back requirements, and so on.

State Support Agreement

The implementation of a BOT agreement requires extensive and continued support and grant of certain rights and authorities by the Government. For this purpose, the State support agreement is signed and included in the contract documents.

Substitution Agreement

The contractor generally obtains the large amount of finance required for the project from financial institutions/banks who are called 'lenders'. The lender proposes a person called 'selectee' who is approved by the employer and who is empowered to substitute the contractor for the balance period, in the event of the contractor's failure to cure the default leading to suspension of the concession in the contractor's favour. An agreement signed by the employer, the contractor and the financial institutions to this effect also forms part of the contract agreement.

Hand-back Requirements

This lists the minimum requirements and item-wise inventory of all assets to be handed to the employer by the contractor at the end of the concession period.

Advantages

The main advantage is that the state or public authority is not required to finance the project immediately, and the project is made to generate the funds and be self-financing to the extent required. The developmental works which otherwise would get delayed can be undertaken forthwith.

Disadvantages

The main drawback of this form of contract is that it is not suitable for each and every type of work. In particular, works which are not likely to generate capital or be self-financing, are not suitable for this contract. For example, rural water supply project, village roads, city roads, and so on.

Suitability

Ideal for highways and express ways, bridges, tunnels that save the cost of operation and/or time of travel for which users will willingly pay reasonable toll charges.

Questions

1. Explain the nature of the relationship between the owner and the contractor.
   (**Ans.:** Section 6.2)
2. Describe the (a) main features, (b) advantages and (c) drawbacks of the following types of contract:
   (i) lump sum contract
   (ii) item rate contract
   (iii) percentage rate contract
   (iv) cost plus percentage contract
   (v) cost plus fixed fee contract   (**Ans.:** Sections 6.4 to 6.7.3)
3. State how a variation in a lump sum contract may be made and priced.   (**Ans.:** Section 6.4)
4. What sorts of works are unsuitable for the lump sum type of contract? Why?
   (**Ans.:** Section 6.4)
5. Under what circumstances may an owner prefer to have his project built under a lump sum contract instead of an item rate contract?   (**Ans.:** Section 6.4)
6. Distinguish between the percentage rate contract and the cost plus percentage contract.
   (**Ans.:** Section 6.7)
7. The cost plus percentage type contract should, as far as possible, be avoided. Discuss.
   (**Ans.:** Section 6.7)
8. Under what circumstances are the cost plus percentage type contracts suitable?
   (**Ans.:** Section 6.7)

9. In what way is a target type contract preferable to other forms of cost plus contracts? Illustrate your answer with a suitable example. **(Ans.:** Section 6.7.4)

10. Write short notes on:
    (a) all in contract
    (b) labour contract **(Ans.:** Section 6.3 and 6.8)

11. What is a subcontract? What are its special features? **(Ans.:** Section 6.9)

12. Write a detailed note on 'BOT form of contract' and its variations highlighting nature of agreement, mode of payment, advantages, drawbacks and suitability for execution of projects. **(Ans.:** Section 6.11)

Conditions of Contract

7.1 GENERAL

The conditions of any contract virtually amount to the law laid down by the parties for governing their respective conduct and fulfilling their contractual duties and responsibilities. The law of contract differs from any other branch of the law inasmuch as it does not lay down rules for the parties to abide by but gives freedom to the parties to make their own rules so long as the rules are within the four corners of the law of contract. An engineering contract is no exception to this basic rule. The conditions of contract form part of the contract documents as mentioned in Section 4.5, which may be referred to as a prelude to this chapter.

The clauses to be included in the conditions of contract will be governed by the character of the proposed work. Since the majority of massive projects are executed for the public by government departments and public sector undertakings through well-established contracting companies, it is in the interest of all concerned to have a common set of conditions generally called standard form contract conditions. This will avoid the trouble of reading all the conditions afresh, every time a contract is to be entered into. All the concerned parties will be familiar with the standard conditions and only special conditions incorporated for each project need to be read and understood.

In India almost every state government department and departments of central government have their own standard forms. For example, the State Public Works Department has Form No. B-2 for item rate contracts, the Central Public Works and other departments too have forms identified by the name of the department using them such as the CPWD form, MES (Military Engineering Services) form, Railways form, and so on. A number of attempts were made to prepare and adopt one uniform set of conditions by the Planning Commission of India. However, the attempts have not succeeded. For massive construction projects for which tenders are invited from international bidders, the FIDIC form has been extensively used in India. An attempt is made to acquaint the reader with the provisions in the FIDIC form and how the said conditions differ from other standard forms.

Not all the clauses of a standard set of conditions are required in every contract, whereas some contracts of a special nature require the drafting of special clauses. In this chapter it is proposed to discuss those clauses of a standard set of conditions, which are most frequently found in ordinary civil engineering contracts. These clauses relate to:

1. definitions of the terms used,
2. security deposit,
3. planning and scheduling of the project,
4. general obligations of the parties to the contract,
5. materials, plant and workmanship,
6. suspension,
7. time for completion, delay in completion,
8. liquidated damages and bonus,
9. measurement and bills and issue of interim payment certificates,
10. variations, additions and alterations,
11. defects: improper work; substandard material, and so on,
12. subletting,
13. breach of contract,
14. settlement of final accounts,
15. procedure for claims, and
16. arbitration.

7.2 DEFINITIONS

For the sake of clarity and in order to avoid disputes regarding the meaning of important terms frequently used in the contract documents, a clause is usually incorporated in the conditions of contract to define various terms. Illustrative definitions of some important terms are considered as follows:

1. *Employer/Owner*
   'Owner' shall mean the person named as such in the Appendix to the conditions and the legal successors in title to such person. The FIDIC provides further that the term will not include an assignee unless consented to by the contractor.
2. *Contractor*
   'Contractor' shall mean person or persons, corporation, company or firm whose tender has been accepted by the owner and shall include the contractor's personal representatives, successors, and permitted (with the consent of the employer) assignees.
3. *Engineer*
   'Engineer' shall mean the person appointed from time to time by the owner and notified in writing to the contractor to act as engineer for the purpose of the contract and notified in appendix to the contract.
4. *Work*
   'Work' shall mean the work or works contracted to be executed under or in virtue of the contract, whether temporary or permanent and whether original, altered, substituted or

additional. The FIDIC conditions include definitions of permanent works and temporary works separately.

5. *Site*

   'Site' shall mean the lands and other places on, under, in or through which the works are to be carried out and shall include any other lands or places provided by the employer/owner for the purpose of the contract.

6. *Drawings*

   'Drawings' shall mean all the drawings referred to in the agreement, special conditions, calculations, maintenance manuals and other technical information supplied to the contractor by the engineer or approved by the engineer and shall include any modifications of such drawings or any additional drawings approved by the engineer.

7. *Specifications*

   'Specifications' mean the specifications of the works included in the contract and any modifications thereof or addition thereto made by the Engineer and supplied to the contractor or submitted by the contractor and approved by the Engineer.

8. *Contract Price*

   'Contract price' means the sum stated in the letter of acceptance for execution of the contract works and remedying of any defects if any, and payable to the contractor in terms with the agreement.

Many other terms may be defined and included in the contract under definitions such as time limit for completion, retention money, day, foreign currency, and so on.

7.3 PERFORMANCE SECURITY/SECURITY DEPOSIT

The contractor is required to deposit with the owner a sum stated as a percentage of the cost of the work in order to safeguard the interests of the owner in the event of improper performance of the contract. The relevant clause should specify the time limit within which the payment has to be made and the mode of payment. It should also state the action to be taken if the contractor fails to comply with the provisions of the clause. The other points to be mentioned are whether or not the amount will be an interest bearing deposit and the period after which the security deposit will be refunded to the contractor. The following clause shows fairly standard wording used to describe the above-mentioned points:

'The Contractor shall (within ten days of the receipt by him of the notification of the acceptance of his tender) deposit with the Owner cash ₹ _______________[1] and permit the Owner while making any payment due, to deduct 10% of all moneys so payable until the full amount of the security deposit specified in the tender is made up. The security deposit referred to when paid in cash, may, at the cost of the Contractor, be converted into interest bearing securities provided that the Contractor has expressly desired this in writing. If the amount of security deposit to be paid in lump sum within the period specified above is not

---

1. Usually the amount is 1% of the cost of the work if in cash and full security deposit in case of bank guarantee.

paid, the tender already accepted shall be considered as cancelled and legal steps taken against the Contractor for the recovery of damages. The security deposit lodged by the Contractor shall be refunded to him after the expiry of three months from the date on which the final bill is paid, or after the expiry of the period of guarantee, whichever is later.'

However, in big projects the security is taken in the form of bank guarantees. The reason is obvious. If the contractor is required to deposit or pay cash he is bound to include the interest on the amount deposited in his tendered rates or cost. The bank guarantee costs the contractor a small percentage, seldom exceeding two per cent per annum of the amount guaranteed with small margin money. Accepting the security in terms of the bank guarantee is thus to the advantage of the employer.

7.4   PROGRAMME TO BE SUBMITTED AND OTHER GENERAL OBLIGATIONS

It has now become customary to ask the contractor to submit his planning and scheduling of the work with details of deployment of machineries and so on to be deployed, to be submitted along with the tender or within time stipulated after the issue of the work order. The engineer and the contractor discuss and finalise the programme. The progress is watched by holding periodic review meetings, generally once every month and the programme revised, if necessary.

Miscellaneous Clauses/Provisions

In addition to the main clauses discussed hereinafter, a contract may include several miscellaneous clauses in order to define clearly and properly the general obligation, responsibilities and liabilities of the contractor. Some such clauses, which are self-explanatory, are as follows:

1. *Setting out of work*
    The contractor shall supply without charge the requisite number of workmen with the means and materials necessary for the purpose of setting out works and counting, weighing and assisting in the measurement or examination of the work or materials at any time or from time to time during the performance of the contract.
2. *Materials, plant and workmanship*
    All materials, plant and workmanship shall be of the respective kind and quality described in the contract and in accordance with the instructions of the engineer issued from time to time. All samples shall be supplied by the contractor at his own cost. The cost of conducting any tests shall be borne by the contractor.
3. *Provision of lights, fencing and related items*
    The contractor shall provide all necessary equipment such as fencing, lights and related items required to protect the public from accident and shall indemnify the owner in respect of any damages arising from non-provision of such equipment.

4. *Damage done inside or outside the work area*

   The contractor shall bear all the expenses of defending any action or legal proceedings that may be brought by any person and also pay any damages and cost that may be awarded by the court in respect of all damage done intentionally or unintentionally by the contractor's labourers inside or outside the work area.

5. *Labour welfare*

   (i) The contractor shall not employ any person who is under the age of twelve (12) years.

   (ii) The engineer shall have the power to remove from work any person who, in his opinion, does not satisfy the conditions of contract and no responsibility shall be accepted by the owner for any delay caused in the completion of the work by such removal.

   (iii) The contractor shall pay fair and reasonable wages to the labourers employed by him for the performance of the contract. If there is any dispute between him and the workmen in this respect, the same shall be referred to the engineer without delay. The engineer shall decide the issue and his decision shall be final and binding. Such a decision shall not entitle the contractor to any revision of the agreed rates. In case of government works, the clause further states that the contractor should employ:

   (a) the labourers registered in the nearest employment exchange and

   (b) the labourers affected by famine.

6. *Work on Sundays or at night*

   No work shall be carried out on Sundays and at night without the written permission of the engineer.

7. *Entire agreement—Acceptance of all conditions*

   Acceptance of all the conditions of contract is compulsory before the parties sign the contract. The conditions normally include a provision reading somewhat as follows:

   This Agreement and the Schedules together constitute a complete and exclusive statement of the terms of the agreement between the parties on the subject hereof, and no amendment or modification hereof shall be valid and effective unless such modification or amendment is agreed to in writing by the parties and duly executed by persons especially empowered in this behalf by the respective parties. All prior written or oral understandings, offers or other communications of every kind pertaining to this agreement are abrogated and withdrawn.

7.5 SUSPENSION OF WORK

The conditions of contract invariably include a clause empowering the owner to suspend the work, such suspension not amounting to a breach of contract on his part. The reason behind the inclusion of such a clause in the conditions is to protect the owner's interest in the event of an emergency or the happening of an unexpected event. The clause relating to suspension of the work may be worded as follows:

'If at any time the Engineer shall, for any reason whatsoever, require that whole or any part of the work to be suspended for any period, he shall give notice in writing (ordering such suspension of the work) to the Contractor who shall thereupon suspend or stop the work totally or partially as the case may be. However, upon suspending the work, the Engineer shall take all appropriate steps to minimise the duration of such suspension, and the Contractor shall be entitled to reasonable compensation for the resultant unavoidable expenses to him on account of the suspension of the work. He shall also be entitled to reasonable extension of time to complete the work.'

To illustrate the necessity of such a clause, assume that an owner has let a contract to build the first storey to an already existing ground-floor house. The construction of the first storey is partially completed when the walls of the old structure develop cracks. Under such circumstances, the engineer, on behalf of the owner, should have the power to suspend the work until he can determine what is to be done. The investigations regarding the causes of development of cracks and the remedial measures may cause considerable delay so that several weeks may elapse before he is sure that he can advise the owner to go ahead with the work. In connection with the above clause, the following points may be noted.

1. The right to suspend the work is not to be given to the contractor.
2. The owner or the engineer should make judicious use of the authority that he gets under the provision of this clause.
3. When a contract specifies a definite time limit for completion, an extension of time should be granted to the contractor to make up for the time lost due to suspension. Such an extension shall be a liberal one, allowing sufficient margin for resumption of the work at the speed at which it was progressing when suspension was ordered for.

7.6   TIME LIMIT FOR COMPLETION

Usually, civil engineering contracts provide for the completion of the work within a specified period. The period for the completion of the work is specified in any one of the following ways:

1. on a given date, or
2. in a specified number of calendar months, or
3. in a given number of working days.

The clause should state the time allowed for the completion of the entire work in one of the above-mentioned ways and also the date from which it shall be reckoned.

In addition, if the time limit for the completion is to be strictly enforced, the clause should stipulate that time is the essence of the contract.

Here is the wording of a typical clause prepared to apply to the above-mentioned matter:

'The time allowed for carrying out the work shall be 24 (twenty-four) calendar months and shall be reckoned from the date fifteen days after the date on which the order to commence the work is received by the Contractor. (Time is deemed to be of the essence to the Contract and shall be strictly observed by the Contractor.)'

Delays and extension of time: Delays in the completion of a work may be caused by the owner, the contractor, or acts of god (natural calamities). The owner is generally assumed to have an implied obligation to provide site of work, all drawings, designs and other necessary information needed to enable the contractor to complete his contract within the specified time. As such, the contractor should generally be entitled to recover the losses sustained because of delays caused by the owner or his engineer in addition to a suitable extension of time for completion of the works. In spite of this, many civil engineering contracts contain substantially the following provision:

'The Contractor shall not be entitled to claim any compensation from the Owner for the loss suffered by him on account of delay caused by the Owner'. This clause, though primarily included to safeguard the interest of the owner, is in reality of no benefit to him. If such a clause appears in the contract, the contractor will most probably add to the contingency item in his tender an amount that in his opinion will cover such eventualities. This means that the owner is going to pay more for the work because of the increased tender price even though there may be no delay in completion.

A better way to handle this problem of delays would be to include provisions reading somewhat as follows:

'In the event of delays caused by … (list of the causes of delay), the Engineer shall, after due consultation with the employer and the Contractor, determine (a) the amount of extension to which the contractor is entitled, and (b) the amount of such costs, which shall be added to the contract price and shall notify the same to the Contractor with a copy to the Employer....'

Sometimes the causes of delay are continuing in nature and in such events the engineer is empowered to determine an interim extension of time.

The provision for compensation in addition to the extension of time can be worded as follows:

'If, in the opinion of the Engineer, any of the salaried men or plant and equipment used by the Contractor at the site in performance of the Contract are necessarily kept idle, on account of any act or omission of the Owner or Engineer and without the fault of the Contractor, the Contractor shall be entitled to a reasonable compensation. This compensation shall be an amount equal to that which he may pay during such period of idleness for the services of such idle salaried persons and expenses connected therewith, plus a sum that he may have to pay in the form of rent for such idle plant and equipment. No compensation shall be paid to the Contractor unless the Contractor shall have submitted a claim with full details in writing to the Engineer within one month of the cause of such claim occurring.'

FIDIC standard conditions provide for an addition to the contract sum, of such sum as may compensate the contractor for the loss incurred due to compensatory events which include delay in giving possession of site, drawings, instructions, and so on.

The other reasons which may cause delay and on account of which the contractor becomes entitled for a reasonable extension of time may include the following:

1. hostile acts of any public enemy,
2. acts of the government, or
3. acts of another contractor over whom this contractor has no control.

Here is a suggested clause dealing with extension of time:

'If the Contractor be delayed at any time in the progress of the work by an act or omission of the Owner or Engineer, or by acts of god, or by the act of another Contractor employed by the Owner, or by acts of the government, or by hostile acts of enemies, or by fire, unusual delay in transportation, or unavoidable casualties, or any causes beyond the control of the Contractor, then the time of completion shall be extended by such reasonable time as the Engineer may decide. The Contractor shall not be entitled to extension of time unless he has submitted to the Engineer full and detailed particulars of any claim to extension of time to which he may consider himself entitled, within one month of the occurrence of the event giving rise to the claim'.

FIDIC conditions stipulate 28 days as the time limit for making an application seeking an extension of time from the date of occurrence of the event causing the delay. The reason for stipulating the time limit is to enable the engineer to verify the causes of delay and their impact on the progress of the work, which may be difficult to verify with the passage of time.

7.7 LIQUIDATED DAMAGES AND BONUS FOR EARLY COMPLETION

Delays that are caused by the contractor himself are naturally his own responsibility. And in order that he should make every effort to prevent such delays, a suitable provision has to be made in the conditions of the contract. Delays that are caused as a result of strikes, bad weather, or storms are generally considered as *force majeure or vis majeure* (the acts of god) for which neither party may be responsible. Delays caused by acts of god (for example, earthquake) are supposedly not the fault of either party and any loss on account of such events must remain where it falls. However, reasonable extension in the time limit should be granted. The condition usually found in engineering contracts provide for extension of time on all these counts.

If there is delay by the contractor resulting in loss to the employer, the contractor has to compensate the employer by paying a sum called penalty or liquidated damages. The liquidated damages are usually stated in the form of a specific percentage of the estimated cost of the whole work for each day or week beyond the date specified in the contract—that the actual completion of the work exceeds. Here is a sample clause dealing with the liquidated damages:

'The Contractor shall pay as compensation an amount equal to one per cent of the estimated cost of the whole work as shown by the tender for every week that the work remains unfinished after the proper date.'

The liquidated damages are generally a pre-determined amount of loss likely to be caused by the delay in completion. Generally, the contract provisions limit the maximum liquidated damages to be recovered to 10% of the contract sum.

Bonus

Many contracts include provision for payment of bonus to the contractor if the contractor achieves earlier completion than the time stipulated for completion by the contract. The amount of bonus is also specified as a per cent of contract sum payable for a maximum of say 10 per cent or if a fixed sum per week or month is specified, the maximum period for which the bonus will be paid is specified such as four months or six months if the time limit specified is in excess of 24 months.

7.8   MEASUREMENT AND PAYMENT TO CONTRACTOR

It is a good practice to make periodical payments to the contractor during the progress of a work. It is unfair to expect a contractor to provide finance from his own funds for all the work, particularly when the project is large and extends over a period of several months to a year or two. If he is not paid at intervals during the progress of the work, he will generally have to borrow money. The contractor will recover the interest on such borrowed money by a suitable increase in the sum tendered by him as the cost of the work or by an increase in unit rates. Thus, it is ultimately the owner who will have to bear the interest on the capital borrowed by the contractor.

To avoid the above-mentioned situation, most of the contracts include a provision of partial payments to the contractor, proportionate to the work done by him, generally once every month, during the progress of the work. This applies to all types of contracts except the BOT type. Of course, in case of the cost plus contracts, the work is financed by the owner, but even in these cases, the fee payable to the contractor as profit and overheads may be paid in instalments. A clause to be inserted in the conditions of contract should clearly specify the procedure in respect of the following:

1.   methods or mode of measurements,
2.   mode of payments,
3.   payment at reduced rates,
4.   final payment, and
5.   payment to subcontractor.

1.   *Mode of Measurements*
     There are many methods of measurements of work and there are wide variations amongst them. It is therefore advisable that the clause should classify the procedure that will be

adopted in order to avoid future misunderstandings. A measurement contract might well contain a clause such as the following:

'Unless otherwise specified in detailed Specifications, the work under different items shall be measured in accordance with the Standard Mode of Measurement of Building Works (I.S. 1200).'

2. *Mode of Payment*

Some ideal requirements in connection with the procedure to be followed for release of payment in case of an item rate contract are illustrated by the following clause:

The contractor shall submit a bill on the printed form on or before the twentieth (20th) of each month for all work executed in the previous month. The charges to be made in the bill shall always be entered at the rates specified in the tender or in the case of any extra work not provided for in the tender at the mutually agreed rates for such work.

If the Contractor has submitted a bill for payment as above, the engineer shall, not later than the first (1st) of each month, issue a certificate for payment to the Contractor. The amount payable shall be ninety (90) per cent of the sum obtained by applying tendered rates to actually measured quantities of the approved work completed by the Contractor during the preceding month. Else he shall state in writing his reasons for withholding the certificate.

Should the Owner fail to pay the sum stated in any certificate for payment issued by the Engineer or in award by arbitration, upon demand when due, the Contractor shall be entitled to receive, in addition to the sum named in the certificate, interest thereon at the rate specified in the appendix to the Conditions.

No interim certificate for partial payment nor partial or entire use or occupancy of the work by the Owner shall be deemed to amount to an acceptance of any work or materials not in accordance with this Contract.

The following comments relate to the above clause:

(i) *Retention money:* The engineer has the right to withhold about 10 per cent of the estimated monthly payments due to the contractor. This gives the owner a sort of leverage over the contractor, as the latter will be encouraged not to abandon the work when it is nearing completion. Also, the employer can use the amount so retained, in addition to the security deposit, to make good losses that he may suffer, in the event of improper performance by the contractor.

(ii) *Delay in payment:* If the clause does not make a provision to grant an interest on the amount which is due to the contractor, but not paid to him in reasonable time, he is bound to increase his tendered sum to account for likely delayed payments and loss of interest thereon. Thus, the cost of the work to the owner may go up in the absence of such a provision.

(iii) *Interim payment to be regarded as advances:* The above-mentioned clause specifies that though the contractor is paid for the work carried out by him it does not, in any way, mean that his work has been accepted. Making good defects, if any, noted at future dates until the expiry of the maintenance period, is the responsibility of the contractor. For this reason the interim payments are to be regarded as payments by way of advances against the final payment which will be due after the expiry of the defects liability period.

3. *Payment at Reduced Rates*

   If the contractor fails to carry out the work strictly in accordance with the drawings and specifications, and, if in the opinion of the engineer, the work is structurally sound, he may accept the work, but the contractor will be paid at reduced rates which will be decided by the engineer. By way of an illustration, assume that the specification for teak wood panelled windows and doors mentions that the thickness of the shutters should be 4 cm and if the shutters actually provided by the contractor are only 3.5 cm thick, the engineer may accept the work and reduce the rate per square metre of the shutters by an amount equal to the reduction in the cost of the materials used in the work. The clause inserted in the conditions in case of an item rate contract may read as follows:

   The rates for several items of work agreed to within shall be valid only when the item concerned is accepted as having been completed fully in accordance with the contract Specifications. In cases where the items of work are not accepted as so completed, the Engineer may make payment on account of such items at such reduced rates as he may consider reasonable.

7.9 ADDITIONS AND ALTERATIONS OR VARIATIONS AND DEVIATIONS

It is a rare occurrence, especially in case of large projects, that they are carried through to completion exactly in accordance with the original plans and specifications on which the tender and contract were based. Additions and alterations are rendered desirable or advisable during the progress of the work due to unforeseen difficulties. For this reason the conditions should always contain a clause giving the engineer power to make additions, alterations, omissions and variations without invalidating the contract in any way.

Extras

The following are some of the reasons on account of which extra costs may be incurred:

1. variations in the quality of work involved,
2. change in the drawings, designs and specifications, and
3. unforeseen happenings may render the work difficult to perform.

1. *Extras on Account of Variations in the Quantity of the Work*

   Civil engineering contracts usually provide for the adjustment of errors in bills of quantities when these are included in the contract documents. Such errors, it is mentioned in the conditions, are not to vitiate the contract but are to be adjusted. In case of the measurement contract the provision made is as follows:

Quantities shown in the Bill of Quantities are approximate and no claim shall be entertained for quantities of work carried out being more or less than those entered in the Bill of Quantities.

The above clause stipulates that if the quantities are more than those mentioned in the Bill of Quantities, the contractor will have to execute them at the agreed rates. In case actual quantities are less than those entered in the Bill of Quantities, the contractor will not be paid for any loss of profit caused thereby. However, it must be remembered that considerable increase or reduction in the quantity of work does result in increase in the cost and the loss of profit and the portion of allowance for the overheads. It will be unfair not to compensate the contractor for such loss. Minor changes may not justify the extra payment. The best method to deal with this situation is to add a provision in the conditions of contract whereby the owner's or engineer's power to order for extra work and variations is limited, for which the contractor's claim for extra payment will not be considered. FIDIC[2] provides for a new rate for any item, of which the quantity increases by more than 10% of the original quantity, provided the cost of the change in quantity is not less than 0.01% of the contract sum. If on completion of the work the cost due to all variations is found to exceed 15% of the contract sum, the contractor is entitled to additional sum to make good his loss. Most of the public works contracts, however, provide plus/minus 25% as the limit of variation in quantities of an item of work for which tendered rate alone will apply and for variation in excess of the said limit, a new rate will be mutually agreed upon.

2. *Extras on Account of Change in Drawings, Designs and Specifications*

In cases where extras result from changes in the drawings, designs and from the quality of materials and workmanship with respect to that specified in the contract, the extra cost resulting therefrom is the responsibility of the owner. Here, the contractor will have to be adequately compensated for such extra work. The extra cost of the work may be computed by using the agreed rates or, where it is not possible to construe the extra rates from the existing agreement; the rates should be worked out and mutually agreed to. It is important to mention in the clause that the contractor is bound to execute the work in accordance with the revised drawings, designs and instructions and his claim for the extension of time will be allowed.

The following is one suggested method of phrasing a clause dealing with extras:

'The Engineer, without invalidating the contract, may order extra work or make changes by alterations in or additions to the specifications, drawings and designs. The contract sum shall be adjusted for the extra cost at the rates as are specified in the tender or at mutually agreed rates when alterations and deviations include any class of work for which no rate is specified in the contract. All such work shall be executed under the conditions of the original contract, except that claim for extension of time on

---

2. FIDIC Conditions – Ed. 1999

account of extras may be accepted and the contractor granted a reasonable extension of time.'

3. *Extras on Account of Additional Difficulty of Performance*
    Sometimes, the contractor has to carry out the work under difficult conditions which could not have been predicted from the drawings and details supplied to him. This may be a just cause for the contractor to claim for extras. For example, if the drawings and details supplied to the contractor did not show the presence of sandy strata and if it is encountered in pile driving, it will cause the contractor considerable additional difficulty and expense of use of jetting water to facilitate driving of piles. In such cases, the owner has to pay an additional sum to compensate the contractor for the additional expense incurred by him in executing the work.

7.10 EXECUTION OF WORK: DEFECTS, MAINTENANCE AND IMPROPER WORK

Every contract generally includes a clause such as:

The Contractor shall execute the work in the most substantial and workman-like manner conforming exactly, fully, and faithfully to the Designs, Drawings and Specifications.'

The condition further provides that the Contractor must appoint a full-time qualified Engineer to get the work done according to the Contract.

Defects, Maintenance and Improper Work

To check whether the contractor's performance is satisfactory, the work is required to be supervised and inspected by the owner's engineer. It is always preferable that, on large projects, the engineer appoint a resident engineer and the supervisors who will be continuously looking after the construction work. The resident engineer along with a team of supervisors is expected to check the work as it proceeds and point out any deviations to the contractor, who will rectify the same. The owner's engineer or architect will generally conduct a periodical inspection of the works. In cases where the extent and type of work do not demand the appointment of a full-time resident engineer, the inspection of the work is limited to the periodical visits of the engineer. In such cases, it is unlikely that the engineer will be able to detect all defects in the materials and workmanship.

Even when there is a resident engineer with a team of assistants, there is a possibility of defects being overlooked. For this reason, contracts include a clause known as the 'defects and maintenance clause'. Under the provision of this clause, the contractor is bound to make good defective work and replace improper materials as and when noticed during the progress of the work and also for a specified period (known as the maintenance period) after the completion of the work. Here is an example of a clause that deals with bad work, improper materials, and so on:

'The Contractor shall promptly:

1. remove from the site any materials which in the opinion of the Engineer are not in accordance with the Contract and shall replace them with proper and suitable materials, and
2. replace and re-execute his own work, which is condemned by the Engineer and the work of the other contractors which is damaged by such removal or replacement, at his risk and expense. In the event of default on the part of the Contractor in carrying out the written order of the Engineer, within the period specified by the Engineer in his order, the Owner may remove the materials and the bad work by employing other persons and all expenses consequent thereon or incidental thereto shall be borne by the Contractor. If the Contractor fails to pay the expenses for such removal and replacement of materials and work within ten days' time thereafter, the Owner may deduct the same from the Security Deposit or from payment then or thereafter due to the Contractor, or may, upon ten days' written notice, sell such materials at an auction and shall account for net proceeds thereof, after deducting all the costs and expenses that are payable by the Contractor to the Owner.'

Defects and Maintenance Clause

A typical defects and maintenance clause may read as follows:

The Contractor shall remedy all defects due to faulty materials and workmanship which shall appear within one year from the grant of the final certificate of completion. He shall also pay for any damages to other work resulting while remedying such defects. The Owner shall give written notice of observed defects with reasonable promptness. If the Contractor shall fail to carry out such work, the Owner shall be entitled to carry out such work by his own workmen or by employing another contractor and the cost of remedying such defects observed during the maintenance period shall be recovered from the Contractor's Security Deposit, or any payment that is due or shall be due to the Contractor.

The provision of the defects and maintenance clause is advantageous to all the parties. From the owner's standpoint, the clause gives him the right of action for breach of the covenants with regard to defects or maintenance at any time during the period agreed upon in the clause. On the other hand, the advantage that accrues to the contractor is the ending of his responsibility and liability at a definite date unless very special circumstances exist.

7.11 SUBLETTING

Subletting means the letting by the main contractor to another, work he has himself contracted to perform. Such transfer of the whole or part of the contract is, in general, inadvisable and the conditions of contract generally prohibit or allow it, subject to the owner's approval. The main reason for preventing subletting is that the contract is made upon the qualifications and responsibility of the contractor and its transfer to anyone in whom the owner does not have equal confidence is likely to be opposed by the latter. Secondly, the transfer of a contract is also accompanied by a lot of problems in respect to settling of the accounts, picking up of the work by the new contractor, division of responsibility, and so on, all resulting in delay in completion. The clause preventing subletting may read as follows:

The Contract shall not be assigned or sublet without the written approval of the Owner. If the Contractor shall assign or sublet his Contract, or attempt to do so, the Owner may, by notice in writing rescind the Contract. In the event of the Contract being rescinded, the Security Deposit of the Contractor shall thereupon stand forfeited and be absolutely at the disposal of the Owner. The Owner may, without prejudice to any other right or remedy, take possession of the premises and of all materials, tools and appliances thereon and finish the work by whatever method he may deem expedient. The Contractor shall not be entitled to receive any payment until the work is completed. If the unpaid balance of the Contract price shall exceed the expense of completing the work such excess shall be paid to the Contractor. In case the expenses of finishing the work exceed such unpaid balance, the Contractor shall pay the difference to the owner. The Engineer's certificate as to the expense of finishing the work shall be final.

Though subletting of the contract is in general prohibited, there may be circumstances under which it may be desirable for the owner to allow subletting. For instance:

1. subletting a part of the work to a specialist firm, which is in a better position to carry out the same than the general contractor,
2. sickness of the contractor,
3. death of some of his key personnel,
4. financial difficulties faced by the contractor, and
5. strained relationship between the labour and the contractor preventing the smooth performance of the job.

Payment to Subcontractors

The certificates issued by the engineer include the cost of work executed or materials supplied by the subcontractors. The subcontractors are to be paid by the principal contractor. The main contractor should make prompt payments to the subcontractors and avoid the possibility of disputes arising between him and the subcontractors over the issue of payments. Any such disputes invariably result in hampering the progress of the work. Should the principal contractor fail to comply with the above requirement, the engineer is empowered (by inserting a suitable clause in the conditions) to order direct payments to the subcontractors and deduct such payment from the next payment due to the main contractor.

7.12 BREACH OF CONTRACT

It is essential that every contract contain a provision to safeguard the interests and rights of either party, in the event of a breach of contract.

Breach on the Part of the Contractor

In case a contractor fails to carry out the contractual obligations, the owner must have a right to claim damages for breach of contract. However, such damages will not always be an adequate remedy and provision must be made in the contract conditions, empowering the owner to have alternative arrangements made in order to get the work completed.

The standard forms of contracts contain clauses that give the owner an express power to determine the contractor's employment and expel him from the site. The following clause is one such example:

In case of abandonment of the work owing to bankruptcy or liquidation, serious illness or death of the Contractor or any other cause, the Owner shall have power to adopt any of the following courses as he may deem best suited to his interests:

1. *To Rescind the Contract*
2. *To Take Over and Complete the Work as an Agent of the Contractor*
   The Owner shall have the right to employ labour and supply materials to carry out the work, or any part thereof, debiting the Contractor with the cost of the labour and the price of the materials and crediting him with the value of the work done, at the same rates as if it had been carried out by the Contractor under the terms of the Contract. The Owner shall also have the right to take possession of and use or permit to use any or all tools, plant, materials, stores, provided by the Contractor for the execution of the work, paying or allowing for the same in accounts at the Contract rates or in case the Contract rate is not applicable, at the current market rates. The Contractor shall not remove tools, plant, materials, and stores from the site without the written permission of the Engineer. The Engineer shall determine the cost of all the work done by the Owner acting as an agent for the Contractor; and in that case the certificate of the Engineer regarding the value of the work done shall be final and conclusive against the Contractor. Any difference between actual cost above the Contract price shall be payable to the Owner by the Contractor.
3. *To Employ Another Contractor*
   To order that the work of the Contractor be measured up and to take the unexecuted part of the work out of his hands and to give it to another Contractor to complete; in which case, expenses which may be incurred in excess of the Contract price shall be borne and paid by the original Contractor. In the event of any of the above courses being adopted by the Owner, the Contractor shall have no claim to compensation for any loss sustained by him.

The wording of the above clause clearly indicates that any action taken under the provisions of the clause determines the employment of the contractor under the contract and not the contract itself; the contract remains in existence and both parties continue to have liabilities and obligations under it.

Breach on the Part of the Owner

When the breach is on the part of the owner, the contractor should be given the right to abandon the contract and bring an action for damages. If the breach of the contract does not go to the root of the contract, it is preferable that the contractor should complete the work first and then sue for the contract price and damages, if any, in addition. A suitable clause drafted for this purpose should make provisions in respect of the following:

1. *Bankruptcy of the owner:* On the bankruptcy of the owner, the contractor is entitled to abandon the contract, but unless any special provision is made in the contract, he can only claim according to the bankruptcy procedure as an ordinary creditor.
2. *Non-payment for the value of the work done:* If the owner does not pay the contractor the value of the work done as certified by the engineer, the contractor, after giving written notice and waiting for the specified period, is entitled to abandon the contract at any juncture that this state of affairs arises and claim for the cost of the work. However, the contractor has to take this step very cautiously because he has to prove that the owner does not intend to make the payment, or is incapable of paying. If he fails to establish this and if the non-payment is proved to be due to some temporary difficulty faced by the owner, which he may shortly overcome, the breach may be only partial and the contractor's action of determining the contract may be adjudged as premature.
3. *Abnormal delay in giving possession of the site, supply of drawings and so on:* When there is a definite obligation to hand over the site on a definite date, and the owner fails or refuses to do so, or if in the absence of any fixed date, he fails to hand over the site or the drawings and so on, within a 'reasonable' time, his conduct amounts to a breach of the contract and the contractor is entitled to recover damages for the loss sustained by him.
4. *Rescission by the owner:* If the owner orders the contractor to stop the work, that would constitute a rescission by the owner. Such rescission will entitle the contractor to seek settlement of accounts including claims for compensation for the loss, if any, caused by such termination of the contract.

7.13   SETTLEMENT OF ACCOUNTS OR FINAL PAYMENT

The contract may contain the following clause, which is self-explanatory:

Prior to final payment the Contractor shall obtain and furnish to the Owner the final certificate issued to him by the Engineer. No such certificate shall be given by the Engineer nor shall the work be considered to be complete until the Contractor shall have removed all his rubbish, tools, scaffolding and surplus materials from and about the work. In case of dispute the Owner may remove the rubbish and charge the cost to the Contractor. The contractor shall submit the final bill, or statement of accounts, six copies of draft final statement with supporting documents showing in detail the cost of the work done and further sums which the contractor may consider due to him, not later than two months after the issue of the Defects Liability Certificate by the engineer.

The provision further stipulates scrutiny by the engineer, discussions between the contractor and the engineer resulting in the submission of the final statement by the contractor. The engineer thereafter issues a certificate of undisputed amount as the final certificate if no dispute remains or an interim certificate if there is a dispute in respect of the final sum payable.

Discharge
Upon submission of the final statement by the contractor, the contractor is expected to give to the employer a written discharge stating that the final payment shall release the owner from all claims and liability to the contractor in the performance of the contract.

7.14   DISPUTES ARISING OUT OF CONTRACT

This heading, given what follows, doesn't appear strange. Why? It is our everyday experience. However, let us reframe it by using synonyms for a couple of words. 'Dispute' means disagreement. It is well-known that every contract is an agreement but the converse, 'every agreement is a contract', is not true. An agreement foreseeable by law alone is a contract. Reframing by using alternate words the phrase reads: '*disagreements arising out of the agreement*'.

Now does it not sound somewhat strange? At least to the author, who has devoted nearly 50 years of his active life to study and analyse thousands of court cases to understand how and why this happens, it is not strange. Are seeds of disagreement sown in the agreement, knowingly or unknowingly, by the parties to their agreement? Construction contracts provide a fertile soil for the growth of disputes between the parties. They provide elaborately for resolution of such disputes by making the construction contract *two in one*; the main contract for execution of the construction work invariably incorporates an agreement for the resolution of disputes by way of naming a dispute resolution Expert (DRE) or Dispute Resolution Board (DRB), and by conciliation and/or arbitration. It is well established that the arbitration clause in a construction contract is an independent contract. Thus, even if the main agreement is found invalid or terminated, the arbitration clause may survive to resolve dispute(s).

Forums for Dispute Adjudication

A party to a dispute arising out of their contract has four judicial forums available to resolve the dispute. The first jurisdiction exercised by a Court of law is compulsory, in the sense that a party invoking the said jurisdiction does not need the consent to do so from the opposite party or parties. Conciliation, the second forum, on the other hand, can be said to be voluntary inasmuch as to have recourse to it for the resolution of a dispute, the disputing parties must have a written agreement to refer the dispute(s) for conciliation by one or if the parties so agree, two or three conciliators. In a pending case, the Court can also direct the parties to take recourse to conciliation. If successful, conciliation results in a Settlement Agreement. signed by the parties and witnessed by the conciliator(s).The Settlement Agreement is given the status of an award which is treated as a decree of the Court.

Arbitration

The last forum arbitration is more commonly used in construction contracts. Since the law of arbitration embodied in The Arbitration and Conciliation Act 1996, has been extensively amended in the years 2015, 2019 and 2020, it is considered desirable to give the reader an overview of the Arbitration and Conciliation Act 1996 as amended.[3]

'Conciliate' means to make calm and amenable, pacify. Conciliation is a process in which conciliatory measures are used to bring about an amicable settlement between or amongst the parties to a dispute. It is in effect a negotiated settlement by and between the parties but with the intervention of a third person, or persons called conciliator(s). The process is in many respects similar to arbitration but with a major deviation inasmuch as the conciliator does not give a decision but persuades the parties to reach an amicable settlement. The settlement so reached is put into writing, signed by the parties and witnessed by the conciliator(s), The settlement agreement reached in conciliation process has the advantage over privately negotiated settlements in as much as the said settlement agreement countersigned by the conciliator(s) is given the status of an award, which is enforceable as a decree of a Court of the law. The process has many advantages and is becoming popular in a number of countries. In India though, the process is given a statutory recognition, and yet it is not very popular.

The law applicable to arbitration and conciliation is embodied in the Arbitration and Conciliation (Amendment) Act 1996, as amended in 2015, 2019 and 2020. Salient features of some important provisions of the said Act are discussed in what follows.

The essential requirement to attract provisions of the Arbitration Act is that there must be an arbitration agreement. What is an arbitration agreement? This question assumes importance and is discussed at length because an arbitration agreement is the foundation on which the jurisdiction of an arbitrator rests. If there is no valid arbitration agreement between the parties, an award made by an arbitrator is invalid and has no force. The Act defines an arbitration agreement in such a comprehensive manner that all the case laws developed over the past many years, under the Act of 1940, are embodied in it.

Section 2(1) (b):
Definitions *'Arbitration agreement'* means, an agreement referred to in Section 7 reading as follows:

Section 7: Arbitration Agreement
(1) In this Part 'arbitration agreement' means an agreement by the parties to submit to arbitration all or certain disputes which have arisen or which may arise between them in respect of a defined legal relationship, whether contractual or not.
(2) An arbitration agreement may be in the form of an arbitration clause in a contract or in the form of a separate agreement.
(3) An arbitration agreement shall be in writing.
(4) An arbitration agreement is in writing if it is contained in:

---

3. For further discussion see the author's book: *Law of Arbitration and Conciliation*, 6th Ed., B S Patil and S P Woolhouse.

    (a)   a document signed by the parties,

    (b)   an exchange of letters, telex, telegrams or other means of telecommunication *including communication through electronic means* which provide a record of agreement, or

    (c)   an exchange of statements of claim and defence in which the existence of the agreement is alleged by one party and not denied by the other.

(5)   The reference in a contract to a document containing an arbitration clause constitutes an arbitration agreement if the contract is in writing and the reference is such as to make that arbitration clause part of the contract.

'In this part' restricts the definition to Part I, Arbitration. It will not apply to Part II: Enforcement of certain foreign awards.

Section 8 empowers a judicial authority before which an action is brought (such as a civil suit) and which is the subject matter of the arbitration agreement to refer the parties to arbitration unless the authority finds that *prima facie* no valid arbitration agreement exists. In short, once the parties have agreed to refer the dispute to arbitration, the parties should abide by their own agreement and any breach of the arbitration agreement will result in a court staying its own proceedings to allow the arbitration to proceed. However, a party can always seek the help of the Court for interim relief as provided for in Section 9.

Section 9: Interim measures etc. by court

(1)   A party may, before or during arbitral proceedings or at any time after the making of the arbitral award but before it becomes a decree of a Court, apply to a Court

   (i)   for the appointment of a guardian for a minor or a person of unsound mind for the purposes of arbitral proceedings, or

   (ii)   for an interim measure of protection in respect of any of the following matters; namely,

      (a)   the preservation, interim custody or sale of any goods which are the subject matter of the arbitration agreement,

      (b)   securing the amount in dispute in the arbitration,

      (c)   the detention, preservation or inspection of any property or thing which is the subject matter of the dispute in arbitration, or as to which any question may arise therein and authorising for any of the aforesaid purposes, any person to enter upon any land or building in the possession of any party, or authorising any samples to be taken or any observation to be made, or experiment to be tried, which may be necessary or expedient for the purpose of obtaining full information or evidence,

      (d)   interim injunction or the appointment of the receiver, or

      (e)   such other interim measures of protection as may appear to the Court to be just and convenient, and the Court shall have the same power of making orders as it has for the purpose of, and in relation to, any proceedings before it.

(2)   Where, before the commencement of the arbitral proceedings, a Court passes an order for any interim measure of protection under sub-section (1), the arbitral proceedings shall be commenced within a period of ninety days from the date of such an order or within such further time as the Court may determine.

(3)   Once the arbitral tribunal has been constituted, the Court shall not entertain an application under sub-section (1), unless the Court finds that circumstances exist which may not render the remedy provided under section 17 efficacious.

The amendment was necessary to avoid overlapping of the provisions of S.9 and S.17 of the Act prior to its amendment. Prior to its amendment, S.17 did not provide for any measure for enforcement of an order by the tribunal granting interim measure, and parties preferred to apply to the Court for interim measures not only prior to but even after the constitution of the arbitral tribunals. The amendment will help to reduce the burden on the Courts to deal with applications under S.9 of the Act.

Section 9 is followed by Chapter III with the heading: **Composition of Arbitral Tribunal**, containing Sections 10 to 15.

Section 10 gives the parties freedom to determine the number of arbitrators, provided that such a number shall not be an even number. If the parties fail to agree, the reference shall be to a sole arbitrator. With each party appointing its own arbitrator, experience showed that the tribunal's decision would result in a tie necessitating a reference to an umpire and all proceedings had to be repeated before the umpire. The amendment provides, if the parties fail to agree upon the number and/or procedure, each party shall appoint one arbitrator and the two arbitrators nominated by the parties shall appoint the third arbitrator (presiding arbitrator) and decision of the majority would become the award of the tribunal.

The Section further provides the procedure for filling the vacancy by a party, through an application to the arbitral institution nominated by the Supreme Court for international commercial arbitration or by the High Court in other than international commercial arbitration. The Section provides a sixty-day time limit for the disposal of an application for filling a vacancy. Such a vacancy may arise from the death or illness of an arbitrator, for example.

Section 12 asserts that it is the solemn duty of an arbitrator to disclose in writing, circumstances likely to give rise to justifiable doubts as to his independence or impartiality. It further incorporates grounds on which an arbitrator's appointment can be challenged. The provisions of this Section assume great importance in respect of arbitrations pursuant to arbitration the standard form contracts for public works and other public sector undertakings which empower the persona designata to appoint a sole arbitrator who may be an officer of the department. The amended Act of 2015 has changed this scenario to some extent. Section 12 as amended reads as follows.

Section 12: Grounds for challenge

Sub-section (2) states that an arbitrator, from the time of his appointment and throughout the arbitral proceedings, shall, without delay, disclose to the parties in writing any circumstances referred to in sub-section (1) unless they have already been informed of them by him.

Subsection (1) reads

(1)   When a person is approached in connection with his possible appointment as an arbitrator, he shall disclose in writing any circumstances,

    (a)   such as the existence either direct or indirect, of any past or present relationship with or interest in any of the parties or in relation to the subject-matter in dispute, whether financial, business, professional or other kind, which is likely to give rise to justifiable doubts as to his independence or impartiality, and

    (b)   which are likely to affect his ability to devote sufficient time to the arbitration and in particular his ability to complete the entire arbitration within a period of twelve months.

Explanation 1: The grounds stated in the Fifth Schedule shall guide in determining whether circumstances exist which give rise to justifiable doubts as to the independence or impartiality of an arbitrator.

Explanation 2: The disclosure shall be made by such a person in the form specified in the Sixth Schedule.

(3)   An arbitrator may be challenged only if

    (a)   circumstances exist that give rise to justifiable doubts as to his independence or impartiality, or

    (b)   he does not possess the qualifications agreed to by the parties.

(4)   A party may challenge an arbitrator appointed by him, or in whose appointment he has participated, only for reasons of which he becomes aware after the appointment has been made.

(5)   Notwithstanding any prior agreement to the contrary, any person whose relationship, with the parties or counsel or the subject-matter of the dispute, falls under any of the categories specified in the Seventh Schedule shall be ineligible to be appointed as an arbitrator:

    Provided that parties may, subsequent to disputes having arisen between them, waive the applicability of this sub-section by an express agreement in writing.

Section 13 challenge procedure

The law gives freedom to the parties to agree to challenge procedure. Failing such agreement, the law provides the procedure be to be adopted. If a challenged arbitrator withdraws from the office or the other party agrees to the challenge, the arbitral tribunal shall continue with proceedings and make an award. Such an award will be open to objection and subject to an appeal.

Sections 14 and 15 deal with the termination of the mandate and substitution of an arbitrator.

The remaining Chapters of Part I, II, III and IV of the Act deal with the conduct of the arbitration proceedings, the making of an award and the challenges to it. Part II deals with international commercial arbitration awards made outside India which may be enforced in India under the New York Convention 1958.[4] The award is made by a majority if all three arbitrators cannot agree. A limited challenge is available to the court. The basic principle is that the law will respect party autonomy in choosing a forum other than the court for the resolution of their disputes.

7.15 PROCEDURE FOR CLAIMS

It is in the best interests of the parties to the contract that disputes are avoided and that the work is carried out to satisfactory completion in an atmosphere of mutual confidence and harmony. To avoid disputes, the parties should take care to: avoid oral agreements and prepare written agreements clearly expressing the full intention of both parties, and avoid oral instructions for alterations and additions. However, disputes are likely for various reasons, not least a deliberate breach of contract by a party or simply incompetence and the provision for the settlement of such disputes is essential.

Modern conditions of contract include an elaborate procedure for lodging and settlement of claims by either party to the contract and in particular by the contractor. The procedure specifies that the contractor must give early warning notices to the engineer or employer of situations likely to give rise to his claims. The idea is to give an opportunity to avoid a potential dispute or to minimise damages. If however, the occasion arises to make a claim for an extra sum, a further notice within four weeks of the cause of action should be given. The engineer is expected to scrutinise the claims and call for full particulars, supporting documents and so on, and if satisfied, certify the extra payment. If the contractor is not happy with the decision given by the engineer, he is entitled to refer the disputes to the Dispute Review Expert (generally one person) or Dispute Resolution Board (DRB) (generally three persons one each appointed by the two parties and the third selected by the two members nominated by the parties). The DRE or DRB is constituted at the time of signing of the agreement or soon thereafter. Some conditions name the DRB as the Steering Committee. This is to avoid arbitration and litigation, if possible. The decision of the DRE or DRB or Steering Committee is to be given within the stipulated time. The contract generally provides the decision to be provisional in nature. Either party not satisfied by the decision so given can demand a reference to arbitration within the time allowed to do so failing which the decision of the DRE or DRB becomes final and binding on the parties.

---

4. The detailed discussion of the balance Act is beyond the scope of this book.

7.16 COMMON DISPUTES

Disputes in case of civil engineering contracts are mainly with reference to:

1.   the interpretation of drawings and specifications,
2.   quality of materials and workmanship,
3.   measurement and payment,
4.   claims for extras,
5.   delay in completion, either extra items or excess quantities and liquidated damages,
6.   issue of certificates,
7.   defects and maintenance, and
8.   subletting of the contract.

In respect of items (1), (2) and (8) above, the engineer is given the power to settle disputes and his decisions are quite often made final and binding on the parties to the contract. In the event of disputes arising in reference to the remaining points, the contract usually provides for referring such disputes to an independent arbitrator. In no case can the engineer be made the sole arbitrator in all matters between the owner and the contractor. Appointing him as the only arbitrator is to expect a man to be a 'judge of his own cause', which is a clear violation of the principles of natural justice. The number of arbitrators has to be uneven. General practice is to appoint a sole arbitrator for small works and a panel of three, one is appointed by the employer, one by the contractor and the third by the two arbitrators for larger works. Where three arbitrators are provided, or by another authority named by the parties. The law does not provide for the choice of arbitrators except that they are to be neutral and independent. It is up to the parties to select a suitable person in the first instance. An effort should be made to select someone of good standing, with the knowledge of the area of the dispute and any experience of arbitration is helpful. The arbitration tribunal need not comprise lawyers or retired judges as the parties do not have to follow the formalities of the court procedure.

An arbitration clause in a civil engineering contract may read as follows:

If any dispute or difference of any kind shall arise between the Owner or the Engineer and the Contractor either during the progress of the work or after completion or abandonment, determination, or breach of the Contract it shall be referred to an Arbitral Tribunal consisting of a sole Member to be agreed upon between the parties and if that is not possible, to three members one to be appointed by each of the party and the third—Presiding Arbitrator—to be appointed by the two members of the tribunal. The Arbitral Tribunal shall have the full power to open up and review any certificate, opinion, decision, requisition or notice and to determine all matters in dispute which shall be submitted to the Tribunal. Notice of the demand for arbitration shall be filed in writing with the other party within thirty (30) days after the dispute has arisen and or decision given by the DRE/DRB/Steering Committee.

It happens in rare cases, particularly in public works, that the parties abide by the decisions of DRE/DRB/Steering Committee; usually a reference to arbitration results. Though the arbitral tribunal's decision is made final and binding upon the parties to arbitration, experience shows that awards are invariably challenged on the grounds permitted by the Arbitration and Conciliation Act, 1996, resulting in further litigation in the courts of law. Settlement of disputes by conciliation, a mode incorporated in the Act of 1996, really recommends itself. It involves the settlement of disputes by an agreement between the parties through the intervention of a third person conciliator. This usually involves an informal and confidential process of arriving at an agreement on all or at least some disputes. Where two parties may be reluctant to talk to each other, the conciliator can speak to each, in confidence, and facilitate a settlement. The reluctance of public authorities to sign such settlement agreements, though the process of conciliation is declared confidential by the law, has put this otherwise excellent mode of dispute resolution virtually in cold storage.

Labour Laws

Construction contracts invariably include a provision that the contractor shall comply with all relevant labour laws. For example, FIDIC Conditions of Contracts for Construction Clause 6.4 reads:

6.4 **Labour Laws:** The contractor shall comply with all relevant labour laws applicable to the contractor's personnel, including laws relating their employment (including wages and working hours), health, safety, welfare, immigration and emigration, and shall allow them all their legal rights….)

The Contractor shall require his personnel to obey all applicable laws.

The important labour laws applicable to construction industry include

(1)   Employees's Compensation Act 1923, and
(2)   The Employees Minimum Wages Act, 1948.

Some important provisions of the above two Acts are highlighted in what follows, the contents are self-explanatory. The objective of the 1923 Act is to ensure that injured employees or in the case of those who die in work-related accidents, their dependents, are paid timely compensation. Examples of possible incidents could be a worker's fall from the scaffolding or a ladder, accident caused by a crane, collapse of a part-constructed wall or a bridge, debris falling from a height, and so on. The Act provides for medical assistance to those injured and sets out clear principles for the calculation of compensation. Any self-inflicted injuries (example, from intoxication at work) are excluded. Employers ought to and usually do buy an insurance to cover such unforeseen accidents.

Every Act lays down in brief the subject matter and purpose of the enactment in its Preamble which in the case of the first of the two Act reads:

Employees' Compensation Act
PREAMBLE

An Act to provide for the payment by certain classes of employers to their Employees of compensation for injury by accident. Whereas it is expedient to provide for the payment by certain classes of employers to their workmen of compensation for injury by accident; It is hereby enacted as follows:

Section 2 comprises important definitions.

2. Definitions...

(b) **'Commissioner'** means a Commissioner for employee's Compensation appointed under Section 20.

Section 20 reads:

Appointment of Commissioners (1) The State Government may, by notification in the Official Gazette, appoint any person *[who is or has been a member of a State Judicial Service for a period of not less than five years or is or has been for not less than five years an advocate or a pleader or is or has been a Gazetted officer for not less than five years having educational qualifications and experience in personnel management, human resource development and industrial relations] to be a Commissioner for employees' Compensation for such area as may be specified in the notification.

Employer

Section 2

(e) 'Employer' includes anybody or persons whether incorporated or not and any managing agent of an employer and the legal representative of a deceased employer, and, when the services of an employee are temporarily lent or let on hire to another person by the person with whom the employee has entered into a contract of service or apprenticeship, means such other person while the employee is working for him.

(c) **'Compensation'** means compensation as provided for by this Act.

'**Dependent**' means any of the following relatives of the deceased employee, namely:

(i) a widow, a minor legitimate or adopted son, an unmarried legitimate or adopted daughter or a widowed mother, and

(ii) if wholly dependent on the earnings of the employee at the time of his death, a son or a daughter who has attained the age of 18 years and who is infirm;

(iii) if wholly or in part dependent on the earnings of the employee at the time of his death, includes

(a) a widower,

(b) a parent other than a widowed mother,

The list includes eight relations who can be dependents.

'**Employee**' means a person, who is

(i) a railway servant as defined in clause (34) of Section 2 of the Railways Act, 1989 (24 of 1989), not permanently employed in any administrative district or sub-divisional office of a railway and not employed in any such capacity as is specified in Schedule II; or

(ii) (a) a master, seaman or other members of the crew of a ship,

   (b) a captain or other member of the crew of an aircraft,

   (c) a person recruited as driver, helper, mechanic, cleaner or in any other capacity in connection with a motor vehicle,

   (d) a person recruited for work abroad by a company, and who is employed outside India in any such capacity as is specified in Schedule II and the ship, aircraft or motor vehicle, or company, as the case may be, is registered in India; or

(iii) employed in any such capacity as is specified in Schedule II, whether the contract of employment was made before or after the passing of this Act and whether such a contract is expressed or implied, oral or 2 in writing; but does not include any person working in the capacity of a member of the Armed Forces of the Union; and any reference to any employee who has been injured shall, where the employee is dead, include a reference to his dependents or any of them;]

Construction and railway industries are among those listed in Schedule 2 to the Act.

   (e) **'Partial disablement'** means, where the disablement is of a temporary nature, such disablement as reduces the earning capacity of an employee in any employment in which he was engaged at the time of the accident resulting in the disablement, and, where the disablement is of a permanent nature, such disablement as reduces his earning capacity in every employment which he was capable of undertaking at that time: provided that every injury specified [in Part II of Schedule I] shall be deemed to result in permanent partial disablement;

   (f) **'Prescribed'** means prescribed by rules made under this Act.

   (g) **'qualified medical practitioner'** means any person registered under any Central Act, Provincial Act, or an Act of the Legislature of a State providing for the maintenance of a register of medical practitioners, or, in any area where no such last-mentioned Act is in force, any person declared by the State Government, by notification in the Official Gazette, to be a qualified medical practitioner for the purposes of this Act;

Section 3: Employer's liability for compensation

According to Section 3, employers are liable to pay compensation to any employees who may have suffered personal injury due to an accident in the course of their employment. It is a no-fault system in that the employee does not have to establish negligence of the employer. However, there are some exceptions as is clear from the wording of section 3 quoted below.

(1) If personal injury is caused to an employee by accident arising out of and in the course of his employment, his employer shall be liable to pay compensation in accordance with the provisions of this Chapter: Provided that the employer shall not be so liable,

   (a) in respect of any injury which does not result in the total or partial disablement of the employee for a period exceeding three days,

(b)   in respect of any injury, not resulting in death or permanent total disablement caused by an accident which is directly attributable to—(i) the employee having been at the time thereof under the influence of drink or drugs, or (ii) the wilful disobedience of the employee to an order expressly given, or to a rule expressly framed, for the purpose of securing the safety of employees, or (iii) the wilful removal or disregard by the employee of any safety guard or other device which he knew to have been provided for the purpose of securing the safety of employee,

(2)   Section 3(2) also makes a provision for the employer to pay a compensation to employees for occupational diseases, that is, any illness caused or made worse by the work environment. Schedule III lists all the occupational diseases some of which are more relevant to the construction sector than others. For example, diseases caused by inhalation of dust or working toxic compounds or compressed air.

According to Section 3(4):

(4)   Save as provided by sub-sections (2), (2A)] and (3) no compensation shall be payable to a employee in respect of any disease unless the disease is directly attributable to a specific injury by accident arising out of and in the course of his employment.

According to Section 3(5):

(5)   Nothing herein contained shall be deemed to confer any right to compensation on an employee in respect of any injury if he has instituted in a Civil Court a suit for damages in respect of the injury against the employer or any other person; and no suit for damages shall be maintainable by an employee in any Court of law in respect of any injury,

(a)   if he has instituted a claim to compensation in respect of the injury before a Commissioner, or

(b)   if an agreement has been come to between the employee and his employer providing for the payment of compensation in respect of the injury in accordance with the provisions of this Act.

Thus, section 3(5) ensures that there is no double dipping. The employee cannot make a claim under this Act or settle a claim with the employer and also make a claim in the court for the same injury.

1.   **Amount of compensation** (1) Subject to the provisions of this Act, the amount of compensation shall be as follows, namely,

(a)   where death results from the injury: an amount equal to fifty per cent of the monthly wages of the deceased employee multiplied by the relevant factor; or an amount of *[one lakh and twenty thousand rupees], whichever is more,

(b)   where permanent total disablement results from the injury, an amount equal to sixty per cent of the monthly wages of the injured employee multiplied by the relevant factor; *[one lakh and twenty thousand rupees], whichever is more.

*[Provided that the Central Government may, by notification in the Official Gazette, from time to time, enhance the amount of compensation mentioned in clauses (a) and (b).] The relevant factor is determined according to the age of the injured employee and is provided on the basis of a table set out in Schedule 4 of the Act.

4A. Compensation to be paid when due and penalty for default

(1) Compensation under Section 4 shall be paid as soon as it falls due.

(2) In cases where the employer does not accept the liability for compensation to the extent claimed, he shall be bound to make provisional payment based on the extent of liability which he accepts, and, such payment shall be deposited with the Commissioner or made to the employee, as the case may be, without prejudice to the right of the employee to make any further claim.

(3) Where any employer is in default in paying the compensation due under this Act within one month from the date it fell due, the Commissioner shall-- (a) direct that the employer shall, in addition to the amount of the arrears, pay simple interest thereon at the rate of twelve per cent per annum or at such higher rate not exceeding the maximum of the lending rates of any scheduled bank as may be specified by the Central Government, by notification in the Official Gazette, on the amount due; and

(b) ...

5. Method of calculating wages: In this Act and for the purposes thereof the expression 'monthly wages', means the amount of wages deemed to be payable for a month's service (whether the wages are payable by the month or by whatever other period or at piece rates), and calculated as follows, namely,

(a) where the employee has, during a continuous period of not less than twelve months immediately preceding the accident, been in the service of the employer who is liable to pay compensation, the monthly wages of the employee shall be one-twelfth of the total wages which have fallen due for payment to him by the employer in the last twelve months of that period.

6. Commutation of half-monthly payments: Any right to receive half-monthly payments may, by agreement between the parties or, if the parties cannot agree and the payments have been continued for not less than six months, on the application of either party to the Commissioner, be redeemed by the payment of a lump sum of such amount as may be agreed to by the parties or determined by the Commissioner, as the case may be.

7. Distribution of compensation: (1) No payment of compensation in respect of an employee whose injury has resulted in death, and no payment of a lump sum as compensation to a woman or a person under a legal disability, shall be made otherwise than by deposit with the Commissioner, and no such payment made directly by an employer shall be deemed to be a payment of compensation; provided that, in the case of a deceased employee, an employer may make to any dependent, advances on account of compensation of an amount equal to three months' wages of such employee and so much of such amount as does not exceed the compensation payable to that dependent shall be deducted by the Commissioner from such compensation and repaid to the employer.

9. Compensation not to be assigned, attached or charged: Save as provided by this Act no lump sum or half-monthly payment payable under this Act shall in any way be capable of being assigned or charged or be liable to attachment or pass to any person other than the employee by operation of law nor shall any claim be set off against the same.

10. Notice and claim: (1) No claim for compensation shall be entertained by a Commissioner unless notice of the accident has been given in the manner hereinafter provided as soon as practicable after the happening thereof and unless the claim is preferred before him within two years of the occurrence of the accident or in case of death within two years from the date of death: Provided that where the accident is the contracting of a disease in respect of which the provisions of sub-section (2) of section 3 are applicable, the accident shall be deemed to have occurred on the first ofthe days during which the employee was continuously absent from work in consequence of the disablement caused by the disease: Provided further that in case of partial disablement due to the contracting of any such disease and which does not force the employee to absent himself from work, the period of two years shall be counted from the day the employee gives notice of the disablement to his employer: Provided further that if an employee who, having been employed in an employment for a continuous period, specified under sub-section (2) of section 3 in respect of that employment, ceases to be so employed and develops symptoms of an occupational disease peculiar to that employment within two years of the cessation of employment, the accident shall be deemed to have occurred on the day on which the symptoms were first detected:

30. Appeals: (1) An appeal shall lie to the High Court from the following orders of a Commissioner, namely, (a) an order awarding as compensation a lump sum whether by way of redemption of a half-monthly payment or otherwise or disallowing a claim in full or in part for a lump sum; (aa) an order awarding interest or penalty under Section 4A; (b) an order refusing to allow redemption of a half-monthly payment; (c) an order providing for the distribution of compensation among the dependents of a deceased employee, or disallowing any claim of a person alleging himself to be such a dependent; (d) an order allowing or disallowing any claim for the amount of an indemnity under the provisions of sub-section (2) of section 12; or...

31. Recovery: The Commissioner may recover as an arrear of land-revenue any amount payable by any person under this Act, whether under an agreement for the payment of compensation or otherwise, and the Commissioner shall be deemed to be a public officer within the meaning of Section 5 of the Revenue Recovery Act, 1890 (1 of 1890).

2) THE MINIMUM WAGES ACT, 1948

The 1948 Act is intended to avoid exploitation of labour so that at least a minimum wage is guaranteed to the employees in the construction industry to enable them to meet their basic needs. The Act is not specifically designed for the construction industry but it is an industry with potential for exploitation of workers. Different categories of workers within the industry will have their minimum wages set by either the state or the central government. Construction is a 'scheduled' employment under this Act and different rates are set for the skilled, semi-skilled and unskilled labourers. These are enforced by inspections and penalties. Employers need to keep due records for this purpose. There is no difference in the minimum wages by gender, only by skill. Payments for overtime are also a right under this Act. It is important to ascertain the applicable minimum wages when estimating the cost labour. They will vary for different trades and by region. They may include the wages and a Dearness Allowance. Some of the important sections are reproduced below:

Preamble:

An Act to provide for fixing minimum rates of wages in certain employments WHEREAS it is expedient to provide for fixing minimum rates of wages in certain employments;

It is hereby enacted as follows:

1. **Short title and extent:** (1) This Act may be called The Minimum Wages Act, 1948.
    (3)   It extends to the whole of India.

2. **Interpretation:** In this Act, unless there is anything repugnant in the subject or context,
    (a)   'adolescent' means a person who has completed his fourteenth year of age but has not completed his eighteenth year;
    (aa)   'adult' means a person who has completed his eighteenth year of age
    (b)   'appropriate Government' means,
        (i)   in relation to any scheduled employment carried on by or under the authority of the Central Government or a railway administration or in relation to a mine, oil field or major port, or any corporation established by a Central Act, the Central Government; and
        (ii)   in relation to any other scheduled employment by the State Government;
    (bb)   'child' means a person who has not completed his fourteenth year of age;
    (b)   'competent authority' means the authority appointed by the appropriate Government by notification in its Official Gazette to ascertain from time to time the cost of living index number applicable to the employees employed in the scheduled employments specified in such notification;
    (c)   'cost of living index number' in relation to employees in any scheduled employment in respect of which minimum rates of wages have been fixed, means the index number ascertained and declared by the competent authority

by notification in Official Gazette to be the cost of living index number applicable to employees in such employment;

(d)   ...

(e)   'employer' means any person who employs, whether directly or through another person, or whether on behalf of himself or any other person, one or more employees in any scheduled employment in respect of which minimum rates of wages have been fixed under this Act, and includes, except in sub-section (3) of Section 26,

    (i)   in a factory where there is carried on any scheduled employment in respect of which minimum rates of wages have been fixed under this Act, any person named under clause (f) of sub- section (1) of section 7 of the Factories Act, 1948 (63 of 1948) as manager of the factory,

    (ii)   in any scheduled employment under the control of any Government in India in respect of which minimum rates of wages have been fixed under this Act, the person or authority appointed by such Government for the supervision and control of employees or where no person or authority is so appointed, the head of the Department,

    (iii)   in any scheduled employment under any local authority in respect of which minimum rates of wages have been fixed under this Act, the person appointed by such authority for the supervision and control of employees or where no person is so appointed the Chief Executive Officer of the local authority,

    (iv)   in any other case where there is carried on any scheduled employment in respect of which minimum rates of wages have been fixed under this Act, any person responsible to the owner for the supervision and control of the employees or for the payment of wages,

(f)   'prescribed' means prescribed by rules made under this Act;

(g)   'scheduled employment' means an employment specified in the schedule, or any process or branch of work forming part of such employment;

(h)   'wages' means all remuneration, capable of being expressed in terms of money which would if the terms of the contract of employment express or implied, were fulfilled, be payable to a person employed in respect of his employment or of work done in such employment and includes house rent allowance but does not include-

(i)   the value of -

    (a)   any house-accommodation, supply of light, water, medical attendance; or

    (b)   any other amenity or any service excluded by general or special order of the appropriate Government;

    (ii)   any contribution paid by the employer to any Pension Fund or Provident Fund or under any scheme of social insurance;

(iii)   any travelling allowance or the value of any travelling concession;

(iv)   any sum paid to the person employed to defray special expenses entailed on him by the nature of his employment; or

(v)   any gratuity payable on discharge.

(j)   'employee' means any person who is employed for hire or reward to do any work skilled or unskilled, manual or clerical, in a scheduled employment in respect of which minimum rates of wages have been fixed; and includes an out-worker to whom any articles or materials are given out by another person, to be made up, cleaned, washed, altered, ornamented, finished, repaired, adapted or otherwise processed for sale for the purposes of the trade or business of that other person where the process is to be carried out either in the home of the out-worker or in some other premises not being premises under the control and management of that other person; and also includes an employee declared to be an employee by the appropriate Government; but does not include any member of the Armed Forces of the [Union].

3. Fixing of minimum rates of wages:

(1)   The appropriate Government shall, in the manner hereinafter provided,

(a)   fix the minimum rates or wages payable to employees employed in an employment specified in Part I or Part II of the Schedule and in an employment added to either part by notification under section 27:

Provided that the appropriate Government may, in respect of employees employed in an unemployment specified in Part II of the Schedule, instead of fixing minimum rates of wages under this clause for the whole State, fixing such rates for a part of the State or for any specified class or classes of such employment in the whole State or any part thereof];

(b)   review at such intervals as it may think fit, such intervals not exceeding five years, the minimum rates of wages so fixed and revise the minimum rates, if necessary:

[Provided that, where for any reason the appropriate Government has not reviewed the minimum rates of wages fixed by it in respect of any scheduled employment within any interval of five years, nothing contained in this clause shall be deemed to prevent it from reviewing the minimum rates after the expiry of the said period of five years and revising them, if necessary, and until they are so revised the minimum rates in force immediately before the expiry of the said period of five years shall continue in force.]

[(IA) Notwithstanding anything contained in sub-section (1), the appropriate Government may refrain from fixing minimum rates of wages in respect of any scheduled employment in which there are in the whole State less than one thousand employees. Subs. by the Adaptation of Laws Order, 1950.

engaged in such employment, but if at any time, the appropriate Government comes to a finding after such inquiry as it may make or cause to be made in this behalf that the number of employees in any scheduled employment in respect of which it has refrained from fixing minimum rates of wages has risen to one thousand or more, it shall fix minimum rates of wages payable to employees in such employment as soon as may be after such finding.

(2) The appropriate Government may fix,

    (a)   a minimum rate of wages for time work (hereinafter referred to as 'a minimum time rate');

    (b)   a minimum rate of wages for piece work (hereinafter referred to as 'a minimum piece rate');

    (c)   a minimum rate of remuneration to apply in the case of employees employed on piece work for the purpose of securing to such employees a minimum rate of wages on a time work basis (hereinafter referred to as 'a guaranteed time rate').

    (d)   a minimum rate (whether a time rate or a piece rate) to apply in substitution for the minimum rate which would otherwise be applicable in respect of overtime work done by employees (hereinafter referred to as 'overtime rate').

[2(2-A) Where in respect of an industrial dispute relating to the rate of wages payable to any of the employees employed in a scheduled employment any proceeding is pending before a Tribunal or National Tribunal under the Industrial Disputes Act, 1947 (14 of 1947), or before any like authority under any other law for the time being in force or an award made by any Tribunal, National Tribunal or such authority is in operation, and a notification fixing or revising the minimum rates of wages in respect of the scheduled employment is issued during the pendency of such proceeding or the operation of the award; then, notwithstanding anything contained in this Act, the minimum rates of wages so fixed or so revised shall not apply to those employees during the period in which the proceeding is pending and the award made therein is in operation or, as the case may be, where the notification is issued during the period of operation of any award, during that period; and where such proceeding or award relates to the rates of wages payable to all the employees, in the scheduled employment, no minimum rates of wages shall be fixed or revised in respect of that employment during the said period.]

(3) In fixing or revising minimum rates of wages under this section-

    (a)   different minimum rates of wages may be fixed for-

        (i)   different scheduled employment;

        (ii)   different classes of work in the same scheduled employments;

        (iii)   adults, adolescents, children and apprentices;

        (iv)   different localities;

    [(b) minimum rates of wages may be fixed by any one or more of the following wage periods, namely,

    (i) by the hour,
   (ii) by the day,
  (iii) by the month, or
  (iv) by such other larger wage period as may be prescribed and where such rates are fixed by the day or by the month, the manner of calculating wages for a month or for a day, as the case may be, indicated]:

      Provided that where any wage periods have been fixed under Section 4 of the Payment of Wages Act, 1936 (4 of 1936), minimum wages shall be fixed in accordance therewith.

4. Minimum rate of wages:

(1)   Any minimum rate of wages fixed or revised by the appropriate Government in respect of scheduled employments under Sec. 3 may consist of,

    (i) a basic rate of wages and a special allowance at a rate to be adjusted, at such intervals and in such manner as the appropriate Government may direct, to accord as nearly as practicable with the variation in the cost of living index number applicable to such workers (hereinafter referred to as the 'cost of living allowance'); or

   (ii) a basic rate of wages with or without the cost of living allowance and the cash value of the concessions in respect of supplies of essential commodities at concessional rates, where so authorised; or

  (iii) an all inclusive rate allowing for the basic rate, the cost of living allowance and the cash value of the concessions, if any.

(2)   The cost of living allowance and the cash value of the concessions in respect of supplies of essential commodities at concessional rates shall be computed by the competent authority at such intervals and in accordance with such directions as may be specified or given by the appropriate Government.

5. Procedure for fixing and revising minimum wages:

(1)   In fixing minimum rates of wages in respect of any scheduled employment for the first time under this Act or in revising minimum rates of wages so fixed, the appropriate Government shall either,

   (a) appoint as many committees and sub-committees as it considers necessary to hold enquiries and advise it in respect of such fixation or revision, as the case may be, or

   (b) by notification in the Official Gazette, publish its proposals for the information of persons likely to be affected thereby and specify a date, not less than two months from the date of the notification, on which the proposals will be taken into consideration.

(2)   After considering the advice of the committee or committees, appointed under clause (a) of sub-section (1), or as the case may be, all representationsreceived by it before the

date specified in the notification under clause (b) of that sub-section, the appropriate Government shall, by notification in the Official Gazette, fix, or, as the case may be, revise the minimum rates of wages in respect of each scheduled employment, and unless such notification otherwise provides, it shall come into force on the expiry of three months from the date of its issue:

Provided that where the appropriate Government proposes to revise the minimum rates of wages by the mode specified in clause (b) of sub-section (1) the appropriate Government shall consult the Advisory Board also.]

11. Claims:

(1)   The appropriate Government may, by notification in the Official Gazette, appoint 1[any Commissioner for workmen's compensation or any officer of the Central Government exercising functions as a Labour Commissioner for any region, or any officer of the State Government not below the rank of Labour Commissioner or any] other officer with experience as a judge of a Civil Court or as a stipendiary Magistrate to be the Authority to hear and decide for any specified area all claims arising out of the payment of less than the minimum rates of wages 2[or in respect of the payment of remuneration for days of rest for work done on such days under clause (b) or clause (c) of sub-section (1) of section 13 or of wages at the overtime rate under section 14] to employees employed or paid in that area.

(2)   [Where an employee has any claim of the nature referred to in sub-section (1)], the employee himself, or any legal practitioner or any official of a registered trade union authorised in writing to act on his behalf, or any Inspector, or any person acting with the permission of the Authority appointed under sub-section (1), may apply to such Authority for a direction under sub-section (3):

Provided that every such application shall be presented within six months from the date on which the minimum wages [or other amounts] became payable:

Provided further that any application [may be admitted after the said period of six months when the applicant satisfies the Authority that he had sufficient cause for not making the application within such period.

[(3)   When any application under sub-section (2) is entertained the Authority shall hear the applicant and the employer, or give them an opportunity of being heard, and after such further inquiry, if any, as it may consider necessary, may, without prejudice to any other penalty to which the employer may be liable under this Act, direct,

    (i)   in the case of a claim arising out of payment of less than the minimum rates of wages, the payment to the employee of the amount by which the minimum wages payable to him exceed the amount actually paid, together with the payment of such compensation as the Authority may think fit, not exceeding ten times the amount of such excess;

    (ii)   in any other case, the payment of the amount due to the employee together with the payment of such compensation as the Authority may think fit, not exceeding ten rupees; and the Authority may direct payment of such compensation in cases where the excess or the amount due is paid by the employer to the employee before the disposal of the application.

22. Penalties for certain offences: Any employer who—

(a)   Pays to any employee less than the minimum rates of wages fixed for that employee's class of work, or less than the amount due to him under the provisions of this Act or

(b)   Contravenes any rule or order made under section 13

shall be punishable with imprisonment for a term which may extend to six months or with fine which may extend to five hundred rupees or with both:

Provided that in imposing any fine for an offence under this section the Court shall take into consideration the amount of any compensation already awarded against the accused in any proceedings taken under section 20.

22A. General provision for punishment of other offences: Any employer who contravenes any provision of this Act or of any rule or of order made thereunder shall if no other penalty is provided for such contravention by this Act, be punishable with fine which may extend to five hundred rupees.

22B. Cognizance of Offences:

(1)   No Court shall take cognizance of a complaint against any person for an offence-

(a)   under clause (a) of section 22 unless an application in respect of the facts constituting such offence has been presented under section 20 and has been granted wholly or in part, and the appropriate Government or an officer authorised by it in this behalf has sanctioned the making of the complaint;

(b)   under clause (b) of section 22 or under section 22-A, except on a complaint made by, or with the sanction of, an Inspector.

(2)   No Court shall take cognizance of an offence,

(a)   under clause (a) or clause (b) of section 22, unless complaint thereof is made within one month of the grant of sanction under this section;

(b)   under section 22-A, unless the complaint thereof is made within six months of the date on which the offence is alleged to have been committed.

22C. Offences by companies.– (1) If the person committing any offence under this Act is a company, every person who at the time the offence was committed was in charge of, and was responsible, to the company for the conduct of the business of the company as well as the company shall be deemed to be guilty of the offence and shall be liable to be proceeded against and punished accordingly;[5]

---

5.   The details, analysis and disscusion are beyond scope of this book.

Questions

1. Define the following terms as used in the conditions of contract: (a) owner; (b) contractor; (c) engineer; (d) site; (e) work.   (**Ans.:** Section 7.2)
2. Explain the purpose of the security deposit. What are the provisions made in the clause in respect of security deposit?   (**Ans.:** Section 7.3)
3. Should a contractor be entitled to damages on account of delays caused by the owner? Why?   (**Ans.:** Section 7.5)
4. Explain why delays caused by the contractor may entitle the owner to damages.   (**Ans.:** Section 7.6)
5. Explain the purpose and operation of liquidated damages.   (**Ans.:** Section 7.7)
6. Under what conditions would the contractor's claim for an extension of time be justified?   (**Ans.:** Section 7.6)
7. State the purpose and procedure adopted in respect to partial payments made to the contractor in case of an item rate contract.   (**Ans.:** Section 7.6)
8. Can the principle of partial payment be applied in case of (a) lump sum contracts and (b) cost plus type contracts? If yes, how may these payments be computed?   (**Ans.:** Section 7.6)
9. What is the purpose of' retention money'?   (**Ans.:** Section 7.8)
10. Why are interim payments to be regarded as advances?   (**Ans.:** Section 7.8)
11. When may payments be made at reduced rate? How and by whom are the reduced rates determined?   (**Ans.:** Section 7.8)
12. When are final payments due and what may delay their execution?   (**Ans.:** Section 7.13)
13. Should an engineer be given the authority to make direct payments to subcontractors? Why?   (**Ans.:** Section 7.11)
14. Explain when and why the contractor may claim extra compensation (on account of additions and alterations). In which of these cases may the contractor's claim for compensation be justified?   (**Ans.:** Section 7.9)
15. State the provision made in the conditions of contract in respect to the following:
    (a)   bad work
    (b)   improper materials   (**Ans.:** Section 7.10)
16. Why is subletting of civil engineering contracts prohibited? Under what circumstances may it be desirable to permit subletting?   (**Ans.:** Section 7.11)
17. Why is it desirable to appoint an independent arbitrator?   (**Ans.:** Section 7.15)
18. What qualifications should the arbitrator possess?   (**Ans.:** Section 7.15)
19. Why should the owner be given the right to suspend the work? Why should the contractor be denied such a right?   (**Ans.:** Section 7.5)
20. Why should the conditions of contract provide for labour welfare when the labour is employed by the contractor?   (**Ans.:** Section 7.4)
21. What action will you recommend, if you were to be the engineer, in respect to the following?
    (a)   The contractor fails to pay the security deposit within the specified time limit.
    (b)   The contractor fails to start the work as per schedule.
    (c)   Progress of the work is slow and likely to cause delay.
    (d)   When half the work is finished, the contractor abandons it.
    (e)   When the work is nearing completion, the contractor abandons it.
    (f)   In the course of performance of the contract, the contractor declares himself bankrupt.

(g) The contractor, without permission, sublets the work to another firm.

(h) During the progress of the work the contractor becomes seriously ill and suspends the work.

(i) During inspection a certain portion of the brickwork completed is found defective.

(j) The materials brought on the site by the contractor are not according to the specifications.

(k) On completion of the work, the contractor fails to remove the scaffolding and debris and also to clear the site.

(l) Certain defects in the work are noted after completion, but before the maintenance period is over.

22. Draft suitable clauses to be incorporated in the conditions of contract in respect to the following: (a) subletting; (b) arbitration; (c) extension of time; (d) extras; (e) final payment; (f) suspension of contract; (g) breach of contract.

23. Write a detailed note on provisions made in the standard form contracts in respect to claims and settlement of disputes.   (**Ans.:** Section 7.14)

8 | Specifications

8.1 GENERAL

A specification is a statement of particulars. Construction specifications may be defined as written instructions distinguishing (or limiting) and describing in detail the construction work to be undertaken. Although conditions of contract and specifications have been discussed in Chapter 4 as two separate contract documents, the former are, in effect, the preliminary clauses of a specification and the two should be read together.

8.2 PURPOSE OF SPECIFICATIONS

The specifications are generally written to supplement the information shown on drawings. As such, they cover those features of the work, which can best be described in words. The specifications, when prepared for and included in the contract documents of a project, serve the following purposes.

1. *Contract document:* Specifications serve as a contract document between the owner and the contractor, limiting and describing their responsibilities.
2. *Guide to bidders:* Specifications enable the estimators of the contractors to arrive at a fair price for the work involved.
3. *Guide to supervisors:* Specifications serve as a guide for the fabrication and installation of materials and equipment.

8.3 TYPES OF SPECIFICATIONS

Depending upon the purpose served, specifications are grouped as follows:

1. *Contract specifications:* The specifications developed for a particular construction project to accompany the working drawings are called contract specifications or project specifications. These are further divided into two groups:
    (i)   general specifications, and
    (ii) particular specifications.

These have been discussed in Chapter 4.[1]

2. *Guide specifications:* These are sample specifications prepared to serve as a guide in the preparation of project specifications.

3. *Standard specifications:* These are specifications prepared to cover a specific material, or a group of materials, used by a specific trade or a segment of the construction industry. These specifications invariably, include methods of manufacture, installation, application, and so on. For example, the Indian Standards Institution has prepared volumes of specifications for a wide variety of materials. These specifications are the result of long and intensive study by experts in the field concerned and are generally accepted as authoritative. Standard specifications are usually made project specifications by reference.

4. *Manufacturer's specifications:* Manufacturers of building products or materials publish specifications of their products or materials. These are in the form of guide specifications, standard specifications or installation instructions shipped with the products or materials to the work site.

Experience shows that while drafting project specifications, a specification writer is tempted to copy from specifications of some other similar project or standard specifications, and so on. Many times it results in incorporation of material not relevant or contradicting express provision elsewhere such as wording in the Bill of Quantities, creating difficulties for resolution of disputes, if they arise. This is so because one of the rules of interpretation of documents is to give meaning to every part of what is stated, not to reduce to silence any part and to read harmoniously the document as a whole. It is advisable to score out unwanted material of specifications from which project specifications are proposed to be prepared by cut and paste method.

8.4   IMPORTANCE OF SPECIFICATIONS

The conditions of contract often include a statement describing the procedure to be followed in the event of a conflict between the drawings and the specifications. The specimen provision made in this respect is as follows:

1. In the event of conflict within the drawings, the large-scale details shall govern the small-scale details.

2. In the event of conflict between the specifications and the drawings, the specifications shall govern.

This emphasises the importance of the specifications and the responsibility of the specification writer.

---

1. See Section 4.9

8.5 SPECIFICATION WRITING

Proper writing of specifications does not require as much imagination and originality as it requires visualization, research, clear thinking, organisation and a thorough knowledge of materials and construction methods gained through study and experience. The variety of subject matter that may have to be covered by a given set of specifications is often extensive and beyond the scope of a textbook. However, the general principles of specification writing are substantially the same regardless of the subject matter. Some of these are stated as follows:

1. *Specification language:* Simple and clear language should be used. The use of indefinite or ambiguous words or clauses should be avoided. The specification should require no interpretation regarding the meaning. Rules of grammar should be adhered to. The style and the tense should remain the same. The sentences should be so framed that addition, omission or misplacement of a comma does not alter the sense. Repetition of the noun is preferable to the use of relative pronouns.

2. *Brevity:* The specifications should be brief. Long and involved sentences should be avoided. Standard articles should be specified by references to standard, accepted specifications. The specifications are instructions, and the specification writer should not give reasons for what he specifies.

3. *Fairness:* The specifications should be fair to all parties. The contractor should not be required to assume all the risks and responsibilities for possible errors and omissions of the engineer or architect.

4. *Inclusion of all items:* All items affecting the cost of work should be included and described in such detail as to leave no doubt about the requirement. Repetition of information shown or scheduled on the drawings and duplication within the specifications themselves should be avoided. This will eliminate the possibility of contradictions.

5. *Inapplicable text should not be included:* When old or standard specifications are used for the preparation of a new set, care should be taken to delete inapplicable material.

6. *The text of specifications should be separated into paragraphs:* The paragraphs should follow an established sequence. For easy reference, each paragraph should be distinguished with underlined word or words.

7. *Standard sizes and patterns:* While specifying sizes and patterns, commercial sizes should be specified as far as possible. These sizes are easily obtainable from the stock and are less expensive than special items of unusual dimensions.

8. *Cross-references:* Cross-references to sections and paragraphs of the sections should be minimised to the extent possible. When absolutely necessary, titles should be referred to.

9. *Either specify the means or the end:* Specifications should prescribe the quality of materials and workmanship, and also the desired final result, but they should not attempt to instruct the contractor as to how he should go about attaining the result. For example, the specification for cement concrete should state the desired strength, such as M 150, and not the proportions in which the various ingredients have to be mixed to attain that strength.

10. *Do not specify the impossible:* Nothing that is practically impossible to achieve, and not intended to be enforced, should be specified.

(It may be noted that the above is an attempt to write a specification of 'specification writing' and all the rules mentioned therein have been followed.)

8.6   SPECIFICATION PARAGRAPHS

The paragraphs of a particular specification should be arranged in the following sequence:

1. specifications for materials,
2. combination of materials,
3. preliminary work prior to construction or installation of materials,
4. installation of materials,
5. tests, if any,
6. clearing on completion,
7. schedules, if any,
8. mode of measurement.

8.7   TYPICAL SPECIFICATIONS

Some typical specifications relating to civil engineering works are set out in the following sections. It must be clearly understood that these are but examples (set out to illustrate the points mentioned in the foregoing sections) and subject to a great deal of variation according to the circumstances. Not many items can be covered in a textbook of this size.

8.8   SPECIFICATIONS FOR EXCAVATION

1. *Levels:* The contractor shall notify the engineer before starting excavation to enable the engineer to take levels before the ground is disturbed. The average ground level shall be worked out and used as the datum from which all levels shall be measured.
2. *Length, width and depth:* The excavation shall be carried out to the lengths, widths and depths shown on the drawings or to such other dimensions as the engineer may specify. If by the contractor's mistake, the excavation is made deeper than that shown on the drawings or ordered by the engineer, the extra depth shall be made up with concrete of the foundation grade, at the contractor's cost.
3. *Unsound ground:* If any patches of bad ground are found at the bottom of the foundation, the contractor shall bring them to the notice of the engineer and excavate to such further depth as the engineer may direct and fill up such extra depth with concrete or any other material as instructed by the engineer. Any additional excavation and concrete or other fill material so required shall be paid for at the agreed rates.
4. *Disposal of excavated material:* No materials excavated from foundation trenches shall be placed nearer than 1.5 m from the outer edge of the excavation. The rate for excavation

is inclusive of lifting, removing and sorting out of useful materials and stacking them separately within the lead of 50 m and lift of 1.5 m. All excavated materials shall remain the property of the owner.

5. *Water in excavations:* The contractor shall provide good quality and a sufficient number of pumps and appliances to keep the excavations clear of water, and shall pump out all water that may accumulate in the excavation, during the progress of the work. The sump holes shall be in positions approved by the engineer. The pumping operations shall be so conducted as not to endanger the foundations or stability of any adjoining structures. Unless specially provided for as a separate item in the contract, the rate for excavation shall be inclusive of pumping out all water.

6. *Shoring:* The contractor shall provide all timbering and other supports necessary for securing the sides of trenches and excavations and shall be responsible for their safety. The timbering shall be removed after its necessity ceases and trash of any sort shall be cleaned out from the excavation. Any timber left in permanently at the written direction of the engineer shall be paid for at the agreed rates.

7. *Passing of foundation:* No footing shall be laid until the engineer has approved the dimensions of the excavation, the nature of the foundation material and the levels, and measurements are recorded. The contractor shall remove at his own expense any footing laid before such approval.

8. *Backfilling:* All space between the foundation masonry or concrete and the sides of excavations must be refilled to the original surface with approved materials. The filling shall be done in layers of 15 cm thickness, watered and rammed.

9. *Protection:* The excavation shall be strongly fenced and marked with red lights at night.

10. *Mode of measurement:* Measurements shall be taken in accordance with the procedure specified in the latest edition of the Standard Mode of Measurements for Building Works, IS: 1200-1965, for excavation. The unit of measurement shall be the cubic metre.

8.9 SPECIFICATIONS FOR CONCRETE MATERIALS

8.9.1 Cement

Ordinary Portland cement shall be used. The cement shall conform to IS: 269 latest edition. It shall be of make and quality approved by the engineer.

Storage of cement: Cement shall be stored on raised platforms 15 cm above the floor level and 30cm away from the walls under a leak-proof roof. Not more than 15 bags shall be stacked vertically in one pile and the maximum width of the pile shall not be more than 3 m. Cement shall be stored in a way that shall permit the removal and use of cement in the chronological order of receipt. Cement remaining in store for more than 60 days from the date of receipt from the factory shall be subjected to tests as indicated in the IS: 269 latest version and used only if found satisfactory.

A suitable weighing scale shall be provided by the contractor and bags underweight by more than 2% of the normal weight shall be rejected and removed from the site. The cement used for any type of concrete shall always be measured by weight and one cubic metre shall be taken as weighing 1440 kg.

8.9.2 Fine Aggregate

Fine aggregate shall be natural sand obtained from a river bed and shall conform to IS:383-1952 and to the relevant portion of IS: 515-1959. Sand shall be clean, hard, strong, durable and free from organic matter, dust, clay, shale, alkali, salts, soft or flaky particles or other injurious substances. Sand shall be washed if it contains more than 5 (five) per cent of clay, dust or silt. Sand shall be graded within limits given in Table III or Table IV in Paragraph 5.2 of IS: 383-1952. The fineness modulus shall range between 2.6 and 3.6.

8.9.3 Coarse Aggregate

Coarse aggregate shall consist of crushed stone and shall be hard, strong, durable and clean. The coarse aggregate shall be crushed from the stone obtained from the quarries approved by the engineer. The coarse aggregate shall meet the requirement of IS: 383 and IS: 515 latest version. The coarse aggregate shall be graded between 6 mm and 40 mm (20 mm in case of RCC work). Grading tests shall be carried out as indicated in IS: 383 and IS: 456 latest version at the cost of the contractor.

8.9.4 Water

Potable water obtained from the sources approved by the engineer shall be used for mixing and curing concrete.

8.9.5 Proportioning Mix

The mix of the coarse and fine aggregate, cement and water shall be designed by preliminary tests to give dense and workable concrete, requiring a minimum quantity of cement to give_________ kg/cm^2 strength after 28 days of curing. The preliminary tests shall be carried out with the materials and the method of compaction that will be actually used on site.

8.9.6 Mixing

Concrete shall be mixed in a mechanical mixer of the approved pattern at the site of work. The drum shall be rotated for a minimum period of two minutes.

Hand mixing: If the quantity of concrete to be mixed is less than 1 m^3, hand mixing shall be permitted. Mixing shall be carried out on smooth watertight platforms of sufficient size to allow efficient turning of the mix. The platform shall be so arranged that the mixing water shall notflow out and no foreign material shall get mixed with the concrete. Cement and sand shall be thoroughly mixed to give a uniform colour of the mixed cement and sand. Coarse aggregate shall then be added. The mix shall be turned three times in the dry condition and three times after adding water. Water shall be added by a hose.

Batching and mixing in weight batch mixer: The batching plant shall be capable of proportioning the materials by weight, each type of material being weighed separately. The cement from the bulk stock shall be weighed separately from the aggregate. The capacity of batching and mixing plant shall be at least 25 per cent higher than the proposed capacity for the laying arrangements. The batching and mixing shall be carried out preferably in a forced action central batching and mixing plant having necessary automatic controls to ensure accurate proportioning and mixing. The accuracy of the weighing scales of the batching plant shall be within ±2 per cent in the case of aggregates and ±1 per cent in the case of cement and water.

8.9.7   Transporting

Plant mixed lean concrete shall be discharged immediately from the mixer, transported directly to the point where it is to be laid by tippers or dumpers within the travel time allowed. The concrete shall be covered by tarpaulin during transit.[2]

8.9.8   Preparations for Concreting before Laying the Concrete

The final surface of the excavation shall be cleaned and thoroughly wetted.[3] All moulds and centering shall be examined and the insides thoroughly cleaned out. Timber moulds and top surfaces of walls which come in contact with the concrete shall be thoroughly wetted.

8.9.9   Laying Concrete

Concrete shall be placed in position, rammed, vibrated and finished within 30 minutes of adding water to the mix. Concrete must not be dropped from a height greater than 60 cm, but shovelled or tipped gently into its final location.

8.9.10   Compaction of Concrete

Concrete shall be properly compacted by an internal vibrator (or by rodding where the vibrator cannot be used) until the concrete is thoroughly compacted and the slurry begins to appear on the surface. No traffic shall be allowed over the concrete until it has set.

---

2. Recently, use of transit mixers for transporting ready mix concrete (RMC), from a batching plant to the site of the work has become common.

3. For concreting in foundation trenches only.

8.9.11   Stops in Concrete

Where it is found necessary to leave off concreting, the concrete shall be finished off vertically except in case of columns and beams, where the construction joints shall be horizontal. Before starting to deposit new concrete, the surface of the existing concrete shall be hacked, wire-brushed, washed down and a thick grout composed of one part of cement and two parts of fine sand applied before the new concrete is placed against it.

8.9.12   Moulds and Centering

The detailed design of moulds and centering shall be prepared by the contractor and approved by the engineer. Formwork shall be so constructed and braced that it shall be durable, strong and rigid, in order to retain the weight of wet concrete and carry working loads without appreciable movement. All joints shall be watertight. Before placing the reinforcement, the inside face of moulds in contact with the concrete shall be oiled. Moulds for beams shall be fixed in perfect alignment and so constructed that the sides can be removed first. The centering for a slab must be laid perfectly flat or true to the slope shown on the drawings. No formwork shall be removed without the permission of the engineer and, while removing, care shall be taken so as not to jar the structure.

8.9.13   Tests

Tests shall conform to the specifications laid down in IS: 456-1962.

8.9.14   Curing

Fresh concrete shall be protected against the inclemency of weather and covered with saturated sacking or similar material. After final setting the concrete shall be kept continuously wet, by ponding water or by any other suitable means approved by the engineer, for a period not less than 14 days from the date of placement.

8.9.15   Mode of Measurement

Measurements shall be taken in accordance with the Standard Mode of Measurement for Building Works (the latest edition of IS: 1200-1965). The unit of measurement shall be the cubic metre.

8.10   SPECIFICATIONS FOR REINFORCEMENT

1. *Steel:* Mild steel reinforcement shall conform to the specifications of the latest IS: 432, 1139, 1786, 1566, 226 and the relevant part of IS: 456 as the case may be.

2. *Cleaning steel:* All the reinforcements shall be thoroughly cleaned of rust, dust, scale, oil and dirt before being assembled.

3. *Bending:* All steelwork shall be bent cold to the shape and dimensions shown on the drawings by the gradual and uniform application of force.

4. *Placing in position:* Reinforcing steel shall be correctly spaced and secured in position by approved wire and shall not be displaced in any way after being fixed in the moulds. Concrete blocks shall be used to provide the specified cover.

5. *Engineer's approval:* Concreting shall not be commenced until the reinforcement has been inspected and approved by the engineer.

6. *Tests:* The contractor shall produce the test certificate of the manufacturer. If necessary, tests shall be carried out according to IS: 223-1950.

7. *Mode of measurement:* Measurements shall be taken in accordance with the Standard Mode of Measurement of Building Works (IS: 1200-1965). The unit of measurement shall be the quintal.

8.11 SPECIFICATIONS FOR BRICKWORK (FIRST SORT)

1. *Bricks:* Bricks shall be of uniform size (19 cm × 9 cm × 9 cm), moulded on a table with sand and thoroughly well-burnt but not over-burnt. The bricks shall have plane rectangular faces with parallel sides and sharp, straight, right-angled edges. They shall give a clear metallic sound when struck and shall be free from lumps, laminations and cracks. On saturation, the increase in weight shall not exceed 20 per cent of the dry weight of the brick when tested according to the latest IS: 1077. The crushing strength of the brick shall not be less than 40 kg/cm^2 when dry and 30 kg/m^2 when saturated. No bats or half bricks shall be used in work except as closers.

2. *Cement:* As per Section 8.9.1.

3. *Sand:* As per Section 8.9.2.

4. *Water:* Potable water shall be used for mixing mortar.

5. *Proportion:* Cement and sand shall be mixed in the proportion of 1:4 by volume.

6. *Mixing:* Similar to Section 8.9.6 except that coarse aggregate is not to be added.

7. *Wetting bricks:* All bricks shall be soaked in water for at least 2 hours before they are utilised in the work.

8. *Scaffolding:* The scaffolding shall be double, sufficiently strong so that it is safe for all construction operations and must be approved by the engineer.

9. *Old work:* The tops of any walls, where work has been suspended, shall be cleaned and thoroughly wetted before work is recommenced.

10. *Laying:* The bricks shall be laid flush in cement mortar with the frog upward, in English bond, unless otherwise specified. The brickwork shall be true to line and plumb, and radiated or curved as shown in the drawings.

11. *Face work:* Bricks of uniform colour shall be used, when the face work is not to be plastered. Brick faces, which are to be plastered or rendered, shall have their joints raked back at least 12 mm.

12. *Joints:* Joints shall not exceed 8 mm in thickness. Bed joints shall be truly horizontal and side joints truly vertical. Joints shall be neatly struck at the close of the day's work.

13. *Uniform raising:* Walls shall be carried up regularly not leaving any part more than 60 cm lower than another. If absolutely necessary, breaks shall be stepped.

14. *Cleaning:* On completion, brickwork shall be washed and any stains or mortar lumps removed from the face.

15. *Curing:* The brickwork shall be kept well-watered continuously for at least 14 days after laying. At no time during the curing period shall the mortar be allowed to dry.

16. *Mode of measurement:* The brickwork shall be measured in accordance with the latest IS: 1200. The unit of measurement shall be the cubic metre.

8.12 SPECIFICATIONS FOR UNCOURSED RUBBLE MASONRY

1. *Building stone:* Stones to be used shall be of the best quality obtained from the quarries approved by the engineer. They shall be tough, hard, dense, sound, durable and of uniform colour and free from flaws, veins, skin, oil, dirt, imperfections and other defects. The stone, when immersed in water for 24 hours, shall not absorb more than 5 per cent of its dry weight of water. The contractor shall submit samples of stones proposed to be used and the engineer shall retain these samples for comparison with the actual stones supplied.

2. *Cement:* As per Section 8.9.1.

3. *Sand:* As per Section 8.9.2.

4. *Proportion:* As per Section 8.11.4.

5. *Mixing:* As per Section 8.11.5.

6. *Size of stone:* Each stone shall be 15 cm thick and 15 cm wide at its thickest part and no stones shall be less than one and one half times its height in length.

7. *Dressing:* Stones shall be hammer-dressed. Weak corners and edges shall be removed. All face stones shall have their sides dressed in straight lines and these sides shall be in one plane.

8. *Face stones:* Stones with larger sides, having good beds, shall be used for face work. At least 50 per cent of the stones shall be more than 10 litres in volume. The beds and joints shall have a minimum bearing of not less than 2 cm.

9. *Through stones:* Through stones shall be about 0.03 m^2 in face area and shall occupy the full width of the wall for walls up to 60 cm in thickness. If the thickness of the wall is more than 60 cm, a line of headers overlapping each other by at least 15 cm shall be provided. Through stones shall be marked on their faces.

10. *Hearting and backing stones:* Thirty per cent of the stones used in hearting and backing shall exceed 10 litres in volume.

11. *Quoins:* Quoins shall be of 30 cm × 15 cm in size. The faces of the quoins shall be hammer-dressed. Each side of the exposed corner shall be chisel-drafted to a width of 40 mm. The beds and tops of quoins shall be dressed square to the face and rough-tooled to 10 cm from the face. The quoin shall tail into the wall to lengths

of not less than 20 cm, measured at right angles to the shorter and longer faces, respectively.

12. *Scaffolding:* Scaffolding shall be single and sufficiently strong so that it is safe for all construction operations. The ends of poles shall not be placed in the position of header stones. Putlog holes shall be made good by stones to match the face work and holes at the back shall be filled in with PCC of M 85 mix.

13. *Laying stones:* The bed, which is to receive the stone, shall be cleaned, wetted and covered with a layer of fresh mortar. Stratified stones shall be laid on their natural beds. All stones shall be laid full in mortar both in bed and side joints. Face stones shall be laid without any pinning on the exposed faces. Clean chips and spalls shall be wedged into the mortar joints in hearting. No hollows shall be left in the masonry. No dressing or hammering shall be done after the stone is set in place. Quoins shall be laid stretcher and header in alternate layers as seen on each face.

14. *Quantity of mortar:* The volume of mortar used per cubic metre of masonry shall be between 0.30 m^3 and 0.35 m^3.

15. *Bond:* Through stones or line of headers, as the case may be, shall be laid at the rate of two stones per metre, square of face, and shall be evenly distributed. Each stone in a course shall break joint with the stone in course below or above by at least 8 cm. At the junction of the walls, the stones at each alternate course shall be carried into each of the respective walls.

16. *Rate of raising masonry:* Face work and hearting shall be brought up evenly, but the top of each course shall not be levelled. As far as possible, the whole of the masonry in any structure shall be carried up to a uniform level throughout. The rate of raising shall not exceed 60 cm per day. When absolutely necessary, the breaks shall be stepped.

17. *Joints:* No face joint shall exceed 16 mm in thickness. All joints shall be raked to a depth of 16 mm before the mortar has set.

18. *Curing:* All masonry work shall be kept continuously wet for 14 days from the date of laying.

19. *Cleaning:* Exposed faces of masonry shall be washed so as to remove all stains and adhering mortar.

20. *Mode of measurement:* Work shall be measured as specified in the latest Standard Mode of Measurements for Building Works IS: 1200. The unit of measurement shall be the cubic metre.

8.13   SPECIFICATIONS FOR COURSED RUBBLE MASONRY (FIRST SORT)

1. *Stone:* As per Section 8.12.1.
2. *Cement:* As per Section 8.9.1.
3. *Mortar:* As per Section 8.11.3 to 8.11.6.
4. *Size of Stones*

(i) *Khandki:* The height of the *Khandki* stones shall be 15 cm. All *Khandki* stones shall have a larger breadth than height and shall tail back into the masonry 1.5 times the height. Thirty per cent of the face stones shall not be less than 20 litres in volume.

(ii) *Quoins:* These shall be of the same height as that of the course. The length of quoins on the longer face shall not be less than twice their height and on the shorter face not less than their height. The quoin shall tail into the wall to lengths of not less than 20 cm and 10 cm, measured perpendicularly to the shorter and longer sides, respectively.

(iii) *Headers and through stones:* The height of the through stones shall be the full height of the course and the width shall not be less than the height. (Rest as per Section 8.12.9.)

(iv) *Hearting and backing:* As per Section 8.12.10.

5. *Dressing*

(i) *Face stones:* The *Khandki* stones shall be dressed so as to have the vertical and horizontal sides straight and at right angles to the adjacent sides. Bushing on the face of the stones shall not project more than 4 cm. The beds and tops shall be rough-tooled to at least 8 cm from the face and the vertical faces shall be rough-tooled to at least 4 cm from the face.

(ii) *Quoins:* The faces of the quoins shall be rough-tooled and the sides of the exposed corner shall be provided with a chisel draft of about 40 mm. The beds, tops and sides shall be dressed in the same manner as *Khandki* stones.

(iii) *Hearting and backing:* As per Section 8.12.10.

6. *Scaffolding:* Scaffolding shall be double and sufficiently strong to be safe for all construction operations.

7. *Laying:* The stones shall be laid in a horizontal course of 15 cm height. All courses shall be of equal height. The quoins shall be laid header and stretcher in alternate layers as seen on each face of the wall and shall correspond to the arrangement of quoins in the same course. The unexposed faces and those faces which have to be plastered shall conform to UCR Masonry Specifications.

8. *Bond:* As per Section 8.12.15, except that through stones or line of headers (as the case may be) shall be laid 1.5 m apart in each course and shall be staggered in successive courses.

9. *Quantity of mortar:* The volume of mortar used per cubic metre of masonry shall be between 0.25 and 0.3 m^3.

10. *Joints:* The thickness of joint shall not exceed 10 mm. Horizontal joints shall be raked to a depth of 10 mm before the mortar has set.

11. *Rate of raising masonry:* As per Section 8.12.16.

12. *Curing:* As per Section 8.12.18.

13. *Cleaning:* As per Section 8.12.19.

14. *Mode of measurement:* Measurements shall be taken in accordance with the Standard Mode of Measurements for Building Works, IS: 1200-1965. The unit of measurements shall be the cubic metre.

8.14   SPECIFICATIONS FOR CEMENT PLASTER

1. *Cement:* As per Section 8.9.1.
2. *Sand:* As per Section 8.9.2.
3. *Water:* Potable water shall be used for mixing the mortar.
4. *Proportion:* Cement and sand shall be mixed in the proportion of one part of cement to four parts of sand.
5. *Mixing:* Mortar shall be mixed in a mechanical mixer of an approved pattern at the site of work. The drum shall be rotated for a minimum period of 2 minutes. The mortar shall be used within 30 minutes of adding water.
6. *Hand mixing:* Similar to hand mixing for concrete as specified in Section 8.9.6, with the difference that the coarse aggregate shall not be added.
7. *Scaffolding:* Shall be as per Section 8.11.8 or Section 8.12.12, depending upon the type of surface to be plastered.
8. *Preliminary work prior to plastering*
   (i) *In case of old work*—all joints shall be raked out to a depth equal to their width.
   (ii) Smooth surfaces, if any, shall be roughened.
   (iii) All dirt, dust, oil, paint and so on, shall be removed.
   (iv) The surface thus prepared shall be kept wet for at least 6 hours prior to plastering.
9. *Gauges:* Patches of plaster 15 cm × 15 cm shall be put on about 3 m apart as gauges. The surfaces of all such patches shall be in one vertical plane.
10. *Plastering:* The mortar shall be firmly applied on the surface from top downward and well pressed into the joints. The mortar shall then be rubbed and levelled with a flat wooden rule until a perfectly plane and even surface is achieved. All corners shall be finished true to their angles, unless otherwise directed by the engineer. The jambs and reveals of door and window openings shall be finished perpendicular to the sill and lintel bottom. Plastering shall be done in squares or strips as directed by the engineer. Strips or squares shall be so formed that day-to-day breaks are made to coincide with architectural breaks.
11. *Finish:* While the plastered surface is fresh, a thick coat of cement slurry shall be applied and rubbed smooth.
12. *Curing:* All plastered surfaces shall be kept continuously damp for a period of 14 days.
13. *Mode of measurement:* Measurements shall be taken in accordance with the latest Standard Mode of Measurements for Building Works IS: 1200. The unit of measurement shall be the square metre.

8.15   SPECIFICATIONS FOR CEMENT POINTING

1. *Cement mortar:* As per Sections 8.14.1 to 8.14.5.
2. Scaffolding: As per Sections 8.12.8 or 8.12.12, depending upon the type of surface to be pointed.
3. *Preliminary work prior to pointing*

(i) The joints in the masonry shall be raked out to a depth not less than the width of the joint.

(ii) The joints shall then be brushed clean of dust and loose particles.

(iii) The surface shall be washed and the joints thoroughly wetted before pointing is commenced.

4. *Pointing*

(i) The joints shall be filled with mortar of a specified mix and required consistency. The mortar shall then be well pressed and rubbed smooth.

(ii) A 3 mm diameter string shall be stretched on the centre line of the joint and pressed with a trowel until a semi-circular depression is made.

(iii) The depressions so formed shall be immediately rubbed with a *nayla* till they become 6 mm deep and wide, uniformly.

(iv) Care shall be taken to see that at the intersections of the joints the depressed lines do not cross, but just touch each other. The mortar shall be restricted to the joint area only. Any excess mortar spreading on adjoining stones (or bricks, as the case may be) shall be removed with a trowel.

5. *Curing:* The pointed area of masonry (or brickwork) shall be kept continuously wet for 14 days.

6. *Mode of measurement:* Measurements shall be taken in accordance with the latest Standard Mode of Measurements for Building Works, IS. 1200. The unit of measurement shall be the square metre.

8.16   SPECIFICATIONS FOR SHAHABAD STONE FLOORING

1. *Shahabad stone:* Shahabad stones shall be hard, tough, durable, free from soft veins, cracks or flaws and shall be of uniform colour. The thickness of the stones shall be 25 mm and the edges shall be dressed to a depth of 10 mm from the top. The size and shape of the stones should be approved by the engineer.

2. *Concrete:* Plain cement concrete of M 60 strength shall be used in the bedding.

3. *Mortar:* Cement mortar shall consist of one pan of cement mixed with four parts of sand and the minimum quantity of water to give the required consistency.

4. *The plinth filling:* The filling shall be fully watered and rammed to the desired slope prior to placing of the cement concrete.

5. *Plain cement concrete:* Concrete of the specified depth shall be laid to the required level and slope, well rammed and cured for 14 days. The average thickness of the cement concrete shall be 10 cm.

6. *Before spreading mortar:* The top of the concrete bedding shall be cleaned of all dirt, dust and loose material and well wetted. The mortar shall then be evenly and smoothly spread over the base, by the use of screen battens, only over so much area as will be covered with stone slabs within 30 minutes. The average thickness of the mortar bed shall be 25 mm.

7. *Laying Shahabad stones:* Before being laid, the stones shall be wetted in clean water. Cement grout shall be spread over the mortar bed. The stones shall be evenly and firmly bedded to the required level and slope, and gently tapped with a wooden mallet to ensure proper and firm laying without any hollows below.

8. *Joints:* All joints shall be in straight lines and of uniform thickness of 8 mm. The joints shall be raked out to a depth of 8 mm when the mortar is green.

9. *Pointing:* All joints shall be pointed with cement mortar in a 1:3 proportion.

10. *Curing:* The flooring shall be kept wet continuously for 14 days.

11. *Cleaning:* The flooring shall be thoroughly cleaned of all mortar stains and other debris.

12. *Mode of measurement:* Measurements shall be taken in accordance with the Standard Mode of Measurements for Building Works, IS: 1200-1965. The unit of measurement shall be the square metre.

8.17 SPECIFICATIONS FOR DOORS, WINDOWS AND VENTILATORS (FIRST CLASS)

8.17.1 Frames

1. *Timber:* The timber to be used shall be Indian teak wood, well-seasoned, uniform in colour, straight in fibre and free from sapwood, knots, flaws, shakes, warps, cups, twists, bends or defects of any kind. The maximum permissible moisture content for timber shall be twelve (12) per cent.

   Tolerance-, (a) The maximum slope of the grain shall not exceed 1 in 20. (b) No individual live knot shall exceed 25 mm in diameter and the aggregate area of all knots shall not exceed one (1) per cent of the area of the piece.

2. *Hold-fasts:* Hold-fasts shall be made of wrought iron and shall be of 300 mm × 40 mm × 5 mm size with two holes drilled in for fixing screws. The hold-fasts shall be coated with coal tar and fixed to the frame with iron screws. The hold-fasts shall be fully embedded in the mortar joints of the wall.

3. *Workmanship:* All woodwork shall be neatly and truly finished to the exact dimensions as shown on the drawings. Horns in sills and heads shall be 15 cm long.

4. *Joints:* Unless otherwise specified, joints shall be simple tenon and mortise joints with the end of the tenon exposed to view. All joints shall fit truly and fully without wedging or filling.

5. *Engineer's approval:* The woodwork shall be inspected and passed by the engineer before it is erected in position. The primary coat of paint or other treatment shall be given after the engineer has passed the woodwork, but before it is erected in place.

6. *Erection:* The frames shall be erected in position and held truly vertical with strong supports from both sides and built-in masonry. The shutters shall be fixed later.

7. *Painting:* The faces of frames in contract with the masonry shall be given two coats of hot coal tar. The exposed faces of the frame shall be finished with two coats of oil paint

of the approved colour and shade (or with French polishing/varnishing as mentioned in the description of the item).

8. *Mode of measurement:* Measurements shall be taken in accordance with the Standard Mode of Measurements for Building Works, IS: 1200-1965. The unit of measurement shall be the cubic metre.

8.17.2 Shutters

1. *Timber:* As per Section 8.17.1.
2. *Fixtures and fastenings:* All fixtures and fastenings shall be of brass, which shall be new, sound and strong and fixed to the joinery in a secured and efficient manner. Unless otherwise specified, each leaf shall be hung with three brass butt hinges with brass screws. Each window shall have a handle, tower and bottom bolts. The type and quality shall be approved by the engineer.
3. *Workmanship:* As per Section 8.17.1.
4. *Panels:* The doors and windows shall have double leaf shutters. Each panel shall be in a single width piece and shall have beaded edges on both sides.
5. *Glazing*[4] (for doors/windows/ventilators): The sizes of the openings for glazing shall be as detailed on the drawings. Unless otherwise specified, 8.5 kg/m^2 glass, plain or obscured, and conforming to IS: 1861-1960 shall be used. The putty to be used for fixing the glass in frames shall conform to IS: 419-1953. The glass panes shall be properly cut to fit the rebates of the frames and sashes truly with about 1.5 mm minus margin on all sides. Before glazing, the shutters shall be given a primary coat. The rebates shall be puttied first and the glass panes shall then be pressed into position and secured with glazier's springs and firmly back-puttied and the rebates neatly chamfered. After a week, the putty shall be painted along with the shutter.
6. *Mode of measurement:* Measurements shall be taken in accordance with the latest Standard Mode of Measurements for Building Works, IS: 1200. The unit of measurement shall be the square metre.

8.18   SPECIFICATIONS FOR MANGALORE TILE ROOF

1. *Tiles:* Mangalore pattern class A tiles conforming to the latest IS: 654 shall be used.
2. *Battens:* Battens shall be sown true to line, width and thickness, shown on the drawings, from country teak wood. Timber to be used shall be well-seasoned and free from knots, warps, twists, bends or defects of any kind. The exposed surface of the battens shall be oiled with one coat of linseed oil, while the portion of the battens embedded in masonry shall be coated with coal tar.

---

4. Applicable to partly or fully glazed doors, windows and ventilators only.

3. *Iron work:* All mild steel flats, bars, nails, bolts and nuts, and binding wire to be used in the work shall be of approved quality. Iron work to be embedded in the masonry shall be coated with coal tar while the remaining shall be brushed over with linseed oil.

4. *Preliminary work:* Before fixing battens, the lines, levels and slopes of the trusses, ridges, purlins and rafters shall be verified, and defects, if any, shall be rectified so that the tiled surfaces shall be regular plane surfaces, without any unevenness.

5. *Fixing battens:* The battens of specified size shall be fixed to the upper surface of the rafters with iron nails. The spacing between the top edges of the successive battens shall be 31.85 cm. The lowest battens shall have double the ordinary thickness and shall be fixed at a distance of 25 cm from the one immediately above it. The bottom faces of the battens shall be paned smooth and the top edges shall be perfectly straight and parallel to the eaves on each face of the roof slope and to the walls.

6. *Laying tiles:* Tiles shall be laid square and breaking joint, fitting properly with respect to one another and the 'catches' resting fully against the battens. The titles at the hips and valleys shall be cut to the required shapes. The finished top surface of the tiled roof shall be in uniform planes from the ridge to the eaves. All lines of tiles in each surface shall be perfectly straight, horizontal and parallel to each other. The tiles at the gables, eaves or other exposed parts shall be tied with an 18 gauge wire underneath the battens. In addition, a steel flat of 40 mm × 3 mm shall be provided near the eaves with the necessary arrangement for fixing the same through the Mangalore tiles to the common rafters by 12 cm-long screws at every alternate rafter.

7. *Mode of measurement:* Measurement shall be taken in accordance with the latest Standard Mode of Measurements for Building Works, I.S: 1200. The unit of measurement shall be the metre square.

8.19   SPECIFICATION FOR ROAD AND BRIDGE WORKS

A large-scale highway modernisation programme has been undertaken in India, introducing innovative materials and techniques in highway construction and maintenance. The Ministry of Road Transport and Highways of the Government of India has published through the Indian Roads Congress, New Delhi 'Specifications for Road and Bridge Works' which is revised periodically. The agreement specifies the edition to be referred to in a given case.

Most of the contracts for road and bridge works specify the work to be carried out in accordance with the above-said specifications with occasional variations to suit the requirements of a given project. The general practice is to include reference to the specification number in the description of the item of work in the bill of quantities itself. For example: an item of earthwork in excavation is described as follows:

'Earth work in excavation necessary for construction for roadway in all types of soil including all leads and lifts complete as per Technical Specification clause 301.'

Clause 301 under Section 300 describes the work in full under the headings: scope, classification of excavated materials, construction operations methods, tools and equipment, excavation in different types of soil such as rock, de-watering, disposal of excavated materials, backfilling, finishing operations, mode of measurements, rates, and so on.

It is beyond the scope of this book to include detailed specifications and the reader may refer to the standard specifications directly.

Questions

1. What is a specification?   (**Ans.:** Section 8.1)
2. State the different purposes served by specifications.   (**Ans.:** Section 8.2)
3. What are standard specifications? What are the precautions to be taken while using standard specifications?   (**Ans.:** Section 8.3 and 8.5.6)
4. What are the principles of specification writing?   (**Ans.:** Section 8.5)
5. Write detailed specifications for the following items of work:
    (a) excavation in trenches for laying drainage pipes
    (b) plain cement concrete in foundation
    (c) RCC slab
    (d) brickwork-II sort in cement mortar
    (e) UCR masonry in foundation and plinth
    (f) coursed rubble masonry-I sort in cement mortar
    (g) partly panelled and party glazed door shutter
    (h) wood work in doors and windows
    (i) cement plaster to UCR masonry
    (j) cement pointing to CR masonry
    (k) reinforcement for RCC columns, beams and slabs
    (l) Mangalore tile roof on CT wooden battens
6. Write a brief note on Standard Specifications for Road and Bridgeworks.   (**Ans.:** Section 8.19)

Contract Organisation and Management

9.1 GENERAL

It is increasingly being realised by most of the contractors that organisation and management of a construction business differ only in detail from managing any other profit-making enterprise. However, there are many contractors who still fail to use the basic proven methods of organisation and management. This may be so because they fail to see the changes that are taking place in the industry as a whole and in the construction industry in particular. These days, work projects have become so large and complex, so influenced by organised labour rules and speeded up by the introduction of modern machinery that without the knowledge and understanding of management and all that it implies, one will be in danger of facing disaster.

Introduction of scientific management in the construction industry, it is claimed, will result in an increase in building productivity, simplification of work and rise in profits.

'Organisation and management of the construction business' is in itself a subject for a separate volume. What follows is an introduction to the basic ingredients of management and an introduction to a few modern methods of planning and scheduling large construction projects.

9.2 ORGANISATION

The organisation of a business is defined as the arrangement of the people in it so that they act as one body. Building up of the organisational structure (framework in which people act) is one of the major tasks of the chief executive and is of the utmost importance. While appointing the right people in the right places, he has not only to see the present state of the company but also to look ahead and ensure that the present-day staff are tailored and trained to meet future requirements.

The general organisation and distribution of functions within a firm depend to a large degree upon its size and the type of work generally undertaken by it. The following types of organisational structures are commonly encountered:

1. military or line-type organisation,
2. line-and-staff organisation, and
3. functional organisation.

9.3   MILITARY OR LINE-TYPE ORGANISATION

The line-type or military organisation is the oldest and simplest form of organisation. In its purest form, it involves a leader and a few workmen. This is illustrated as follows (Fig. 9.1).

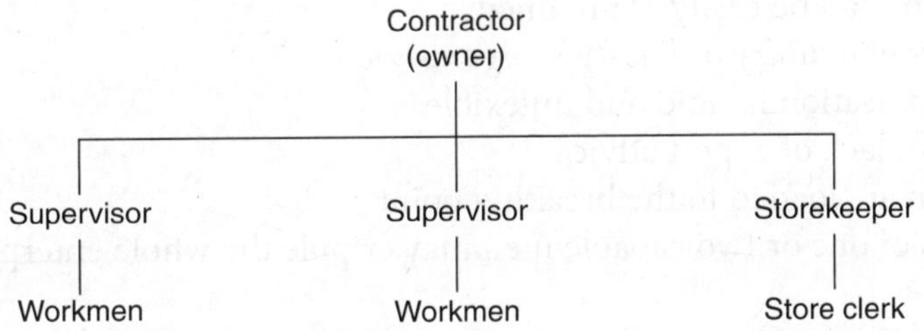

Fig. 9.1   Military or line-type organisation

The leader has authority over the workmen and directs them in respect of what to do, when to do and how to do. The workmen know where exactly they stand in relation to their superior. This helps maintain discipline which is an essential and important factor in any type of organisation.

At present, construction works are so large and complex that one man alone is hardly able to perform the varied duties he is called upon to do and other people have to be brought into the organisation. These additional persons are specialists such as engineers, office managers and general superintendents. The duties of the contractor are partly performed by these men acting as heads of their respective departments and the contractor acts primarily as an administrative head. All details of the different departments have to be cleared through the contractor. These points are illustrated in Fig. 9.2.

The figure shows a typical example of a military or line-type organisation of a medium-sized firm engaged mainly in civil engineering and public works construction.

1. *Advantages of military or line-type organisation*
    (i)   It is simple.
    (ii)   There is a clear-cut division of authority and responsibility.

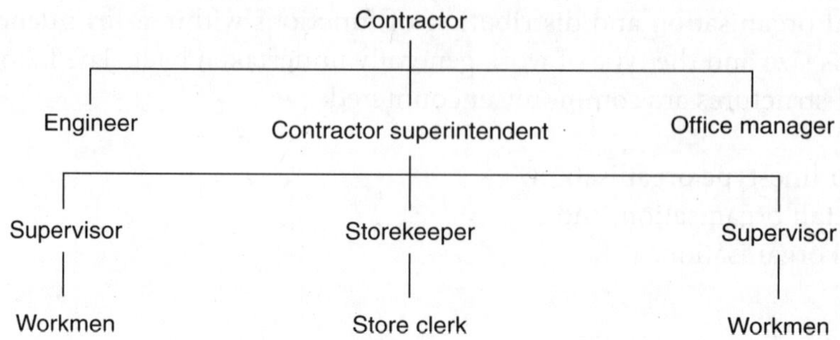

Fig. 9.2   Expanded one-man organisation

   (iii)   It is stable.

   (iv)   Quick action can be taken.

   (v)   Discipline can be easily maintained.

2.   *Disadvantages of military or line-type organisation*

   (i)   The organisation is rigid and inflexible.

   (ii)   There is a lack of expert advice.

   (iii)   Key men are loaded to the breaking point.

   (iv)   The loss of one or two capable men may cripple the whole enterprise.

9.4   LINE-AND-STAFF ORGANISATION

The line-type organisation discussed above is not suitable for firms undertaking the construction work of large and complex projects, where the key men in the organisation have to be assisted, helped and advised by the specialists in different fields. The specialists known as 'staff employees' work under the project manager (who may be a contractor, partner or qualified civil engineer with long-standing experience working directly under the contractor), but unlike in a military organisation, they can communicate directly with each other, or field employees with regard to advice and directions. As they are supposed to give advice and not direct instructions to the line employees, they do not share the responsibility for performance of the line people on the bottom of the chart.

Figure 9.3 shows a typical example of a line-and-staff organisation. The unbroken lines show the authority and the broken lines show contact and communication.

1.   *Advantages of line and staff organisation*

   (i)   The organisation is based on planned specialisation.

   (ii)   It brings expert knowledge to bear upon the management.

   (iii)   Most matters are handled directly by the staff employees leaving the project manager and general superintendent more time to concentrate on getting the work done at the right time and cost.

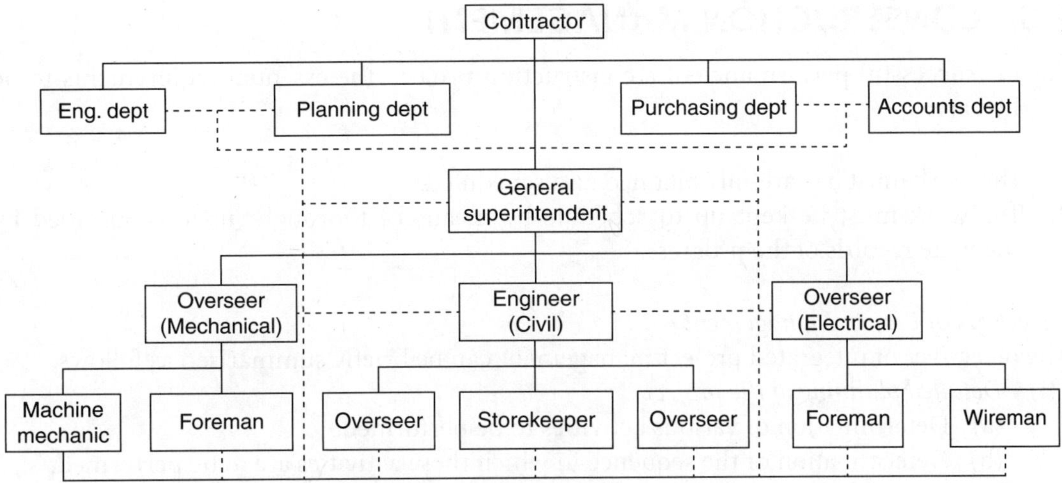

Fig. 9.3   Line-and-staff organisation

(iv) It provides more opportunities for advancement for capable and efficient workers as a greater variety of responsible jobs are available.

2. *Disadvantages of line and staff organisation*
   (i) There may be confusion about the relation between staff and line employees.
   (ii) The staff may be ineffective due to lack of authority to carry out its recommendations.
   (iii) Line employees may resent the activities of staff members, feeling that the prestige and influence of linemen suffer due to the presence of the specialists.

9.5   FUNCTIONAL ORGANISATION

In line-and-staff organisation, the staff departments seldom restrict their activities to advisory or planning capacities. At times, the staff officers may be more influential than some line department heads. Thus, there is disparity between theory and actual practice. This disparity is eliminated in the functional organisation, wherein the authority of staff specialists over the line officials at the lower level is recognised. In all other respects it is similar to the line-and-staff organisation.

1. *Advantage of functional organisation*
    It attempts to remove the drawbacks of the line-and-staff organisation listed above.
2. *Disadvantages of functional organisation*
    (i) It is difficult to define who the 'boss' is.
    (ii) It de-emphasises the position of the line organisation.
    (iii) It may present special difficulties of coordination.

9.6   CONSTRUCTION MANAGEMENT

For the successful performance of a construction project, the essential requirements to be satisfied are:

1. The work must be carefully planned and scheduled.
2. The work must be kept up to schedule by means of thorough supervision aided by adequate records of the progress.

Objectives of Project Management
The objectives of integrated project management can be briefly summarised as follows:
 (i) *Detailed planning of the project*
      (a) Determination of various activities to be performed.
      (b) Determination of the sequence in which these activities are to be performed.
      (c) Determination of the overall requirements of the resources required.
 (ii) *Developing realistic schedules*
      Preparation of a time-table showing the start and finish dates of the various activities involved.
(iii) *Exercising effective control*
      (a) Periodic evaluation of the progress made, and the comparison of the actual progress with that planned.
      (b) Projecting the effect of the current status of the various activities on the project completion time.
      (c) Corrective action to be taken as and when necessary.
           The above cycle is to be repeated at periodic intervals right from the start of the project to its completion. This is achieved by holding monthly progress review (MPR) meetings. Minutes are maintained and an action plan is drawn.
(iv) *Ensuring optimum use of resources*
      The resources are men, materials, machinery and money.

9.7   PLANNING AND SCHEDULING WORK

For a small construction project, an experienced contractor does not have to do much planning because he already knows what to do and when to do it. For a large project, however, planning must be systematically and carefully done. The important stages in planning a project are:

1. Breaking down the project into its various divisions and subdivisions or activities.
2. Deciding the method to be employed in the execution of each activity. While choosing a particular method or a combination of equipment and manpower, alternate methods are considered and the method that involves the least time and expenditure is selected.

3. Once the activities have been separated and the choice of method made, it is possible to find out approximately the time that each activity will take for completion.
4. Planning (i) materials, (ii) labour and (iii) plant.
    (i) *Materials:* Materials of construction will be required on specified dates throughout the progress of the work. The quantities of different materials required and the dates when they will be needed on the site have to be decided in advance.
    (ii) *Labour:* On all construction works, various types of labour would be required for different periods. Unless the labour programme—showing different classes of workmen, their numbers and approximate dates when they will be needed on site— is chalked out in advance, it would become difficult to meet the various demands. The labour cost can be minimised if such a programme is prepared in advance and related exactly to the detailed construction programme.
    (iii) *Plant:* It is essential to know in advance the different types of plants that will be needed on the site and on what dates. Unless this information is obtained in advance, it would be difficult to procure the same on the site, irrespective of the situation whether the contractor owns such plants or has to purchase or hire them.
5. After the project has been properly planned, it is scheduled.
    Scheduling is a mechanical process for formalising planned functions and assigning starting and finishing dates to each part of the work in such a manner that the whole work proceeds in a logical sequence and in an orderly and systematic way. Though every attempt is made to keep the progress of the work as per schedule, some variation is bound to take place on account of factors which could not have been predicted at the time the schedule had been prepared. This necessitates rescheduling of the project during the progress of the work.

    There are several types of schedules which may be designed to suit the needs of the various types of projects. The following methods of scheduling a project are considered here:
    (i) bar charts,
    (ii) critical path method or CPM, and
    (iii) programme evaluation review technique or PERT.

9.8   BAR CHARTS

The breakdown of the project into different activities, their durations, and starting and completion dates can be graphically shown on a chart. There are various types of charts in actual use. In fact, each large firm may have its own special type of a chart. The simplest type of chart is that for the construction of buildings and other structures where the breakdown into different activities usually follows the various items of estimate prepared at the time of submitting the tender.

Figure 9.4 shows one such chart. It is known as the Gantt bar chart. The bar chart represents both the planning and the scheduling of a project. The same chart can also be used to show the progress of the work as indicated in the figure. If on the same chart, both the programme and progress are shown, one can get a clearer mental picture of the actual position (namely, what was planned and what has been achieved) by looking at such a chart rather than by reading a detailed report.

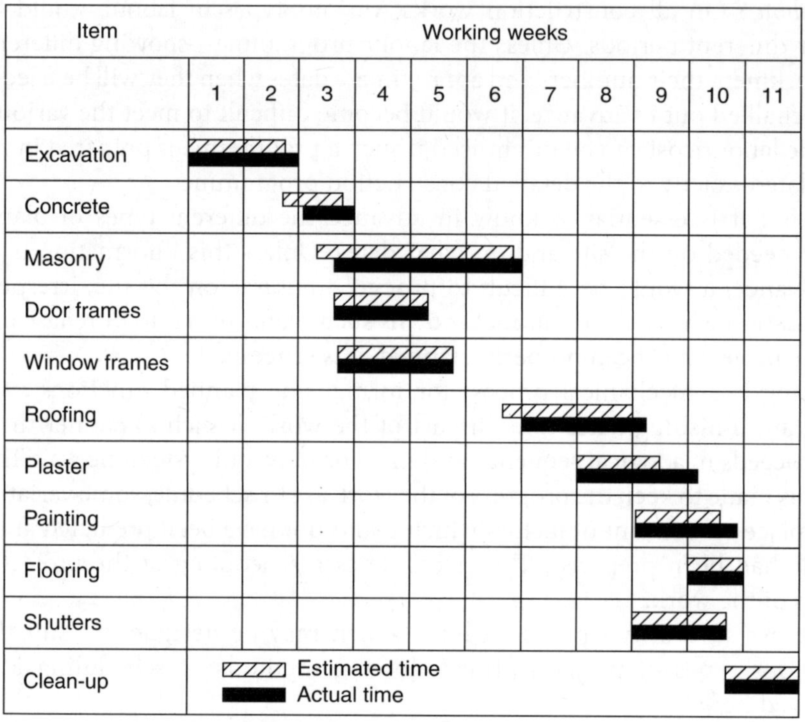

Item	Working weeks

	1	2	3	4	5	6	7	8	9	10	11
Excavation											
Concrete											
Masonry											
Door frames											
Window frames											
Roofing											
Plaster											
Painting											
Flooring											
Shutters											
Clean-up											

Estimated time
Actual time

Fig. 9.4   Gantt bar chart

1. *Technique of Recording Progress*

    The *maistries* or technical assistants, supervising different activities on the site (for example, excavation, brickwork, concreting, and so on) keep notes in their own duplicating notebooks, the originals being torn out at the end of each day and handed over to the supervisor or overseer. On the basis of these notes, the overseer prepares his own report and hands it over to the chief supervisor or site engineer. The chief supervisor may compile the information in the form as in shown Fig. 9.5.

    The amount of work completed periodically (that is, end of period I, period II and so on) is shown on the chart, the final position of which should show graphically the total quantity recorded in writing. For example, on 14.9.2014 in this case, the chart should show 138 m^3 of brickwork completed.

Item _________   Brickwork _________   Sheet No. _________

Estimated quantity ... (500 m^3)

Starting date ...

4 September 2014 _________________

Estimated duration... 30 days

Date	Location description	Workmen	Approximate quantity completed	Total quantity
4.9.2014	Quarter B1 and C2 coloured red on drawing	14	20 m^3	
5.9.2014	—do—	12	18 m^3	
6.9.2014	—do—	15	23 m^3	
	End of period I	41	61m^3	61 m^3
9.9.2014	—do—	12	18 m^3	
10.9.2014	—do—	14	20 m^3	
11.9.2014	—do—	15	24m^3	
12.9.2014	—do—	10	15 m^3	
	End of period II	51	77 m^3	138 m^3

Fig. 9.5   Progress record sheet

The progress on other items which are shown on the chart may be recorded on sheets similar to that shown in Fig. 9.5. The information shown on charts is related to the drawing by colouring. Similarly, different colours can be used advantageously on charts, for example, 'anticipated progress' shown in black, actual progress which is 'on time' shown in green and progress 'behind time' shown in red, and so on.

Only major items should be shown on the charts as too many items would make the task difficult and too few would make it inadequate and useless.

The usefulness and importance of such charts have to be properly understood. If the progress is recorded regularly and the charts maintained up-to-date, they show the following:

(i)   different activities to be performed,

(ii)   the period required for each activity,

(iii)   the dates on which materials, labour and plant will be required so that arrangements can be made to procure them in advance,

(iv)   up-to-date progress on different items, and

(v)   activities which are lagging behind and progress on which if not accelerated may cause delay in the completion time.

It will now be clear that the little extra labour and expense incurred in keeping such records are negligible compared to the usefulness of the charts.

2. *Improved Gantt Bar Chart*

From the above discussion on bar charts, it is clear that this method has certain definite limitations. The various functions required to be performed under each item of work are not separated from one another, but are all shown by a single bar. This does not permit an effective control of the project. As an illustration, let us consider an item of structural steel. This item will be represented on the chart by one single bar, though the following functions control the completion time of such an item:

 (i)   materials ordered,
 (ii) procurement time,
 (iii) erection started,
 (iv) erection completed, and
 (v)   test and approval.

For completion of the item as per schedule, each of the above functions must be allocated a specific time and that time should be properly controlled. The above information can be shown on the bar chart by the use of what are called 'milestones'. These are slidable symbols superimposed upon the bar as shown in Fig. 9.6. The symbols are designated by letters such as A, B and C or by numerals 1, 2, 3, and so on. These symbols may designate the above-mentioned functions, against each of them, or may be used to designate whatever the chart-maker has in mind.

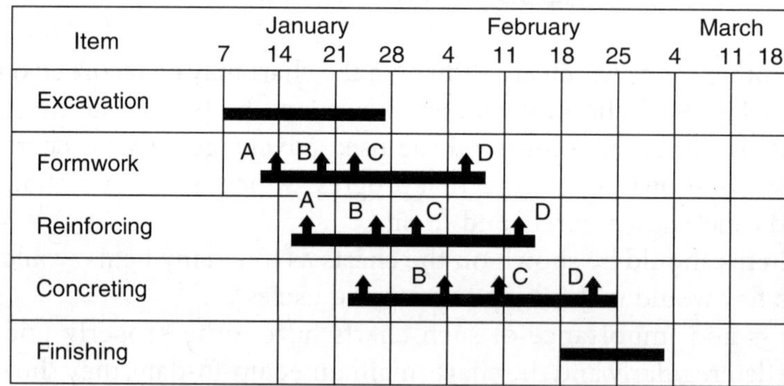

Fig. 9.6   'Milestone' chart

The other serious drawback of the bar chart is that it fails to establish the interdependency of one event on another and, as a result, one fails to identify which activities are critical (in the sense that any delay in the completion of these activities will delay the completion of the whole project) and deserve full attention if the project is to be completed on time.

The above shortcomings of the bar chart have limited their use to small projects only. A search to find out an effective method of construction scheduling gave birth to an entirely different method of scheduling a project. This method is known as the 'Critical Path Method' or CPM.

9.9   CRITICAL PATH METHOD (CPM)

The critical path method is a comparatively recent method of planning and controlling building and construction projects, now coming into general use all over the world. It is basically a 'network diagram' showing the logical sequences in which all operations involved in a project from planning to completion occur.

Any large-scale project consists of a number of activities (that is, jobs to be done) and each activity takes some time before it can be completed. This time taken by an activity for its completion is called the 'duration' of the activity. If the project has to be completed in a specific period, then some activities will get more time than their duration, whereas, in other cases, the maximum time available will be equal to the duration of the activities. If the maximum time available for an activity is equal to its duration, such an activity is called 'critical', because any delay in it will cause a comparable delay in the project completion time. Further, if a project contains critical activities, then it also contains one continuous path of critical activities running through the project diagram from the origin to the terminus. Such a path is called the 'critical path'. The technique of locating these activities and the path is called the 'critical path method'.

Planning of a project by the critical path method involves the following steps, which are mentioned in the order of their occurrence.

Steps for network development and analysis of a project
1. Prepare a complete list of activities of the project.
2. Decide the dependencies, that is, constraints of the activities.
3. Initially draw a rough network and then, by making improvements in the arrangements of the various activity arrows, draw the fair network.
4. Number the events ensuring that event no. (j) is greater than event no. (i) for all activities.
5. Decide the unit of time to be adopted and estimate the durations of all activities.
6. Carry out the FORWARD PASS for obtaining early event times and the project duration.
7. Carry out the BACKWARD PASS to obtain late event times.
8. Identify critical events.
9. Determine the critical activities and hence find the critical path.
10. Prepare the activity times computation table showing earliest start time (EST), earliest finish time (EFT), latest start time (LST), latest finish time (LFT) and total float of all activities (float indicates additional time available for completing an activity without delaying the overall completion).

11. Prepare the project calendar based on the proposed date of commencement of the project.
12. Prepare the activity calendar showing the early and late start and finish dates of the activities of the project.

Method of preparation of an activity arrow diagram
A simple method of illustrating the basic form of the arrow diagram would be to develop it from the conventional bar chart. This is shown in Fig. 9.7.

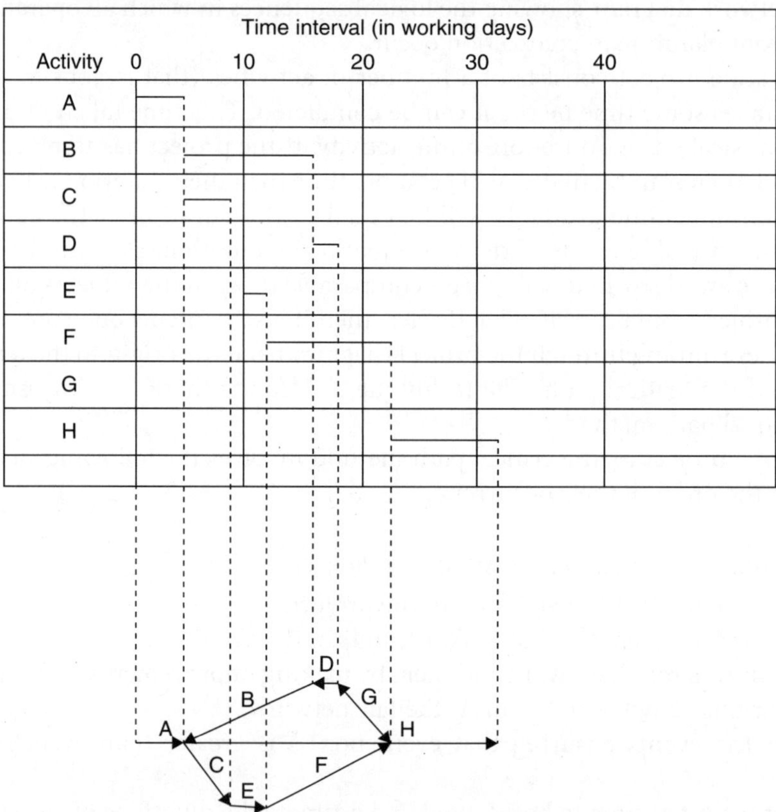

Fig. 9.7   Activity arrow diagram

The development of a network from the Gantt bar chart is simply for the sake of illustration. It is not necessary at all to have a bar chart ready to prepare a network diagram; on the other hand, quite often, the network is converted into a time-phased bar chart for ease of understanding.

The activity arrow diagram can be drawn without the aid of the Gantt bar chart as now explained.

The project is first divided into the different activities in the sequence that they have to be performed, namely, A, B, C, D, E, F, G and H. Then, while drawing the arrow diagram, the basic questions to be answered are:

1. What immediately precedes a particular activity?
2. What can be concurrent with this activity?
3. What immediately follows this activity?

For example, consider activity A. There is no preceding activity. Therefore, this activity will be the starting of the network diagram. As there is no other activity concurrent with A, draw a straight line of convenient length with an arrow at its head to show the direction of progress. Now, the activity immediately following A is B and the activity concurrent with B is C. Further, both B and C cannot be started unless A is complete. Draw two lines as shown starting from the arrow marking the end of activity A (refer also to Fig. 9.8).

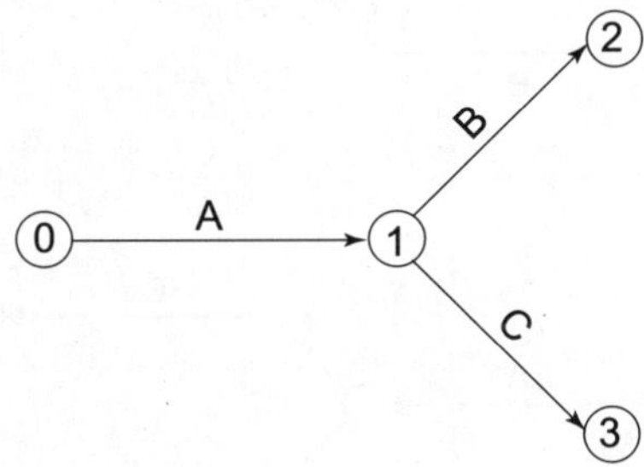

Fig. 9.8 Arrow diagram depicting an event

The meeting point of the activities is known as an 'event' (Fig. 9.8) which is considered to be a point in time at which all activities coming into it have been completed and these activities must have been completed before any activities starting from that event can begin. However, the following activities (B and C) need not start at once.

The diagram in Fig. 9.8 shows the relationship amongst the three activities A, B and C which could exist in a number of different ways as follows:

1. A precedes B and C.
2. B and C preceded by A.
3. B and C follow A.
4. A followed by B and C.
5. A controls B and C.

6. B and C controlled by A.
7. B and C depend on A.
8. B and C cannot start until A is completed.
9. There is a constraint of A on B and C.
10. B and C are constrained by A.
11. A must be completed before B and C can start.
12. Ground is set for B and C on completion of A.

Illustrative examples

(a) B is controlled by A:

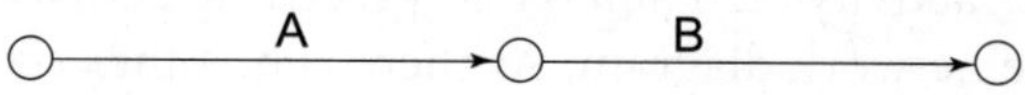

(b) C is controlled by A and B:

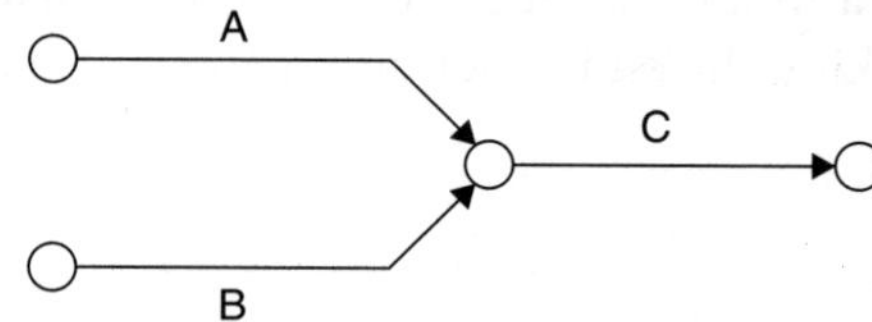

(c) A control B and C:

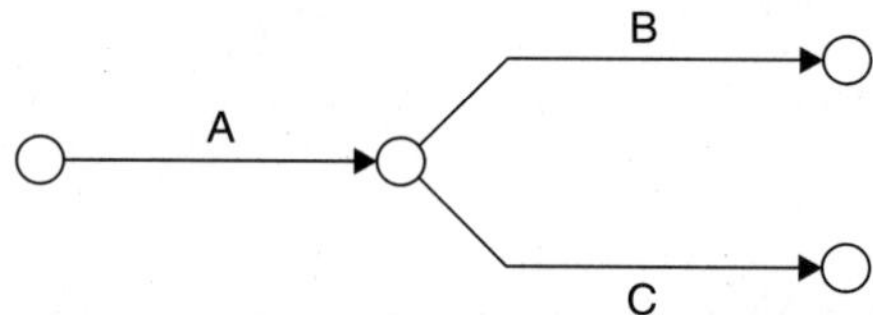

(d) L and M are controlled by A and B:

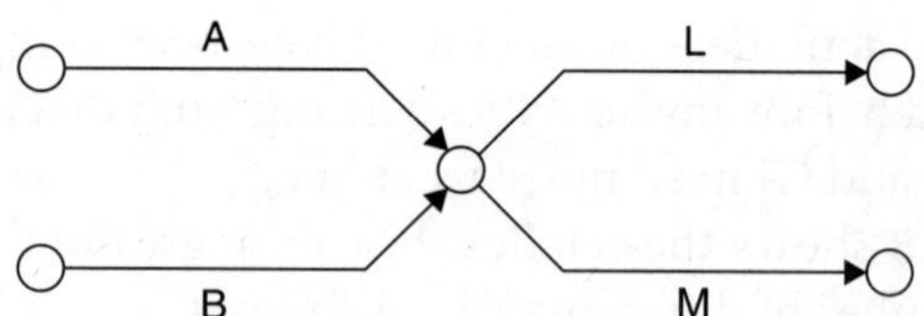

(e) L is controlled by A and B; M is controlled by B:

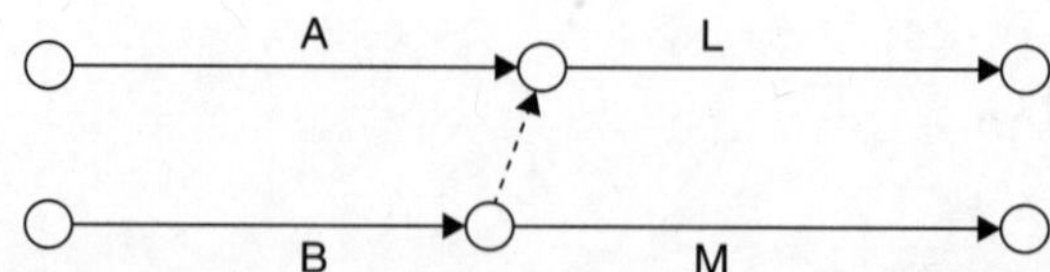

(f) A and B control L; whilst B and C control M:

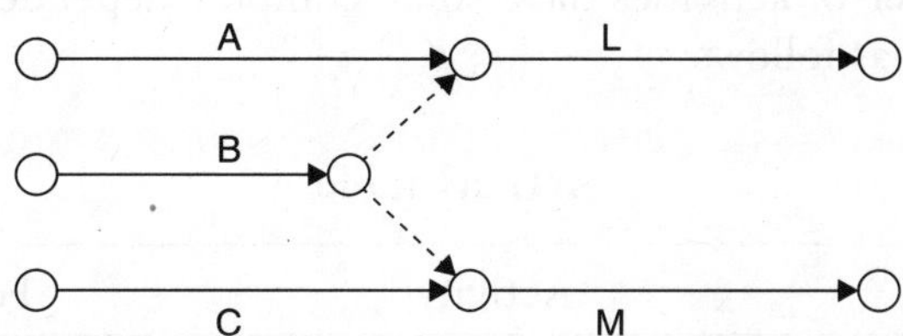

(g) A and B control L; B controls M, B and C controlling N:

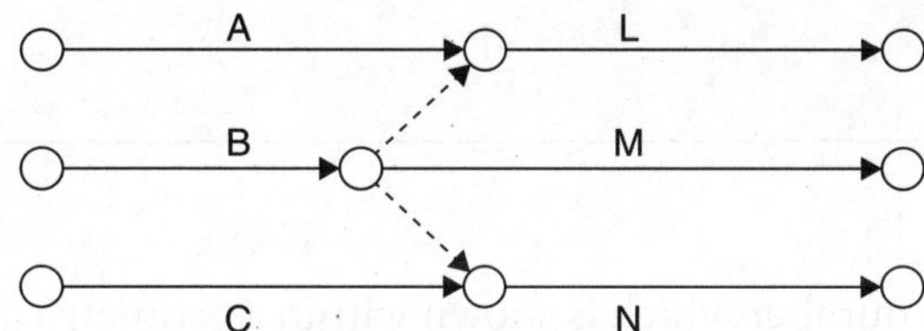

(h) A controls L and M; whilst C controls M and N:

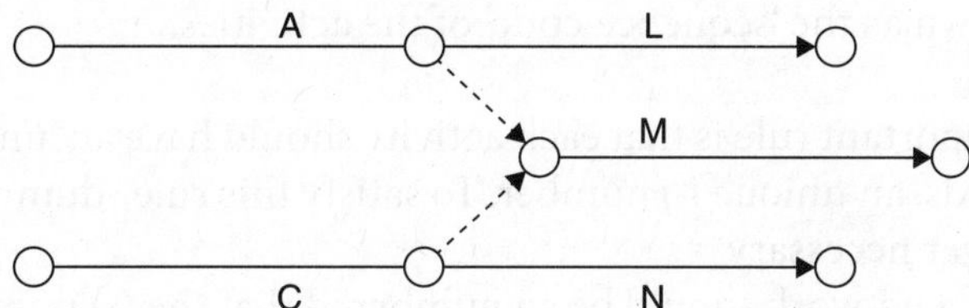

(i) A controls L and M; B and C control M; and N depends on C:

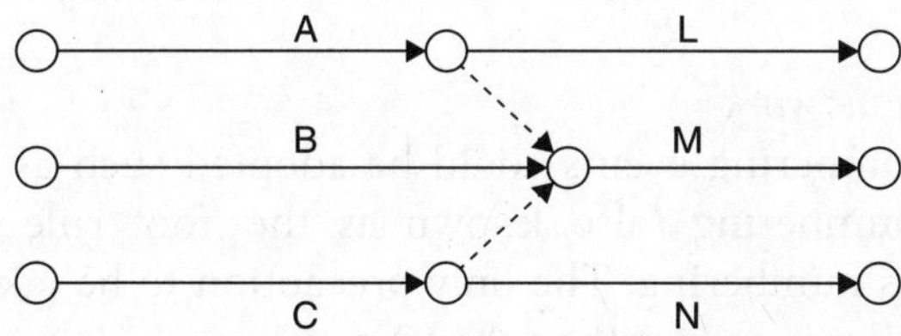

(j) L, M and N are controlled by A; M and N are controlled by B; and N is controlled by C:

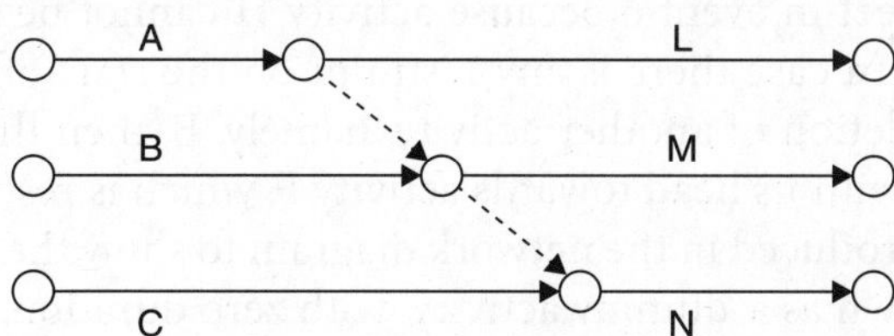

Activity table
In cases where a number of activities have some complex dependencies, it is advisable to prepare an activity table as follows:

Activity table

Preceding activities	Activity	Following activities

Events
Every event is assigned a number which is shown within a geometrical figure, such as a circle. As every activity is bounded by two events, these event numbers could be used to identify the activity. The initial event of the activity is known as the (i) event, whereas the end event of the activity is known as the (j) event. Thus, every activity can be identified by i-j event numbers. This is known as the 'sequence code' of the activities.

Rule 1:   The most important rule is that each activity should have an 'unique sequence code'; in other words, an unique i-j number. To satisfy this rule, 'dummy' arrows should be used wherever necessary.
Rule 2:   The event of a network should be so numbered that the (j) number of every activity is greater than the (i) number of the same activity.
Rule 3:   A network must have only one starting and only one ending event.
Rule 4:   Obviously, event numbers should not be duplicated in a network.

Numbering of events in a network
Different methods of numbering events could be adopted such as horizontal numbering, vertically downwards numbering (also known as the 'foot rule method') or vertically downwards and upwards numbering. The only precaution to be taken is to ensure that for each activity the (j) number is greater than the (i) number.

Following the above procedure, the network of an arrow diagram is depicted in Fig. 9.9 showing interdependencies of activities A, B, C, D, E, F, G and H.

Activities G and F meet in event 6 because activity H cannot be started unless activities G and F are completed. In case there is any restraint on the start of the activity F, which is dependent on the completion of another activity, namely, B, then this restraint is shown by a dotted line and arrow with its head towards activity F, which is restrained. This dotted line with an arrowhead is introduced in the network diagram to show the desired sequence of the real activities and is known as a 'dummy activity' with zero duration.

The lines have brief job descriptions written alongside to identify the activities and the numerals appearing along them indicate the duration of each activity in terms of working days. The completed network diagram is shown in Fig. 9.9.

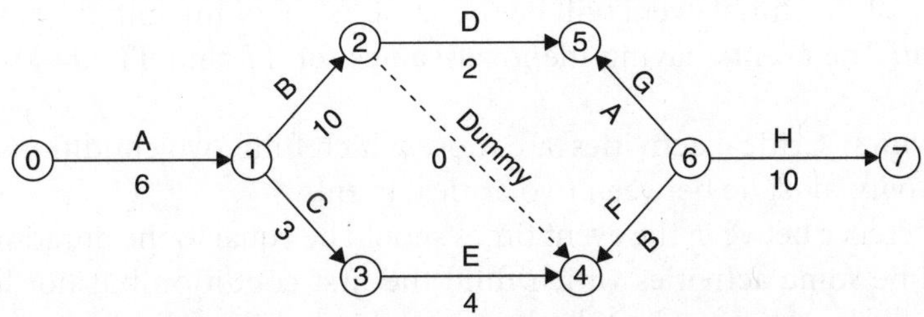

Fig. 9.9 Network of an arrow diagram

Duration of the Activities

1. *Duration and description*: The duration of the activity is written below the arrow while its description is written above the arrow. These durations indicate the actual time, that is, the working time required to execute the activities excluding holidays or non-working days.

2. *Unit of time*: Any convenient unit of time may be adopted, depending on the nature of the project, namely, days, weeks, months or shifts or hours. However, it should be ensured that:

    (i)   The durations of all the activities are indicated in full units of time only.

    (ii) The durations of all the activities in a network are shown in the same unit of time.

3. *Estimation of duration*: Duration can be estimated by any one or more of the following methods:

    (i)   Based on the following data:

        (a)   total quantity of work to be done,

        (b)   number of units of the controlling resource (that is, men, materials, machinery or money) proposed or expected to be employed/available, and

        (c)   estimated output per unit of controlling resource per unit of time.

    (ii) Based on past experience with respect to activities such as 'obtaining administrative approval' or 'evaluation of tenders' and so on.

    (iii) Based on the best judgement.

Event Times Computation

1. *Early event time (TE)*: The 'early event time' is the earliest possible time in which an event can be reached. The event is said to be 'reached' only when *all* the activities leading to it are completed. In the case of a MERGE event, the largest value is adopted. The TE of the initial event is taken as zero, and the TE of the final event gives the 'minimum project duration'. The operation of finding TEs of all the events, beginning with the initial event and proceeding to the final event, is known as the FORWARD PASS.

2. *Latest allowable event time (TL):* The 'latest allowable event time' is the latest time at which the event must be reached without delaying the project. In other words, TL = TE for the final event. We work backwards to find out the TL values of all the events, right from the final event to the initial event, and this operation is known as the BACKWARD PASS. The TL value of the initial event will be the same as TE of the initial event, that is, zero.
3. *Critical event:* The events having identical values for TE and TL are known as 'critical events'.
4. *Critical activities:* 'Critical activities' are those which fulfil two conditions:
    (i)   The activity must lie between two critical events.
    (ii)   The difference between the event times should be equal to the duration of the activity. There could be some activities which fulfil the first condition but not the second one. Such activities are not critical. Only those activities which fulfil *both* the conditions are called critical.
5. *Critical path:* 'Critical path' is a continuous chain of critical activities starting from the initial event and reaching the final event of the project. This is the *longest chain* of activities in the project. If we add the durations of the critical activities in series, the total will be equal to the project duration. In a network, there must be *at least one* critical path. However, there may be more than one critical path in a network. There could also be one critical chain, which in the beginning may diverge into two or more critical chains which may further converge into one critical chain and so on.

Determination of the Earliest Start and Finish Times

After representing the different activities by an arrow diagram and assigning the duration to each activity, the earliest start and finish times can be determined. The activities, their sequence code and durations are first listed in a table (refer to Table 9.1).

The activities are so arranged that a number of activities ending at the same event are in a sequential order, for example, 3–4, 2–4, 4–6, 5–6. With this arrangement, it is easy to select the earliest event occurrence time as the activities controlling the event are near one another in the table. Time may be given either as working days or calendar days. (Working days have been used in this illustration.)

Table 9.1

Activity	Sequence code	Duration	Earliest		Latest		Float		Critical activity
			Start	Finish	Start	Finish	Total	Free	
1	2	3	4	5	6	7	8	9	10
A	0–1	6	0	6	0	6	0	0	*
B	1–2	10	6	16	6	16	0	0	*
C	1–3	3	6	9	9	12	3	0	

(continued)

Table 9.1   (Continued)

Activity	Sequence code	Duration	Earliest		Latest		Float		Critical activity
			Start	Finish	Start	Finish	Total	Free	
1	2	3	4	5	6	7	8	9	10
D	2–5	2	16	18	18	20	2	0	
E	3–4	4	9	13	12	16	3	3	
Dummy	2–4	0	16	16	16	16	0	0	*
F	4–6	8	16	24	16	24	0	0	*
G	5–6	4	18	22	20	24	2	2	
H	6–7	10	24	34	24	34	0	0	*

Hints

1. *Sequence code:* Arrange the activities in the order of their end events within the order of initial events. Dummy activities should also be included.

2. First take the values of Col. (4) and Col. (7) directly from the network — Col. (4) = EST = TE (i) and Col. (7) LFT = TL (i)

3. Then work out values in Col. (5) and Col. (6) — Col. (5) = EFT = EST + D = Col. (4) + Col. (3); Col. (6) = LST = LFT-D = Col. (7) – Col. (3)

4. Then work out values in Col. (8) — Col. (8) = TF = LST – EST = Col. (6) – Col. (4); Also = LST – EFT = Col. (7) – Col. (5)

5. *Remarks:* Activities with zero total float are critical. Remaining activities are non-critical.

The earliest possible starting date for operation A is zero, this being the beginning of the first day of operation. It is obvious that the earliest possible finishing date for operation A is its starting date plus duration, that is, $0 + 6 = 6^{th}$ working day. It follows for both operations B and C that the earliest possible starting date for each of these operations is the sixth working day, since operation A must be completed in its entirety before either B or C can be started. This procedure can be best represented by the following equations:

1. Earliest possible start time for an activity = { Earliest event occurrence with reference to start of project

2. Earliest possible finish time for an activity = { (Earliest start time) + (duration)

To illustrate the use of the above equations, consider activity D (2–5). The earliest event is 2 and the date on which it occurs is the 16th working day, that is, completion of activity B. Therefore, the earliest possible start time for activity D is the 16th working day and the earliest possible finishing date will be $16 + 2 = 18$th working day. In case of activity F (4–6), the earliest event is 4 and the occurrence dates are tabulated as shown in Table 9.2.

Table 9.2

Activity	Event	Occurs on
E (3–4)	4	13th day
Dummy (2–4)	4	16th day
F (4–6)	6	24th day
G (5–6)	6	22nd day

It will be noticed from this table that though event 4 first occurs on the 13th day, operation F cannot start unless operation B (1–2) is complete as the restraint is shown by the dummy activity (2–4). So, for the purpose of computing the values of the earliest start times, occurrence of event 4 will be on the 16th working day. Similarly, in case of activity H (6–7) the earliest event (6) occurrence time is the 24th day and not the 22nd day. To generalise, it may be said that when an event is the junction point of more than two activities, the earliest event occurrence date is the finish date of that activity which ends in the event and is the last to be completed.

Determination of the Latest Start and Finish Times
The procedure for determining the latest start and finish times is very much similar to the procedure followed for the determination of the earliest start and finish times with the difference that the computation is done in reverse, that is, proceeding upward from the bottom of the column. The earliest and latest start and finish times for the last activity are assumed to be the same. The latest start time is then determined by subtracting the operation duration from the latest possible finish time. To put it in the form of equations:

(i) Latest possible finish date for an activity $= \begin{cases} \text{Earliest event occurrence time} \\ \text{with reference to completion of} \\ \text{project} \end{cases}$

(ii) Latest possible start time for an activity $= \begin{cases} \text{Latest finish} \\ \text{time for that} \\ \text{activity} \end{cases} - \begin{cases} \text{Duration} \\ \text{of that} \\ \text{activity} \end{cases}$

For example, the latest finish time for activity F (4-6), the earliest event with reference to completion of the project, is 6 and the time of its occurrence is the 24th working day. Its latest possible starting date would be

$$= \text{latest finish time} - \text{duration}$$

$$= 24 - 8$$

$$= \text{16th working day}$$

When an activity precedes more than one other activity, its latest finish time is determined by those which follow it. For example, activity A precedes activities B and C. The latest starting times of these activities are, respectively, the 6th and 9th working days; the earliest of these late starts is the 6th working day. This then is the latest possible finish time for activity A. The same procedure is adopted for determining the latest possible finish time of activity B.

Calculation of Different Types of Floats
Definition of float: Float is the difference between the *time available* to do the job and the *time required* to do the job. It is thus the time by which activities can be delayed without delaying the completion of the project.

The activities which lie on the critical path have zero float and all the remaining activities have varying degrees of float. Floats are mainly of three types:

1. total float,
2. free float, and
3. interfering float.

1. *Total float:* The total float is the difference between the latest finish time and the earliest finish time or the latest start time and the earliest start time.
    For example, in Table 9.1, for activity D,

$$\text{Total float} = \text{latest finish time} - \text{earliest finish time}$$

$$= 18 - 16 = 2\,\text{days}$$

The significance of the total float of 2 days is that activity D
  (i)   can be started 2 days later, that is, on the 18th day; or
  (ii) can be completed 2 days later, that is, on the 20th day; or
  (iii) can be performed at a slower rate in a 4-day period instead of 2 days as planned.
The total float periods for all the activities are similarly calculated and the results are tabulated as shown in Table 9.1.

The knowledge of the total float periods for different activities is of much value to the construction engineer, who is then able to utilise the resources in a convenient way without delaying the completion of the project.

2. *Free float:* The free float of an activity may be defined as the number of days by which the finish date of the activity can exceed the earliest possible finish date without affecting any other operation. It is thus the difference between the earliest finish time of an activity

and the first of the earliest start times of all activities which immediately follow it. For instance, operation E on the activity arrow diagram (Fig. 9.9) is followed by operation F. The earliest finish time of operation E is the 13th day and the earliest start date of operation F is the 16th day. The difference between these two times is a free float of 3 days for activity E. Free floats for all other activities are determined in the same way and tabulated in Table 9.1.

3. *Interfering float*: The float time for an activity, when its preceding activities have started at their latest start time, is termed as 'interfering float'. It can be calculated as follows:

Interfering float time for an activity

= (earliest finish time – latest start time) – (duration)

In case non-critical activities have been delayed and the delay has exceeded the free float time, but if it is within the interfering float time, some non-critical activity needs revision; such a delay will not necessarily affect the project completion time.

It will be noticed in Table 9.1 that no activity has interfering float time.

Determination of Critical Path

The activities having zero total float are critical activities and the path which moves along them resulting in the largest possible completion time (as can be verified on the network diagram) is called the critical path.

From Table 9.1, it is seen that certain activities, namely, A, B, dummy activity, F and H, are critical. A delay in the completion of any one of these activities would delay the completion of the whole project. On the other hand, a saving of time on any single critical activity would result in a saving of time for the overall completion of the project. This is one of the greatest advantages of the CPM technique. For instance, assume that the completion time of a project has to be reduced by some margin. The network diagram will indicate that only critical activities need to be expedited by resorting to extra labour or by working extra time. In the absence of network analysis, even non-critical activities are likely to be expedited at an increased cost, although the early completion of such activities will be of no use.

Determination of Semi-critical Activities

An activity is considered semi-critical when its degree of float is so small that it may turn critical due to a slight delay. However, the degree of float time which is to be considered to render an activity semi-critical will have a wide variation depending upon the accuracy with which the durations of all activities have been determined. If the durations of all operations are highly reliable, even an activity with a float period less than five (5) per cent of the total project period may be considered as semi-critical. This percentage may be as high as 15 per cent if the durations of the different activities are not reliable. Table 9.3 presents some semi-critical activities.

Table 9.3

10% of project time = 3.4 days	5% of project time = 1.7 days
Activity C	—
Activity D	—
Activity E	—
Activity G	—

From Table 9.3 it is observed that considering 10 per cent of the project time as the value for semi-critical activities, all the slack activities become semi-critical, while with 5 per cent of the project time as the value for semi-critical activities, all the slack activities still remain slack and there is no semi-critical activity. Thus, it is evident that for the CPM to be effective, the estimated duration of each activity must be predicted with a high degree of accuracy.

Illustrative example (for practice)

Let us consider the construction of a single span bridge. Decide the constraints and develop the network. By assuming suitable durations for the activities, identify the critical path. Also prepare the activity time computation table and a project calendar assuming a certain date as the date of commencement of the project. Tabulate the results as follows (see also Fig. 9.10):

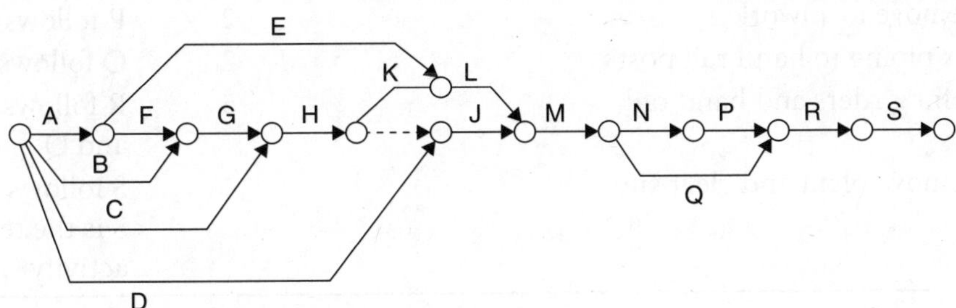

Fig. 9.10   Network for the construction of a single span bridge

Code letter	Activity description	Duration (days)[1]	Constraints/ Dependencies
A	Preliminary arrangements at site including setting out and procuring tools and plant, building materials, and cement and steel reinforcement	10	Starting activity
B	Procure cast steel bearings, including placing order, inspection and delivery at site	60	—do—

(*continued*)

(Continued)

Code letter	Activity description	Duration (days)[1]	Constraints/ Dependencies
C	Fabricate steel girders including procurement of steel and transport at site	30	—do—
D	Manufacture hand rail including procurement of pipes and transport at site	10	—do—
E	Bar bending for deck slab	6	E follows A
F	Construction of abutments including return walls and bed blocks	45	F follows A
G	Install bearings	2	G follows B and F
H	Erection of girders	4	H follows C and G
J	Weld hand rail posts to girders	4	J follows D and H
K	Erect formwork for deck	8	K follows H
L	Place reinforcement for deck	2	L follows K and E
M	Laying concrete in deck	1	M follows L and J
N	Curing	21	N follows M
P	Remove formwork	2	P follows N
Q	Fix piping to hand rail posts	2	Q follows M
R	Paint girders and hand rail	5	R follows P and Q
S	Remove plant and clear site	4	S follows R S is the terminal activity

1. Assumed. *Hint:* See Fig. 9.10.

9.10   PROJECT CONTROL THROUGH CPM

For effective control, the management must be informed about the delays occurring in the project and also about those that are likely to occur in future, so that timely action can be taken to remedy the situation. For this purpose, a suitable system of reporting has to be evolved. Figure 9.11 shows one suitable form for submitting periodic reports. The interval of reporting should be at least one half of the time required for the semi-critical activities to turn critical.

If such reports for each activity are preserved, they will prove valuable in the preparation of the critical path network analysis for future projects.

Report

Name of the project _________________________ Network no. ___________________________

Activity _________________________________ Subnetwork no. ________________________

Critical/semi-critical/likely to turn semi-critical

Description	**Sequence code**	**Delay/probable delay**	**Suggested action**

Fig. 9.11 Periodic report

9.11 PROGRAMME EVALUATION REVIEW TECHNIQUE (PERT)

PERT was conceived by the US Navy's *Polaris* submarine management team. This new technique had to be developed as the conventional methods of management and control of the project to be dealt with (which involved some 11,000 subcontractors) proved entirely inadequate. This method is very much similar to the CPM technique. The most basic difference between the two methods of advance scheduling pertains to the procedure for estimating the durations of various activities. In the PERT system, probabilistic time is used.

The procedure for the determination of the probabilistic time estimate is as follows.

The planner first consults the person responsible for the completion of the activity, whose duration is to be determined, and with his help establishes the following:

1. The optimistic time, that is, the shortest probable time.
2. The most likely or probable time.
3. The pessimistic time, that is, the longest probable time needed to complete the activity.

With this information, the planner estimates the probabilistic time by using the formula:

$$\text{Te} = \frac{A + 4M + B}{6}$$

where Te: expected time by the PERT method, A: the shortest time, M: the most likely or probable time and B: the longest time.

For example, assume that the optimistic, most likely and pessimistic time estimates for an activity are, respectively, 6, 16 and 22 days. The probabilistic time estimate would then be:

$$\text{expected time by PERT} = \frac{6 + (4 \times 16) + 22}{6}$$
$$= 15.33 \text{ days}$$

It is clear from the above formula that while determining the probable time, the most likely time is given four times the weightage given to the optimistic and the pessimistic time estimates.

However, the present trend is towards the use of the deterministic method of estimating durations as used in the CPM technique. The use of the PERT system has so far been confined to construction works with regard to which there has been little or no previous experience.

9.12   TIME–COST RELATIONSHIP

Any project consists of a number of activities. All activities incur a certain cost, and the cost of the project varies with the time taken to perform the activities. The estimated time required to complete a given job is of three types:

1.   normal time,
2.   slow time, or
3.   crash time.

Normal time is the standard time that an estimator would usually allow for an activity. *Slow time* is that time which results in the least possible direct cost. This is based on the assumption that the smaller the crew, the lower would be the cost of construction.

Crash time is that time which the estimator would allot, when faced with a seemingly stiff completion time. This time is the opposite of slow time and results in the highest possible cost known as 'crash cost'.

9.12.1   Activity-Cost Slope

The activity-cost slope is the extra cost required per unit of time reduction in the duration of the activity, and is worked out as follows:

$$\text{Activity-cost slope} = \frac{\text{extra cost required}}{\text{reduction in duration}}$$

$$= \frac{\text{crash cost} - \text{normal cost}}{\text{normal duration} - \text{crash duration}}$$

9.12.2   Network Compression

The technique of reducing the project duration by the sequential crashing of the critical activities in the cost-slope order is known as 'network compression'. The crashing of the activities may be limited by the following methods:

1. The critical activity being fully crashed.
2. The total float of the parallel non-critical activity being reduced to zero, making it critical.
3. If there is more than one parallel critical chain, equal amount of crashing on all the chains is required.
4. When all the critical activities on any one critical chain are fully crashed.

9.12.3   Project Cost Analysis

1. The *normal solution* is the one corresponding to the normal durations of all the activities giving the minimum direct cost for the project.
2. The *crash solution* is the one giving the minimum direct cost corresponding to the minimum project duration.
3. The *all-crash solution* is the one corresponding to crash durations of all the activities.
4. The *optimal solution* is the one giving the *minimum total project cost,* that is, the sum of the direct and indirect costs. As compared to the normal solution, this solution leads to a reduction in the project duration as well as a saving in the total project cost.
5. The *minimum net cost solution* is the one giving the minimum net cost, that is, total project cost minus the value of *additional* benefits as compared to the normal solution.

9.12.4   Reduction in Total Project Duration

Reduction in total project duration can be accomplished in various ways as now mentioned:

1. Re-examination of the activity durations of the critical activities and reducing these durations to the extent possible.
2. Performing concurrently as many activities as possible.
3. Application of the 'ladder technique', that is, splitting the cyclic jobs into a number of parts.

4. Transfer of resources from non-critical activities to critical activities, wherever possible, with a view to expedite them.
5. Lowering the performance requirements of the activities, that is, reducing the specifications, if possible, in a manner that is consistent with quality, technical requirements and safety.
6. Application of the network compression technique.

9.12.5 Time–cost Curves

The relationship between time and cost is best shown diagrammatically as in Fig. 9.12.

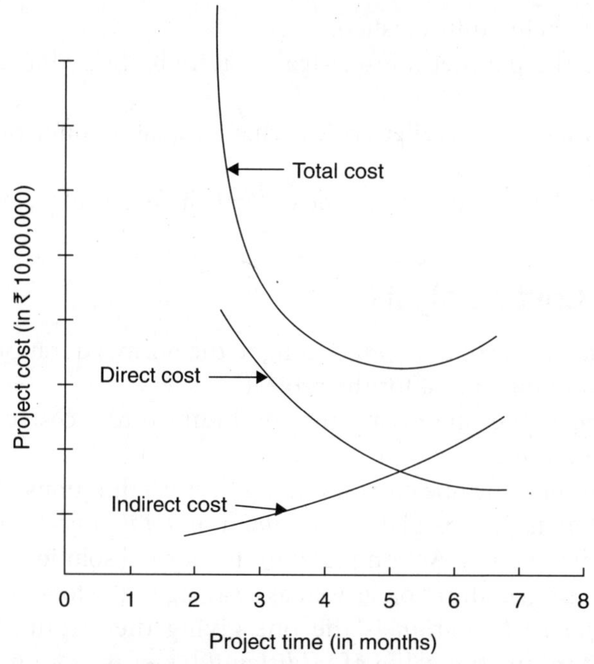

Fig. 9.12   Time–cost Curves

The lowest point on the total cost curve indicates the optimum time for the completion of a project in the sense that the total cost of construction becomes the least. Speeding up the work will result in decreased indirect charges but increased direct charges. If a project has to be completed in less than the optimum time, it will have to be speeded up, which will result in an increase in the total cost. However, each activity has a certain limit beyond which it cannot be shortened any further. This shortest possible duration of an activity is referred to as the 'crash time' and the corresponding cost is called the 'crash cost'. After working out the crash time and crash cost for each operation, the shortest possible project time at the least cost can be determined.

A study of the time-cost curve indicates that the cost of the project increases beyond the optimum time and it is of no advantage to allow the project to linger.

In the past, it was almost standard practice to complete *each and every activity* on a crash time basis if a project were to be completed on this basis. However, the advance scheduling technique (CPM or PERT) has proved that a project generally contains many activities which are non-critical and do not directly affect the project completion time and, as such, need not be placed on a crash time basis. Only the critical activities need to be planned and executed on a crash time basis.

Questions

1. Define 'organisation' and 'organisational structure' of a firm.   (**Ans.:** Section 9.2)
2. What are the different types of organisational structures? Outline in brief the characteristics, advantages and disadvantages of each structure.   (**Ans.:** Sections 9.3 to 9.5)
3. With the help of suitable charts, outline the difference between military organisation and line-and-staff organisation.   (**Ans.:** Sections 9.3 and 9.4)
4. In what respect does the functional organisation differ from the line-and-staff organisation?   (**Ans.:** Section 9.5)
5. State the objectives of integrated project management.   (**Ans.:** Section 9.6)
6. What are the important stages in the planning of a project?   (**Ans.:** Section 9.7)
7. What is meant by 'scheduling of a project'?   (**Ans.:** Section 9.8)
8. How is the Gantt bar chart used for recording the programme and progress of the work? Illustrate.   (**Ans.:** Section 9.8)
9. Why is it important to maintain the charts at the site of a work? What information does a bar chart give?   (**Ans.:** Section 9.8)
10. What are the limitations of a conventional bar chart in the planning and scheduling of a project? How can these be overcome?   (**Ans.:** Section 9.8)
11. Define the following terms as used in the CPM technique:
    (a) activity
    (b) critical activity
    (c) critical path
    (d) event
    (e) float
    (f) semi-critical activity   (**Ans.:** Section 9.9)
12. State the various steps for the network development and analysis of a project in the order of their occurrence.   (**Ans.:** Section 9.9)
13. What are the factors to be considered while dividing a project into its various activities?   (**Ans.:** Section 9.9)
14. Illustrate the CPM technique as used in scheduling a construction project.   (**Ans.:** Section 9.9)
15. Define 'float' and name the different types of floats. Write a full explanatory note on any one type.   (**Ans.:** Section 9.9)

16. What is a semi-critical activity? Illustrate your answer with a suitable example.

(**Ans.:** Section 9.9)

17. Suggest a suitable method for submitting the periodical reports for effective control through CPM.   (**Ans.:** Section 9.10)

18. In what respect does the PERT system differ from the CPM technique?   (**Ans.:** Section 9.11)

19. Illustrate the relationship between 'time' and 'cost'.   (**Ans.:** Section 9.12)

20. Write short notes on:
     (a)   network compression
     (b)   project cost analysis   (**Ans.:** Section 9.12.2)

The PWD Procedure of Executing a Project

10.1 GENERAL

Generally, the procedure and methods in government departments and public bodies are bound by rigid rules, as they involve the administration of public funds. Commercial methods are more flexible and allow a large degree of discretionary power in as much as the funds involved are private property, and a lot can be left to mutual trust.

It is proposed to discuss in this chapter the procedure adopted by the Public Works Department (PWD) in the execution of any project.

10.2 CLASSIFICATION OF WORKS

The works executed by the PWD are generally classified into the following heads as such a classification facilitates execution:

1. major works,
2. minor works,
3. special repairs,
4. maintenance and repair or current repair works, and
5. deposit works.

1. *Major works:* These include all new works and/or additions and alterations to the existing works, the cost of which exceeds a certain stated limit. For instance, the Buildings and Communications Department of a State may classify a works as 'major' when the cost exceeds ₹50,000. In a different category, irrigation projects costing ₹5 crores or more are called 'major' works. The cost criteria may vary from State to State and also from time to time.
2. *Minor works:* New works of a small magnitude and extensions, additions and alterations to the existing works costing not more than the limits mentioned above are grouped

under this heading. In case of irrigation projects estimated to cost between ₹20 lakhs and ₹5 crores, the term 'medium works' is used.

3. *Special repairs:* The work of carrying out repairs to existing structures, in case they are damaged due to extraordinary and unexpected causes such as heavy rains, storms, floods, or fires, is called 'special repairs'. The procedure followed in carrying out such work is similar to the one adopted in the case of new works.

4. *Maintenance and repair works:* The existing roads, buildings, dams and other similar structures, occasionally need minor repairs, which as a matter of regulation are carried out periodically. These are usually of the same magnitude from time to time. An estimate is prepared separately for the maintenance of each work by 15 January of the previous financial year and sanctioned by the competent authority by 15 April of the year concerned.

5. *Deposit works:* Under this heading are grouped the works carried out by the PWD for other bodies like municipalities and airports. The cost of construction is met by the funds of the public body for whom the work is executed. Sometimes, the cost of the work is met partly by the government and partly by local bodies, in which case, the works are called 'contribution works'.

10.3 ADMINISTRATIVE APPROVAL

When any government department proposes to carry out a work to meet its administrative needs, the matter is taken up with the PWD. The office of the divisional engineer, under whose jurisdiction the proposed work lies, prepares a preliminary report in the form of a proposal. This proposal, which includes preliminary drawings (such as the site plan, the layout plan, and so on), approximate estimates, outline specifications, and so on, is then submitted to the proper authority of the department that requires the work to be executed for obtaining the concurrence of that department. If the authorities of the PWD have recommended that the proposals are sound and that the preliminary estimates are reasonable, the proposals are approved by the competent authority of the concerned department. This formal acceptance, which is in effect an order to the PWD to execute the specified work, is called 'administrative approval'.

A similar procedure is followed in the case of works required to meet the administrative need of the PWD and also the works required in the general interests of the public, for example, roads, irrigation projects, bridges, and so on.

For petty works costing less than ₹5,000, the procedure is slightly different. Instead of preliminary plans and approximate estimates, the detailed drawings and estimates are prepared in the first instance and submitted for obtaining the administrative approval.

10.4 TECHNICAL SANCTION

Once the work has been approved administratively, in-depth investigations are carried out and detailed drawings and estimates prepared by the PWD. The detailed estimates are then submitted for sanction to the government responsible for the PWD or to the officer of the Public Works Department to whom the power to accord the sanction has been delegated by the government. Such a sanction is known as 'technical sanction'.

The procedures outlined in Sections 10.3 and 10.4 also apply to modifications of the proposals originally approved in case such modifications are likely to necessitate the eventual submission of a revised estimate. The revised estimate is prepared when the sanctioned estimate is likely to be exceeded by more than five (5) per cent. In some cases, revised administrative approval is taken when

1. the detailed estimates, when prepared, exceed the amount administratively approved by more than 10 per cent, and
2. the expenditure on a work exceeds, or is likely to exceed, the amount administratively approved for it by more than 10 per cent.

10.5   EXECUTION OF WORKS

The method employed in the execution of a project depends to a large extent on the nature of the work. Keeping this fact in mind, any one of the following methods can be used:

1. contracts,
2. employment of daily labour,
3. rate list,
4. piecework, and
5. daywork.

These methods are now individually considered in the subsequent sections.

10.6   CONTRACTS

Generally, the works estimated to cost ₹5,000 and above are prepared on regular contract forms. Tenders are invited publicly in the manner described in Section 3.2. Of the various types of contracts between the owner and the contractor, those commonly used are lump sum, item rate and percentage rate contracts.

Usually, the lowest tender is accepted, unless there is some objection to the capability of the contractor or some other reason outlined in Section 3.19. No officer of a rank lower than that of an executive engineer is permitted to accept any tender other than the lowest. When it is proposed to accept a tender other than the lowest, the decision is not to be taken by an individual officer, but by a committee. This committee generally consists of the following officers:

1. At the level of executive engineer: the executive engineer and the collector concerned.

2. At the level of superintending engineer: the superintending engineer and the commissioner.
3. At the level of chief engineer: secretary (Public Works Department), special secretary (Political and Service Department), and the concerned chief engineer.

The decision of the committee to accept a tender has to be unanimous, otherwise the matter is referred to the higher committee.

The execution of minor works is generally entrusted to the regular subdivisions of the department. Major and medium works, due to their magnitude, are looked after by a special subdivision, or division, or a circle set up for the purpose.

The fieldwork is usually carried out by subdivisions, each subdivision looking after a part or whole of one particular project. A deputy engineer is in charge of one subdivision and is assisted by five to six overseers or junior engineers. These persons are the officers on the field and are primarily responsible for the quality and progress of the work. The sub divisional staff functions under the instructions and guidance of the executive engineer.

Mode of Payment to the Contractor, or Billing
The junior and/or deputy engineer record(s) the measurements of a work carried out by the contractor (in case of lump sum contracts, this is limited to additions and alterations) and a bill is prepared and submitted to the office of the executive engineer. The executive engineer is supposed to check the measurements (5 per cent to 10 per cent) and the quality of the work, before passing the bill for payment to the contractor.

The bills are grouped as under:

1. first and final bill,
2. interim bill, and
3. final bill.

1. *First and final bill:* In case of a small work of short duration, where the preparation of interim bills is not necessary, the contractor is paid the whole of the tender sum in only one instalment, that is, on completion of the work. P. W. Acct. Form No. 24—First and Final Bill—is used for making the payment. This form is also used to make payments to the suppliers, if the whole of the amount is to be paid in one instalment on receiving the goods. Bills of several contractors and suppliers can be prepared on a single form if the payments relate to the same head of account.
2. *Interim payment bills:* These are further classified as under:
    (i)   advance,
    (ii)   secured advance, and
    (iii) on-account payment or running account bill.
    (i)   *Advance:* When a contractor is to be paid an advance, P. W. Acct. Form No. 25 is used. This form is also used if an on-account payment is to be made, but an advance

payment already made for the same work is outstanding. Advance, payments to contractors are generally not favoured and, when made, they are in the form of secured advances.

(ii) *Secured advance:* Sometimes, to assist the contractor financially, the contractor is paid not only for the cost of work actually carried out but also for the materials brought to the site by him for use on the work. The value of such materials is assessed and 75 per cent (maximum) of it is paid to the contractor, on the execution of a special bond whereby the materials become the property of the government.

For secured advances and also for on-account payment where secured advances are already outstanding, P. W. Acct. Form No. 26 is used.

When the materials, on which a secured advance is paid to the contractor, have been used in the construction of a particular item, the amount of the advance payment for that item is not to exceed a sum equivalent to the value of work done less the proportionate amount of the secured advance.

(iii) *On-account payment or running account bills:* When only on-account payments are made both to contractors and suppliers, P. W. Acct. Form No. 27 is used. Payments are ordinarily made monthly. Such payments are treated as payments on account, subject to adjustment in the final bill

A separate running account is maintained for each contract.

3. *Final bill:* On completion of a work, the final bill is prepared on the appropriate form but printed on yellow paper. All the entries in respect of all the items mentioned in the running account bill are repeated in the final bill for the work. While making the final payment, the payee, if he is able to write, has to add in his own handwriting that the payment is made 'in full settlement of all demands'. If the payee is unable to write, these words are filled in by the officer making the payment.

10.7 EMPLOYMENT OF DAILY LABOUR OR THE DEPARTMENTAL METHOD

The main characteristic of this method is the elimination of the middle man—the contractor—between the department and the suppliers of materials and workmen. The department purchases the materials directly from the suppliers and also engages the labour force required for the execution of a project on daily wages. The immediate effect of this will be the elimination of the profit, over and above the actual cost of construction, which will have to be paid to the contractor executing the work, had the work been let out to him on a contract basis. This is in theory. In practice, however, the cost of construction is seldom found to be less than the amount payable to the contractor if the work is executed by him. The main reason for this is, paradoxically enough, the elimination of the contractor! The contractor possesses such qualities as are vital for the efficient and economic construction of works. Lack of these qualities and experience of organisation of large projects on the part of government officials, coupled with the rigid framework of rules within which they have to operate, results in delay in completion, bad quality of work and uneconomic construction. It may be for this reason that the departmental method is not commonly employed in the execution of new works and is restricted to maintenance and repairs of existing facilities.

1. *Purchase of Materials*

   Sealed tenders (or quotations) for the supply of materials delivered to the site of work are invited publicly. Generally, the lowest rates are accepted and the supplier, who has quoted them, is asked to supply the materials. The measurements of the materials received on site are recorded in the measurement book by an officer of the department who is authorised to do so. It is the responsibility of the officer entering the measurements to see that the materials are of specified quality as agreed. A bill is then prepared on an appropriate form and the payment made to the supplier.

2. *Employment of Daily Labour*

   Labourers of different categories, in required numbers, are employed on daily wages. The record of the labourers employed is maintained by the subordinate staff. This record is called the 'nominal muster roll' (NMR).

3. *Muster Roll*

   This is the PWD Acct. Form No. 21. This form lists the names of workers (grouped according to category) and the dates on which they have been employed. Their presence has to be marked daily and the total of the workers of different classes drawn up the same evening by the subordinate maintaining the muster roll. The muster roll is checked by the subdivisional officer once or twice in a month and by the executive engineer whenever he inspects the work. The muster roll is closed and submitted to the subdivisional officer on completion of the work or at the end of the month, if the work lasts for more than one month.

   The muster roll comprises the following particulars:

   Part I of the muster roll: On the front page of its outer sheet, detailed instructions as to how to maintain the muster are given. Additional information shown on it includes the month, name of work, head and subhead of account, subdivision and division by which it is maintained, and so on. The inner part of the outer sheet and inner sheets show the details of labour employed, total cost of labour and passing orders of the officer thereto.

   Part II of the muster roll: On the last page of the outer sheet are written the details of labour not having been present to receive payment when the labour of the muster is paid by the officer and thus shows the arrears of wages due to be paid.

   Part III of the muster roll: This is the abstract of the measurements of the work carried out by the labour and recorded in the measurement book.

   With the help of Part I and Part III of the muster roll, it is possible to find out whether the work has been carried out efficiently and at a reasonable cost. The rates of execution of different items of work are not expected to exceed the rates mentioned in the sanctioned schedule of rates.

4. *Daily Reports*

   To ensure that the subordinate maintaining the muster roll does not change or add to the record of the day's labour, he is required to submit a daily labour report to his superiors. The daily report shows the number of each class of labourers employed on each work or its sub-work with the rate of each category of labourers and the approximate quantity of work done.

   When the muster roll is received by the subdivisional officer, he has to check and see that there is no discrepancy between the daily reports and the muster roll before passing it for payment.

5. *Measurement Book*

   The measurement book is P. W. Acct. Form No. 23. It records the measurement of work done by any approved method, and also the materials received, for which the payments are to be made. Detailed measurements may be dispensed with in the following cases:

   (i) When a work is carried out on a lump sum contract and no additions and alterations are ordered.

   (ii) When the work consists of periodical repairs and quantities which do not change, the measurements are recorded in an efficiently maintained standard measurement book.

   (iii) When on-account payment is to be made and an officer (not below the rank of subdivisional officer) certifies that the quantity of work carried out is not less than the quantity of work paid for, the officer granting such a certificate is held personally responsible for any overpayment, and so on. In no case is the final payment made without detailed measurements.

The muster roll and measurement book are the initial records upon which the accounts of works are based and their importance must be realised by the young and newly recruited engineer, or else he might be drawn into serious trouble. If any error is committed while writing, the erroneous contents should be scored out, in a manner they can be read, and compared with the corrected contents. The correction should be signed.

10.8   RATE LIST

The rate list is one of the recognised systems of carrying out a petty work other than by the employment of daily labour. Petty works costing a few thousand rupees (for example, ₹3,000), are not likely to attract the attention of any contracting firms. As such, invitation of public tenders by advertisement in newspapers is not justified. Generally, petty workers operating in the concerned area come forward to undertake such works. A list of petty workers is maintained in the office of the executive engineer. Whenever any petty work is to be executed, the petty workers are informed about the same. On request, they are supplied with the list of items to be executed with approximate quantities under each item and specifications thereof. The petty workers are asked to quote rates for different items and submit the lists in sealed covers. These covers are opened and, generally, the petty worker who has quoted the lowest rates or cost is selected to execute the work. The rates quoted by him are transferred to the printed form which also shows the estimated rates of the concerned items. The petty worker has to sign or affix his thumb impression to the rate list form in token of his acceptance of the rates. No security deposit is recovered and there is no provision of liquidated damages in case of delays.

More than one petty worker may be employed on one work at the same time provided the cost of the work does not exceed the fixed limit such as ₹3,000.

Measurements of the works done by the petty worker are recorded and he is paid for the work actually executed at the agreed rates.

Original works or special repairs to existing works are normally carried out by this method. The subdivisional officers are authorised to accept and sanction a rate list for works costing up to ₹1,000. For works exceeding this limit, the rate lists are to be approved by the executive engineer concerned.

10.9 PIECEWORK

The piecework system is similar to the rate list system with the difference that the agreement between the department and the petty worker is in respect of rate per unit quantity only without any reference to the total quantity of work or the time limit within which it has to be done. This method is suitable for maintenance and repair works. As and when the repairs have to be carried out, the petty worker may be informed about the same and the work got executed.

The basis of payment to the petty worker is the quantity of work actually carried out as entered in the measurement book and at the agreed rate.

10.10 DAYWORK

Daywork is defined as 'the method of valuing work on the basis of the time spent by the labourers, the materials used and the plant employed'. This is the only suitable method of payment to the contractor where it is impracticable to value the work done at the billed rates. The procedure is as follows.

1. *Materials*

    The selected contractor has to obtain the quotations from the suppliers for the various materials to be used in the work. The engineer-in-charge approves the quotations of a particular supplier, from whom the contractor purchases the materials. The contractor is paid the cost of materials delivered to the site plus a specified percentage as profit. An account of materials used on the site and the vouchers and receipts showing the amounts paid to the supplier form the basis for working out the cost of materials. The cost of the internal haulage of materials is paid for on the basis of labour and plant used.

2. *Labour*

   The contractor maintains a day-to-day list showing names, occupations and times of duty of all workmen employed on the work and submits it to the engineer every day. This list and the agreed rates of wages help in determining the actual amount paid by the contractor to the labourers in the form of wages. The contractor is paid the amount so worked out plus a specified percentage of that amount to cover overheads.

3. *Plant*

   The comprehensive list of items of plant with the agreed hourly, daily or nightly hire rates is used as the basis of making payments to the contractor for the use of his plant on the site.

   The daywork method, as described earlier, is not common in the Public Works Department and its use is restricted to employment of labour through the contractor in the event of an emergency and when the quantity of work done cannot be determined, for example, in the case of urgent repairs to canal breaches. The fee to be paid to the contractor may be included in the wages or expressed as a lump sum or percentage of the net amount of wages.

10.11 STORES

The department obtains large quantities of materials, small tools, implements, machinery and other related equipment required for original works and also for repairs and maintenance of existing structures. All these items taken together are referred to as 'stores'. The care and consumption of the stores are the responsibility of the concerned departmental officers, who also have to maintain correct records and submit periodical returns to the superiors.

The stores of the department are classified as under:

1. stock,
2. tools and plant,
3. road metal, and
4. materials charged directly to the works.

1. *Stock*

   The term 'stock' is used in connection with the consumable materials purchased by the department. Stock may be sent directly to the work site from the source of supply as this saves both time and money. However, it is necessary to maintain a central stock to tide over a temporary shortage or to help bulk purchase of materials which may be needed only in small quantities on individual works. The central stock is maintained at an economic and practical minimum.

   (i) *Stock account:* Stock accounts maintained by the departmental officers are of two kinds:

   (a)   quantity accounts

   (b)   value accounts

   (a)   *Quantity accounts:* The account of quantities received is maintained in the following form:

   | Receipt | Issue | Balance |

   The account shows the heads under which materials are purchased and used, and also the dates of transactions. At the end of every month, the balances are struck, which should correspond to the quantities in stock.

   (b)   *Value accounts:* These accounts show the money value of the materials received in and issued from stock. The payment side of the account shows the value of the materials obtained from different sources, either by cash payments or by book adjustments. The receipt side shows the value of the materials issued from time to time at previously fixed rates. These issue rates are revised by competent authorities to bring them on par with the market rates.

   (ii) *Stock taking:* The executive engineers have to arrange to get the stocks checked throughout their divisions at least once a year. As far as possible, the verification is entrusted to an officer independent of and unconnected with the staff responsible for the custody of stores. In the case of special stores depots or divisions or construction divisions, where there is a large concentration of stores, a continuous and periodical verification of stores by an officer of the audit department is arranged.

   The officer verifying the stock has to set right the book balance, whenever any discrepancy is noticed, treating the surplus as receipt and deficit as issue, with a suitable remark. The result of stock verification is finally to be reported to the divisional officer. Any materials or articles on stock, but which are not likely to be used in the following twelve months, are also to be reported to the executive engineer, who, in consultation with the superintending engineer, will dispose of such materials or articles.

2.   *Tools and Plant*

The tools and plant of a division fall under the following two categories:

   (i)   Minor tools and and plant such as pickaxes, *phawdas* (shovels), iron baskets, and so on, which are required for the general use on the works of a division.

   (ii) Special tools and plant such as concrete mixers, mortar mills, cranes, etc., which are required for special works.

   The costs of supply, repairs and carriage of minor tools and plant are charged to the minor head 'tools and plant', while such costs of special tools and plant are borne by the work concerned.

The account of tools and plant is maintained in the same manner as that for stock articles. The verification procedure is also similar to that of stock verification, with the difference that whenever any articles are found deficient, a note of the deficiency is made in the account of issues, without any correction of the book balance.

3. *Road Metal*

   Materials such as rubble, broken stone, hard and soft *murum,* purchased and stacked in heaps along roads for construction and/or maintenance, are grouped under the general heading 'road metal'. The account of road metal showing receipt, use and balance quantities for every kilometre length of the road in which the metal is lying is maintained. The road metal is periodically verified and examined. If the materials are found to be deteriorated and useless, they are written off.

   The account of road metal submitted periodically to the divisional office is called 'road metal return' (RMR).

4. *Materials Charged Directly to Work*

   When materials are purchased for use in the construction of a particular work, these are generally sent to the site of work. A separate account of such materials known as 'materials-at-site-account' is maintained. It shows the dated transactions, namely, receipt, issue and balance. An abstract of transactions is prepared every month. After every six months (by the end of September and March), these abstracts are submitted to the divisional office in the form of returns.

   Materials issued to a work in excess of requirements, and if in a serviceable condition, may be transferred to other works in progress (if likely to be used on such works within a reasonable period) or brought on to the stock account.

Questions

1. Distinguish between:
   (a) maintenance and repairs and special repairs   (**Ans.:** Section 10.2)
   (b) administrative approval and technical sanction   (**Ans.:** Sections 10.3 and 10.4)
   (c) daily labour and daywork   (**Ans.:** Sections 10.7 and 10.10)
   (d) regular contracts and piecework   (**Ans.:** Sections 10.6 and 10.9)
   (e) stock account and material-at-site-account   (**Ans.:** Section 10.11)
   (f) secured advance and running account bill   (**Ans.:** Section 10.6)
2. What are the different methods of execution of works?   (**Ans.:** Sections 10.5 to 10.10)
3. How is a work executed by the departmental method? What are the advantages and disadvantages of this method?   (**Ans.:** Section 10.7)
4. Write short notes on:
   (a) classification of works   (**Ans.:** Section 10.2)
   (b) secured advance   (**Ans.:** Section 10.6)
   (c) muster roll   (**Ans.:** Section 10.7)
   (d) daily report   (**Ans.:** Section 10.7)
   (e) measurement book   (**Ans.:** Section 10.7)
   (f) rate list   (**Ans.:** Section 10.8)

(g) piecework   (**Ans.:** Section 10.9)
(h) daywork   (**Ans.:** Section 10.10)
(i) stock account   (**Ans.:** Section 10.11)
(j) tools and plant   (**Ans.:** Section 10.11)
(k) road metal return   (**Ans.:** Section 10.11)
(l) material-at-site-account   (**Ans.:** Section 10.11)

Appendix **A**

Invitation to Tender (Limited Competition)

(Address of Architect/Engineer)

....

Dated ....

Gentlemen,

Construction of ....

Estimated cost ....

At ....

For ....

You are invited to tender with others, for the above work.

Bills of quantities and specifications will be supplied on request.

The plans may be seen and copies of bills of quantities obtained at my office at the above address any week day between .... a.m .... and .... p.m.

Tenders are to be delivered at my office as above up to or before .... (time) on .... (date)

The lowest or any other tender will not necessarily be accepted.

I should be obliged if you would inform me as early as possible whether you are able and willing to send in a tender.

Yours faithfully

Signature of Architect/Engineer

To

....

....

(Name and address of Contractor)

Appendix **B**

Tender for a Contract

Name of work ....

At ....

For ....

Dated ....

To

....

....

(Owner or Architect/Engineer)

In compliance with your invitation for tenders dated .... the undersigned hereby undertakes to furnish materials, labour and equipment and perform all work as shown and described in the drawings, specifications and all the contract documents prepared by .... (name) Architect/Engineer for the sum of Rupees .... (Worked out on the basis of the unit rates and quantities shown in the schedule of quantities.)[1]

The undersigned agrees that, upon written acceptance of this tender, he will within .... days of receipt of such notice execute a formal contract agreement with the owner, and that he will pay in cash the security deposit as required.

The undersigned further agrees that, if awarded the contract, he will commence the work within .... days after the date of receipt of written notice to proceed, and that he will complete the work within .... calendar/working days thereafter.

Enclosed is earnest money in the form of a certified cheque payable to .... (owner)

Sd/ ....
(Name of the Contractor or Firm)
Business address ....

---

1. Applicable to measurement contracts.

Note: The persons or firm submitting the tender should see that the rates in the above schedule are filled up by the Engineer on the issue of the form prior to submission of the tender.

Note: All the columns in the schedule should be filled in, in ink, and the total of the entries in the last column should be struck by the Contractor under his signature.

Appendix C

Schedule A

Schedule showing the materials to be supplied by the Owner for the work to be executed and rates at which they are to be charged for

S. No	Approximate quantity	Particulars	Rate at which the materials will be charged to the contractor			Place of delivery
			Unit	₹	P	

Note: The persons or firm submitting the tender should see that the rates in the above schedule are filled up by the Engineer on the issue of the form prior to submission of the tender.

(Signature of Contractor)   (Signature of Owner)

Appendix D

Schedule B

Memorandum showing items of work to be carried out.

Item No.	Estimated quantities	Items of work	Tendered rates		Unit	Total amount according to estimated quantities
			In figures	In words		
			₹ P			

Note: All the columns in the schedule should be filled in, in ink, and the total of the entries in the last column should be struck by the Contractor under his signature.

(Signature of Contractor) (Signature of Owner)

Appendix E

Architect's Interim Payment Certificate

Dated ....

Certificate No ....
Amount of contract ₹ ....
Value of work done to date ₹ ....
Value of materials on site ₹ ....

Total to date ₹

Deduct... % as per conditions
of contract ....   ₹ ....
Previously certified   ₹ ....
Now certified   ₹ ....
Certificate in favour of ....
....
of ....
on account of ....
At ....
under the terms of contract dated ....
20 ....

Amount of ₹ ....

Addressed to

....
....

(Architect's/Engineer's Address)

....
....
Certificate No. .... Dated ....
To ....
I hereby certify that the sum of
₹ ....
is due to .... of ....
on account of works at ....
under the terms of contract dated ....
20 ....
₹ ::::::::::::::::::::::::::

....
Architect/Engineer

Received from ....
the above sum of ₹ ....
Sated 20

(Stamp)

Signature of Contractor

References

[1] Abbett, Robert W., *Engineering Contracts and Specifications,* John Wiley & Sons, New York, Chapman & Hall, London (1948).

[2] *Bombay Public Works Department Manual,* Vol. I, Government of Bombay, Public Works Department, Director, Government Printing, Publication and Stationery, Bombay State, Bombay (1956).

[3] Creshwell, W. T., *The Law Relating to Building and Engineering Contracts,* Sir Issac Pitman & Sons, London (1957).

[4] Deatharage, George E., *Construction Company Organisation and Management,* McGraw-Hill, New York (1964).

[5] Deatharage, George E., *Construction Scheduling and Control,* McGraw-Hill, New York (1965).

[6] Dunham, C.W. and Young, R.D., *Contracts, Specifications and Law for Engineers,* McGraw-Hill, New York (1958).

[7] Griffith Edwards, H., *Specifications,* D. Van Nostrand, New York (1961).

[8] Johnson, A. and King , W. H., *A Manual of Specifications and Quantities for Civil Engineers,* English Universities Press, London (1957).

[9] Marks, R.J., Grant, Alec and Helson, P.W., *Aspects of Civil Engineering Contracts Procedure,* Pergamon Press, London (1965).

[10] Maxwell-Cook, John C., *Civil Engineering Contracts Organisation,* Cleaver-Hume Press, London (1959).

[11] Sansom, R.C., *Organisation of Building Sites,* Her Majesty's Stationery office, London (1959).

[12] *Standard Specifications,* Government of Maharashtra, Buildings and Communications Department, The Government Central Press, Bombay (1965).

[13] *The Indian Contract Act—1872.*

SECTION II

Civil Engineering Estimates

[Including valuation]

Introduction

0.1 GENERAL

An estimate is a calculation of the quantities of various items of work, and the expenses likely to be incurred thereon. The total of these probable expenses to be incurred on the work is known as 'estimated cost' of the work. The estimated cost of a work is a close approximation of its actual cost. The agreement of the estimated cost with the actual cost will depend mainly upon the accurate use of estimating methods and correct visualisation of the work as it will be done. The importance of correct estimating is obvious. Under-estimating may result in the client getting an unpleasant shock when tenders are opened and drastically modifying or abandoning the work at that stage. Over-estimating may lose the engineer his client or his job or in any case his confidence.

Estimating is the most important of the practical aspects of engineering, and the subject deserves the closest attention of one aspiring to a career in the profession. It is a comparatively simple subject to understand; however, as it brings one up against practical work, methods and procedure, a knowledge of it cannot be acquired without close application.

0.2 PURPOSE OF ESTIMATING

1. *To Give a Reasonably Accurate Idea of the Cost*
   An estimate is necessary to give the owner a reasonably accurate idea of the cost to help him decide whether the work can be undertaken as proposed or needs to be curtailed or abandoned, depending upon the availability of funds and prospective direct and indirect benefits. For government works, proper sanction has to be obtained for allocating the required amount. Works are often let out on a 'lump sum' basis, in which case the tenderer must be in a position to know exactly how much expenditure he is going to incur on the work.
2. *Estimating Materials*
   From the estimate of a work it is possible to determine what materials (and in what quantities) will be required for the work so that arrangements to procure them can be made.
3. *Estimating Labour*
   The number and kind of workers of different categories who will have to be employed to complete the work in the specified time can be found out from the estimate.

4. *Estimating Plant*

An estimate will help in determining the amount and kind of equipment needed to complete the work.

5. *Estimating Time*

The estimate of a work and past experience enable one to estimate quite closely the length of time required to complete an item of work or the work as a whole.

Whereas the importance of knowing the probable cost needs no emphasis, it should be stressed that estimating materials, labour, plant and time is also immensely useful in planning and execution of any work.

0.3   TYPES OF ESTIMATES

There are several kinds of estimates. These can be grouped into two main categories:

1. approximate estimates, and
2. detailed estimates.

1. *Approximate Estimates*

An approximate estimate is an approximate or rough estimate prepared to obtain an approximate cost in a short time. For certain purposes the use of such methods is justified. These are dealt with in Chapter 1.

2. *Detailed Estimates*

A detailed estimate of the cost of a project is prepared by determining the quantities and costs of everything that a contractor is required to provide and do for the satisfactory completion of the work. It is the best and most reliable form of estimate. A detailed estimate may be prepared in the following two ways:

(i)   unit-quantity method, and

(ii)   total-quantity method.

(i)   *Unit-quantity method*: In the unit-quantity method, the work is divided into as many operations or items as are required. A unit of measurement is decided. The total quantity of work under each item is taken out in the proper unit of measurement. The total cost per unit quantity of each item is analysed and worked out. Then the total cost for the item is found by multiplying the cost per unit quantity by the number of units. For example, while estimating the cost of a building work, the quantity of brickwork in the building would be measured in cubic metres. The total cost (which includes cost of materials, labour, plant, overheads and profit) per cubic metre of brickwork would be found and then this unit cost multiplied by the number of cubic metres of brickwork in the building would give the estimated cost of brickwork.

This method has the advantage that the unit costs on various jobs can be readily compared and that the total estimate can easily be corrected for variations in quantities.

(ii) *Total-quantity method*: In this method, an item of work is divided into the following five subdivisions:
   (a) materials,
   (b) labour,
   (c) plant,
   (d) overheads, and
   (e) profit.

The total quantities of each kind or class of material or labour are found and multiplied by their individual unit cost. Similarly, the cost of plant, overhead expenses and profit are determined. The costs of all the five subheads are summed up to give the estimated cost of the item of work. For example, the cost of brickwork in a building would be determined as below:

I	(i)	Cost of material at the source			
		Bricks	... ...	₹	
		Sand	... ...	₹	
		Cement	... ...	₹	
		Water	... ...	₹	
	(ii)	Cost of handling and transporting above materials	... ...	₹	
II		Cost of labour both skilled and unskilled	... ...	₹	
III		Cost of plant	... ...	₹	
IV		Overheads	... ...	₹	
V		Profit	... ...	₹	______
Total cost of brickwork:				₹	______

The first method is fully explained in this book and it is also used to frame the example estimates included in the text.

0.4   QUALIFICATIONS OF AN ESTIMATOR

A good estimator should possess the following qualifications:

1. A thorough understanding of the architect's drawings.
2. A sound knowledge of building materials, construction methods and customs prevailing in the trade.
3. A fund of information, collected or gained through experience in construction work, relating to materials required, hourly output of workers and plant, overhead expenses and costs of all kinds.
4. An understanding of a good method of preparing an estimate.
5. A systematic and orderly mind.

6.   Ability to do careful and accurate calculations.
7.   Ability to collect, classify and evaluate data that would be useful in estimating.

Good instruction or careful and thorough study of a standard book will help a beginner to become a good estimator. He must, however, try to develop all the above-mentioned qualities while obtaining practical experience.

0.5   DATA REQUIRED FOR PREPARING AN ESTIMATE

In order to prepare a detailed estimate the estimator must have with him the following data:

1.   Plans, sections and other relevant details of the work.
2.   Specifications indicating the exact nature and class of materials to be used.
3.   The rates at which the different items of work are carried out.

To enable an estimator to take out the quantities accurately, the drawings must themselves be clear, true to fact and scale, complete, and fully dimensioned.

The estimator has also to bear in mind certain principles of taking out quantities. These principles are outlined in Chapter 3.

0.6   STEPS IN PREPARATION OF AN ESTIMATE

There are three clearly defined steps in the preparation of an estimate.

1.   *Taking out Quantities*
     In the first step of taking out quantities, the measurements are taken off from the drawings and entered on measurement sheet or dimension paper. The measurements to be taken out would depend upon the unit of measurement. For example, in the case of stone masonry in superstructures, length, thickness and height of the walls above plinth level would be taken out from the drawings and entered on the measurement sheet, whereas, in the case of plastering only the lengths and height of the walls would be entered. Obviously, the unit of measurement in the first case is cubic metre and that in the second case is square metre.
2.   *Squaring out*
     The second step consists of working out volumes, areas and so on, and casting up their total in recognised units.
3.   *Abstracting*
     In the third step, all the items along with the net results obtained in the second step are transferred from measurement sheets to specially ruled sheets having rate column ready for pricing.

The second and third steps above are known as 'working up'. All calculations in these stages and every entry transferred should be checked by another person to ensure that no mathematical or copying error occurs.

0.7 STANDARD METHOD OF MEASUREMENT OF BUILDING WORKS

The different methods of measuring used by various Central and State Government departments and by construction agencies were found to be a serious difficulty to estimators and a standing cause of disputes. For this reason a unification of the various systems at the technical level had been accepted as very desirable and wanting. The Indian Standards Institution took up for standardisation the method of measurement of building works and published the *Indian Standard Method of Measurement of Building Works* (referred to in this book as SMM of BWIS 1200) in 1958. The standard was republished in 1964 in a revised form after a thorough examination of a large number of suggestions for modifications.

Although the standard has no legal sanction and as such need not be adopted unless it is referred to in the contract, this document, prepared as it is out of an ardent need, is bound to establish itself and would be produced as evidence of custom in the profession. Some universities and technical education boards have already included its study in their courses. It is therefore advisable that students and young engineers in particular study it and follow its recommendations in all cases. This standard has been followed in this book in the measurement of building works.

0.8 PRINCIPLES OF DECIDING UNIT OF MEASUREMENT

A beginner may find it difficult to remember the units of measurement of different items. Memorising of units of measurement would be greatly simplified if the principles kept in view while selecting the units of measurements are known to him. Following are the most important principles of selection of unit of measurement:

1. The unit of measurement should be simple and convenient to measure, record and understand.
2. It should be one which provides for fair payment for the work involved.
3. In the result it should yield quantities which are neither too minute nor too large.
4. The price per unit should not be a very small figure or a very large one, that is, generally costlier items will be measured in smaller units, cheaper ones in larger units.
5. The unit of measurement may sometimes depend upon the unit for the raw material and/or labour and/or important dimensions. For example, stone masonry is measured in cubic metres because raw materials are measured in cubic metres. Plastering or pointing is measured in square metres as the labour is considerable.

Approximate Estimates

1.1 GENERAL

The only safe and sure method of estimating costs is to take out the actual quantities of works under different items and multiply these quantities by prevailing unit rates. This method, however, is laborious, time-consuming and is suitable only when drawings and specifications are already prepared. When the accurate cost of construction is not required and a rough idea of cost would serve the purpose, efforts are made to use shortcuts and approximate methods. The accuracy of costs estimated by such approximate methods depends upon the judgement, skill, and experience of the estimator, and upon the correctness of the prices used.

This chapter includes the description of the methods used in working out approximate costs of the following projects:

1. buildings,
2. irrigation projects,
3. water supply and sanitary schemes,
4. highways, and
5. bridges.

1.2 USES OF APPROXIMATE ESTIMATES

Approximate estimates are prepared to serve the following purposes:

1. *To Give a Rough Idea of the Probable Expenditure*

   At an early stage in the development of a project, it is necessary to ascertain whether the project can at all be financed. For this a rough idea of the probable expenditure has to be obtained, and if this appears feasible, then further details may be considered. At this stage, the information available is of the most meagre nature; such drawings as exist give little, if any, indication of the constructional details to be provided for and the specifications are, generally non-existent. Preparation of an approximate estimate is, thus, the only suitable method to find out the cost and ascertain whether the investment in the project is justified or not.

2. *Administrative Approval*

   In the case of government and other public works, proper sanction has to be obtained for allocating the expenditure required for the detailed investigations and preparation of plans and estimates. This sanction is given on the basis of the approximate cost found out by an approximate method.

3. *Valuation and Rent Fixation*

   Sometimes it is necessary to find out the cost of an existing structure for one or more of the following reasons:

     (i)   sale or purchase,

    (ii)   rent fixation,

   (iii)   framing tax schedules, and

   (iv)   insurance requirement.

   In all these cases the valuators invariably make use of the approximate estimates.

1.3 APPROXIMATE METHODS OF ESTIMATING FOR BUILDINGS

The following six methods of estimating approximate costs of buildings are commonly used:

1. service unit,
2. square metre of floor area,
3. cubic metre of the building,
4. rough grouped quantities or elemental bill method,
5. detailed quantities of a bay, and
6. comparison of costs at relative dates.

The above methods are stated serially with an increasing degree of accuracy. Each of these methods is discussed, in detail, in the subsequent sections.

1.4 SERVICE UNIT METHOD

Each building is constructed to serve some purpose. For example, a hospital is designed to accommodate a certain number of beds. Each bed is then considered as the service unit of a hospital building. Whenever a building is constructed the concerned architect/engineer keeps a record of the place of construction, actual cost of construction, number of service units and the year of construction. From this record, it is possible to work out the cost of construction per service unit, thus,

$$\text{Cost of construction per service unit} = \frac{\text{Total cost of construction}}{\text{Total number of service units}}$$

Such data are accumulated over a number of years' practice. If any structure, similar to one constructed in the past, is proposed to be erected, the architect/engineer will be able to estimate the cost by multiplying the number of service units to be provided in the proposed structure by the rate per service unit of the structure already constructed.

In order to work out a worthwhile figure, care will have to be taken to modify the known rate in the light of the following points.

1. The rate has to be suitably changed by the study and use of cost indices to account for general increase or decrease in the cost of construction during the period between the year of construction of the existing structure and that of the one proposed.
2. Any peculiarities of the proposed building, its location, deviations in the foundation conditions and materials proposed to be used and so on, should reflect in the rate.

The above costs are very approximate and will vary considerably with buildings of different sizes, shapes, kinds of construction, quality, locality and time.

1.5   SQUARE METRE OF FLOOR AREA METHOD

In this method of estimating building costs, the floor area of the proposed structure is worked out by multiplying the external dimensions at the plinth level and at each successive floor. This floor area is then multiplied by a suitable rate. This rate of construction per square metre of the floor area is worked out after dividing the known cost of construction of a similar building by the floor area of that building. However, the rate so worked out needs to be modified to account for any special conditions, variation in the materials, location, time, and so on in respect of the proposed building.

The following are some of the variations of this method:

1. One way is to add the areas of all usable floors excluding roof and basement and to multiply this area by a suitable unit price.
2. Another way is to use a certain unit rate for each floor including the basement and roof.
3. Thirdly, different unit rates for the basement floor, other floors and roof may be used.
4. A fourth method is to have the unit rate for the lower floor include the basement and foundations, and that for the top floor include the roof.

An estimator may use any one of the above methods taking care to state just what is included in the unit rate used in the estimate.

The following illustrative example further clarifies the above methods and shows the effect on unit rate by using the different methods.

Illustrative Example

A building, having 15 m by 10 m external dimensions and consisting of a basement, ground and first floors, is recently constructed at the cost of ₹18,00,000. Compute the cost per square metre of the building, by various methods, for office records and future use.

Solution: The rate of construction per square metre of the building is worked out by each of the above-mentioned methods and in the same serial order.

1.
Ground floor area	=	$15 \times 10 = 150$ m^2
First floor area	=	$15 \times 10 = 150$ m^2
Total area of two floors		300 m^2
Cost per square metre	=	$\dfrac{18,00,000}{300}$
	=	₹6000

2.
Basement area	=	$15 \times 10 = 150$ m^2
Ground floor area	=	$15 \times 10 = 150$ m^2
First floor area	=	$15 \times 10 = 150$ m^2
Roof area (horizontal)	=	$15 \times 10 = 150$ m^2
Total floor area		600 m^2
Cost per square metre	=	$\dfrac{18,00,000}{600}$
	=	₹3000

3. In this method it is assumed that the cost of basement is 60% of the cost of ground floor and that of roof is 50% of the cost of the ground floor. Ground and first floor have equal costs.

Basement area (equivalent)	=	$15 \times 10 \times 0.6 = 90$ m^2
Ground floor area	=	$15 \times 10 \quad = 150$ m^2
First floor area	=	$15 \times 10 \quad = 150$ m^2
Roof area (equivalent)	=	$15 \times 10 \times 0.5 = 75$ m^2
Total equivalent area		465 m^2
Cost per square metre	=	$\dfrac{18,00,000}{465}$
	=	(say) ₹3871

4. With the assumption made in the third method, the equivalent area is worked out thus:

$$
\begin{aligned}
\text{Ground floor (equivalent)} &= 150 \times 1.6 = 240 \text{ m}^2 \\
\text{First floor (equivalent)} &= 150 \times 1.5 = 225 \text{ m}^2 \\
\text{Total equivalent area} &= 465 \text{ m}^2 \\
\text{Cost per square metre} &= \frac{18,00,000}{465} \\
&= ₹3871
\end{aligned}
$$

(1) Cost per square metre of ground floor

$$= 3871 \times \frac{240}{150} = ₹6193.6$$

(2) Cost per square metre of first floor

$$= 3871 \times \frac{225}{150} = ₹5806.5$$

1.6   CUBIC METRE METHOD

The method of estimating building costs by the cubic metre of volume is more accurate, in general, than the square metre of floor area method because the depth of foundation and the height of structure above ground level are the additional dimensions taken into account.

The method consists of finding the total volume of a building in cubic metres and multiplying this volume by a suitable rate per cubic metre. The rate used must be worked out from the record of cost of construction of buildings of the same class and type with the modification in the light of points mentioned in Section. 1.4.

Work out Cubic Contents of a Structure

The dimensions are measured at the outer faces of walls and the floor area is worked out; this floor area is then multiplied by the height of the structure. There are different rules for measuring the height of a building, some of which are mentioned below.

1. *RIBA* and *RICS* have recognised the method in which height is measured from the top of concrete foundations to (i) in the case of ordinary pitched roofs, half the height between the point of intersection of roof with the outer line of the wall and the ridge; or (ii) 60 cm above flat roofs (refer to Fig. 1.1). In the case of mansard roof, the cubic contents of the roof are worked out separately.
2. *American Institute of Architects* (AAA) recommends the method in which the actual cubic space enclosed within the outer surfaces of the outside or enclosing walls and contained between the outer surfaces of the roof and 15 cm below the finished surface of the lowest floor is considered.

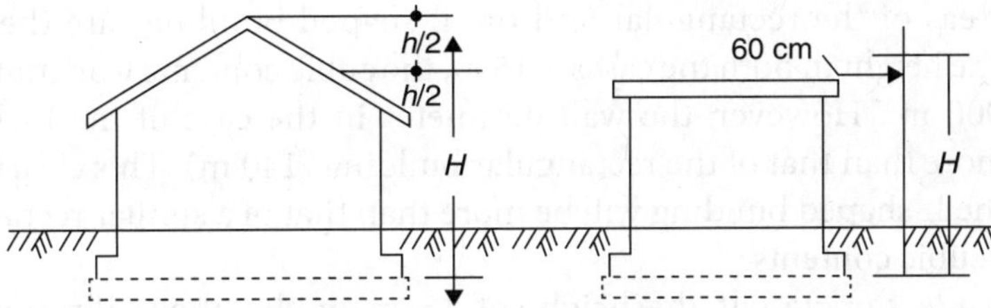

Fig. 1.1   RIBA and RICS method of measuring height or working out cubic contents

3. The method normally used in India is to measure the height from half the depth of foundation to (i) in the case of pitched roof, half way up the roof between the point of intersection of the roof with the outer line of wall and the ridge; or (ii) top in the case of flat roof; or (iii) half the height of parapet (refer to Fig. 1.2).

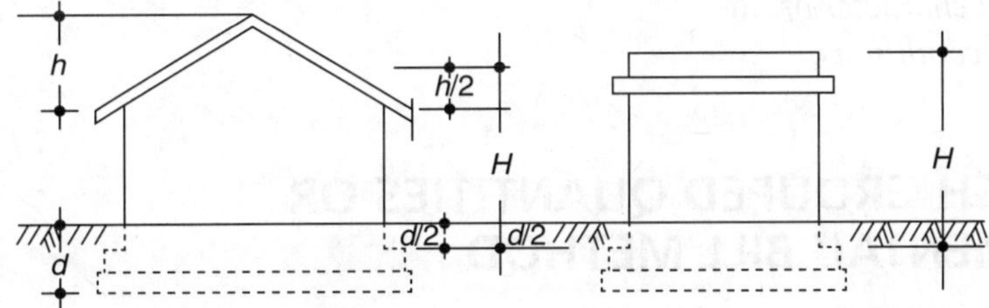

Fig. 1.2   Standard method of measurement of height for calculating cubic contents

The main skill lies in fixing the price per cubic metre in this type of estimate. The assessment must be based on carefully kept records of previously erected buildings of similar type and class.

There are many factors which affect the cube price, some of which are listed below:

1. *Shape*:   The cubic contents of two buildings may be identical and the specifications broadly the same, but the shape of the plan may make a considerable difference in the price per metre cube. To illustrate, consider the sketches in Fig. 1.3.

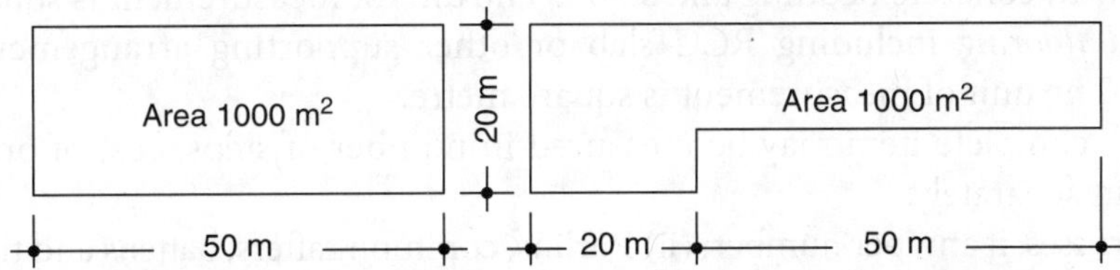

Fig. 1.3   Line plan of rectangular and L-shaped buildings

The floor areas of the rectangular and the L-shaped buildings are the same, namely, 1000 m². If the height in both the cases is 15 m, the cubic contents work out to be identical, namely, 15000 m³. However, the wall perimeter in the case of the L-shaped building (180 m) is more than that of the rectangular building (140 m). This clearly indicates that the cost of the L-shaped building will be more than that of a similar rectangular building of the same cubic contents.

2. *Subdivision into storeys*:   If the height of 15 m in the above illustration is divided into three storeys, this requires an additional floor of 1000 m², which would result in considerable increase in the cost, without any change in the cubic contents.
3. *The amount and type of internal vertical subdivision.*
4. *Magnitude of work*:   Small buildings cost more in proportion than large ones of the same type.
5. *The type of house*:   Bungalows are dearer than the storey houses of the same accommodation.
6. *Relative proportions of voids to solids.*
7. *Position and character of site.*
8. *Foundation conditions.*

1.7   ROUGH GROUPED QUANTITIES OR 'ELEMENTAL' BILL METHOD

In this method, a building is divided into as few items of work as possible and measurements are taken out for the same. The cost per unit quantity of each group is worked out separately and carefully, accounting for all the different operations, which are combined together for simplicity. The quantities are multiplied by these rates and the cost worked out.

The nature of each group or item of work and the total number of items will vary with different kinds of buildings. For a small work—residential building—the different items to be considered may be as follows:

1. *External walls* including excavation, concrete bedding, masonry and brickwork, pointing, internal finishings, doors and windows. Measurements, are taken out in square metre.
2. *Internal walls* including all items mentioned under external walls. The unit of measurement is square metre.
3. *Ground flooring* including surface excavation, hard *murum* and/or rubble filling, floor finishing with concrete bedding and so on. The unit of measurement is square metre.
4. *Suspended flooring* including RCC slab or other supporting arrangement and floor finishing. The unit of measurement is square metre.
5. *Stairs*: The complete item may be measured in number of steps, cost of one step being worked out separately.
6. *Roof*: (i) Trusses, if any, per number; (ii) Purlins, common rafters, battens and tiles—all these are measured together in one item; the unit of measurement is square metre of the roof area.
7. *Water supply and sanitary fittings*: A lump sum may be provided for this work.

8. *Electrical work*: A lump sum provision, based on the past experience, may be made in the estimate.

It is clear from the above list of items that though approximate, this method involves considerable work and takes a longer time; however, the additional details make it more reliable than the methods discussed earlier.

1.8   DETAILED QUANTITIES OF A BAY

This method of approximate costing is particularly suitable for buildings which consist of several identical bays; for example, a workshop building, a cycle stand, a garage and so on. It consists of taking out detailed quantities of all the different items of work involved in a typical bay or slice, and finding out the cost of all such items using suitable rates. This cost, when divided by the length of bay considered, would give the cost per running metre of the building. This cost of construction per running metre is multiplied by the total length of the building and cost of the end walls (worked out by finding out the cost of one wall and multiplying it by their total number) added to it. The cost estimated is generally found to be fairly close to the actual cost of construction.

1.9   COMPARISON OF COSTS

The approximate cost of the proposed structure is found out by modifying the known cost of similar existing structures in accordance with the index figures of costs at the appropriate dates, that is, the dates of construction of the old and proposed structure.

Although this method gives the best results, its use is limited to works which are a repetition of similar works previously constructed and of which the time, year and details of cost of construction are known. For example, construction of an additional ward to a hospital building.

1.10   APPROXIMATE METHODS OF ESTIMATING FOR IRRIGATION PROJECTS

Unlike building works, preparation of approximate estimates in respect to other civil engineering projects involves a considerable amount of field and office work.

The following two methods are normally used to find out the approximate cost of irrigation projects. The first method gives a very rough idea about the cost, whereas the cost worked out by the second method is fairly reliable.

1. *Per Million Cubic Metre of Storage Capacity*
    This is a rough and quick way of estimating the cost of a scheme without carrying out any survey or preparation of drawings. The method of costing involves the following steps:

(i)   After a careful study of the topo-sheet a suitable site for the dam is selected, marked on it, and the catchment area found out.

(ii) Knowing the rainfall data from the rain gauge stations in or near the catchment area, yield in million cubic metres, for a desired dependability percentage is worked out.

   The post-monsoon yield to the extent of ten per cent of the yield so worked out is added to find out the gross yield.

(iii) A percentage of the total available yield (normally 60% to 75%) that would have to be stored is decided upon and the capacity of the reservoir worked out.

(iv) The final and important stage is to decide upon the cost of construction per million cubic metres of the storage with the help of carefully kept records of projects already completed. It is always better to visit the site of the proposed dam in order to ascertain with which of the existing projects, general conditions of the proposed scheme most closely resemble. This helps in deciding a proper cost of construction per million cubic metres of the storage. By multiplying this cost with the total storage capacity, the approximate cost is worked out.

As the storage capacity of a particular dam depends to a great extent upon the characteristics of the reservoir basin and no two sites may have identical characteristics, the cost per million cubic metres of storage of water is subject to a wide variation. The variation may be from ₹1,20,00,000 per million cubic metres of storage of water for large projects to ₹5,50,00,000 per million cubic metres in respect of smaller schemes.

The following illustrative example will clear the above-mentioned points.

Illustrative Example

1.	Catchment area	50 sq. km
2.	Nature of catchment	Good
3.	Dependable rainfall at 60% confidence limit	800 mm
4.	Yield per square kilometre of catchment area (from Strange's table)	0.2234 million cubic metres
5.	Monsoon yield	11.17 million cubic metres
6.	Post-monsoon yield (10% of monsoon yield)	1.117 -do-
7.	Total yield	12.287 -do-
8.	Gross storage capacity (65% of total yield)	8.02 -do-
9.	Approximate cost of the project @ ₹3,00,00,000 per million cubic metre of storage capacity (8.02 × 1,60,00,000)	say ₹24,00,00,000

2. *Elemental Bill Method*

This method resembles the rough grouped quantities method as used in the case of building projects. In order to collect the necessary details that are required in addition to those mentioned in the first method above, the following field work is done:

(i)   Survey of the longitudinal and cross-sections of the channel at the proposed dam site.

(ii)   Exploration of the foundations of the dam to know the depth of impervious stratum such as solid rock.

(iii)   Basin contouring survey of minimum two contours to work out the capacity of the reservoir.

With the help of the above collected data, a rough bill of quantities is prepared by dividing the scheme into the following component parts:

(a)   dam,

(b)   surplus works,

(c)   outlets,

(d)   distribution system,

(e)   de-watering of the foundations and coffer dam,

(f)   land acquisition, and

(h)   overheads, centage and other charges.

(a)   *Dam*: Longitudinal section of the dam showing foundation levels, ground levels, full reservoir level, top of the dam and so on, is prepared. The average height of the dam is worked out. The cross-section of the dam suitable for the height is decided on the basis of type sections. The quantity of masonry, concrete or earthwork, as the case may be, can be found out by multiplying the cross-sectional area by the length of the dam. In the case of earth dams, the total length of a dam may be required to be divided if the cross-section is varying; for example, that portion of the dam where berm at intermediate level is required will be separately considered from the portion where it is not required. Measurements in respect of cut off trench excavation and filling are taken out separately by assuming suitable cross-section and average depth.

A suitable rate per cubic metre of the work is worked out in order to include the cost of all items connected with a dam. For example, in the case of an earth dam the quantities would be grouped into two groups:

- *Cut-off-trench* including excavation and filling complete.
- *Embankment* including drainage, rock toe, hearting, casing, pitching and turfing.

The cost of a dam is worked out by multiplying the total quantity of work with the rate per unit quantity. Rates (2014) for a few items (rough grouped quantities) are given in Table 1.4 which may serve as a rough guide in the preparation of the approximate estimate of an irrigation project.

(b)   *Surplus works*: The cost of excavation is worked out by approximately estimating the quantity and multiplying it by a suitable unit rate. The cost of overflow section of a dam is worked out in a way similar to that of a non-overflow section.

(c)   *Outlets*: Based on past experience and records, a lump sum provision is made in the bill of quantities for the proposed outlets.

(d)   *Distribution system*: The cost of distribution system (such as canals, branches, distributories, regulation works, cross drainage works, and so on) is found out by multiplying the total area likely to be irrigated, expressed in hectares, by a certain rate which is decided on the basis of the records of projects already completed. The rate (2014) normally used is given in Table 1.1.

Table 1.1

Item of work (Rough grouped quantities)	Unit	Rate (2014) ₹
1. Cut-off-trench (excavation and filling)		
(a) soil	m^3	120 to 160
(b) rock	m^3	350 to 500
2. Earthwork (including hearting, casing, rock toe revetment, drains)	m^3	120 to 150
3. U C R masonry in c s m 0.8–0.2:5)	m^3	1800 to 2000
4. Distribution system (including lining)	hectare	30,000 to 50,000

In order to determine the area likely to be irrigated, it is assumed that one million cubic metres of stored water irrigates about 10,000 hectares to 14,000 hectares of area.

(e) *De-watering foundations and coffer dam*: A lump sum provision based on experience of similar projects is made in the bill of quantities.

(f) *Land acquisition*: Land to be acquired for the following purposes is found out and its value worked out on the basis of suitable rates.
- Land likely to go under submergence.
- Land for dam site, approach road, quarries, camp buildings, and for the rehabilitation of villages.

In addition, a lump sum provision for houses, trees, wells and so on, is made.

(g) *Overheads and other charges*
- About 25 per cent of the total estimated cost of works under one to five of the above items is added, over and above the estimated cost, to cover the expenses over miscellaneous items such as preliminary work, buildings, roads, special tools and plant and so on.
- *The estimated cost including* overheads is increased by a percentage (normally 13%) to account for centage charges.
- *Abutment of land revenue*: Normally five per cent of the cost of land to be acquired is added to the total cost thus worked out.

1.11 APPROXIMATE METHOD OF ESTIMATING FOR WATER SUPPLY PROJECTS

Similar to irrigation projects, preparation of an approximate estimate in the case of water supply and sanitary projects includes collection of extensive data, a few design calculations and takes much time. However, considering the large investment both during construction and maintenance, the work of detailed investigations and construction is not undertaken unless the scheme is found to be self-supporting and financially feasible. In order to determine the feasibility of a project the cost of construction and maintenance are approximately worked out. The annual revenue that the project will earn is also approximately estimated. When the annual maintenance burden, including interest on the capital invested, is found to be less than the annual revenue estimated to be raised, the scheme is said to be self-supporting and financially feasible.

The method followed in estimating approximate cost of a water supply project is explained below.

1. *Data to be Collected*
    (i) *Situation*: Name of place, district, nearest railway station, nearest town connected by good road.
    (ii) *Configuration*: Nalla or river flowing nearby, relative elevations of town and nalla or river.
    (iii) *Rainfall*: Maximum, minimum and average rainfall data for as many previous years as are available.
    (iv) *Population*: Population figures for the past few decades, present population as per reports of municipal or panchayat officers, and estimate of prospective population for the next few decades.
    (v) *Existing water supply*: Source, quantity of water available, treatment if any, rate of supply, and so on.
    (vi) *Existing drainage*: Conditions of roads, drainage disposal, and so on.
    (vii) *Necessity of the proposed scheme*: Importance of the place, connecting roads and traffic thereon; railway station requirements if any, and so on.
2. *Estimating the Cost of a Project*
    The entire scheme is divided into the following component parts and the cost of each part is determined separately. The units in which approximate quantities under different component parts or items are worked out and the range of prevailing unit rates (2014) are stated below.

	Item	Unit	Rate (2014)
a.	Preliminary survey	lump sum	
b.	Head works		Based on records of similar projects
	(i) Dam	as per Section 1. 10.2 (i)	
	(ii) Infiltration gallery or trench gallery (450 mm dia)	per metre length	

(continued)

Continued

	Item	Unit	Rate (2014)
c.	Jack well and pump-house specifying diameter and depth	lump sum	Based on the records of similar projects
d.	Inspection well specifying diameter and depth	-do-	-do-
e.	Pumping machinery. Calculating BHP requirement and providing for duplicate set	per BHP	₹35,000 for 75 HP and above to ₹45,000 up to 3 HP
f.	Rising main specifying diameter, length, material	per metre	Depends upon material and dia., for example 20 cm dia. ms pressure pipe ₹3000
g.	Ground service reservoir specifying shape, size and capacity in litres. ₹0.70 for 15 lakh litres and above capacity	per litre of water stored	₹15 for small size reservoirs up to 25,000 litres capacity, to ₹6 for size up to 1.5 lakh litres capacity
h.	Treatment works	million litres per day	Based on records of similar projects
i.	Overhead service reservoir specifying material type and capacity	per litre	₹15 to ₹25
j.	Distribution system, specifying range of dia. of pipes, material, and so on	per capita	From the records of similar projects
k.	Approach road length from plan, specifying width of surface	per kilometre	As per Section. 1.13
l.	Staff quarters	as per Section. 1.4 to 1.9 (*any one method*)	
m.	Land acquisition	per hectare	₹5,00,000 and above depending upon the nature and type of land

Net cost of the scheme is worked out by adding the costs of all the above items. A percentage of the net cost (excluding the cost of land acquisition) is added to cover preparation charges. The cost so worked out is increased by a percentage (about 16%) to account for establishment, tools and plant charges.

1.12 APPROXIMATE METHOD OF ESTIMATING FOR SEWERAGE SCHEMES

1. *Data to be collected* are similar to a water supply project.
2. *Estimating the cost of a scheme*: The project breakdown into different items, the units in which approximate quantity of each item is worked out, and the prevailing unit rates are given as follows:

	Item	Unit	₹
a.	Working survey	lump sum	To be worked out from records of similar projects
b.	(i) Outfall and collecting sewers including excavation, filling, manholes, ventilators and so on		
c.	Branches and laterals	per capita per capita	
d.	Pump house and sump well	similar to water supply project	
e.	Pumping machinery	-do-	
f.	Rising main	-do-	
g.	Sewage treatment and disposal	per million litre per day	-do-
h.	Staff quarters	similar to water supply project	
i.	Land acquisition	-do-	

To the net cost as estimated above, a percentage of it (excluding the cost of land acquisition), normally 17% to 18%, is added for establishment, tools and plant charges. The cost so worked out is increased by a small percentage to account for the preparation charges of the scheme.

1.13 APPROXIMATE METHOD OF ESTIMATING FOR HIGHWAYS

In the case of road works, there is considerable amount of standardisation in respect to land to be acquired, width and type of wearing surface, and other constructional details to be provided for a particular class of a road. It is, therefore, possible to prepare detailed estimates for different types of roads for one kilometre length. Such detailed estimates together with the information gained from the road works already constructed enable one to find out a realistic price of construction of a road per kilometre of its length. With this data at hand it is possible to estimate the approximate cost of a proposed road project, with a fair degree of accuracy, if the type of surface and the total length are known.

The classification of roads (in India) and approximate costs of construction (2014) per/kilometre length are given in Table 1.2.

Table 1.2

S. No.	Classification of roads	Cost per kilometre (2014) ₹
1.	Village roads	₹20 lakhs to ₹30 lakhs
2.	Other district roads	₹70 lakhs to ₹80 lakhs
3.	Major district roads	₹100 lakhs
4.	State highways	₹160 lakhs
5.	National highways	
	2 lane	₹3 crores
6.	4 lane	₹6 to 10 crores
7.	Expressway (6 lane)	₹10 crores to 12 crores

The cost of constructing bridges is found out separately as explained in the next section. Similarly, the cost of providing a special type of wearing surface such as concrete, bituminous, and so on, is also worked out separately.

The common types of wearing surfaces are listed in Table 1.3. The estimated cost of construction (2014) of 3.70 m wide and 1 km long pavement of each type is also given in the same table.

Table 1.3

Type of wearing surface	Cost/kilometre (width 3.70 m) (2014)
10 cm thick metal coat	₹6,00,000
8 cm thick metal coat	₹4,80,000
5 cm thick full grout	₹8,00,000
5 cm thick semigrout	₹6,00,000
4.00 cm thick premix carpet with seal coat	₹16,00,000
2.5 cm thick premix carpet without seal coat	₹10,00,000
2.0 cm -do-	₹8,50,000
Two-coat surface dressing	₹7,50,000
Seal coat	₹3,00,000

1.14   APPROXIMATE METHOD OF ESTIMATING FOR BRIDGES

The approximate cost of a bridge is worked out in the following steps.

1. The linear waterway, during the time of maximum floods at the proposed site, is found out.
2. The known cost of construction of a similar bridge is divided by its linear waterway to find out the rate of construction per metre of linear waterway. This is modified to

account for special features, if any, of the proposed bridge, change in the cost index, and so on.

3. The total linear waterway (in metres) at the proposed site is, then, multiplied by the rate of construction per metre of linear waterway, as worked out in step 2 above.

Table 1.4 gives the classification of bridges (for the purpose of approximate estimating only) and the approximate costs (2014) per linear metre of waterway. In the case of culverts and pipe drains, the cost is expressed per number.

Table 1.4

S. No.	Classification	Unit per (two lane)	Cost in rupees
1.	Minor bridges (that is, bridges with linear waterway less than 30 m)	metre of linear waterway	3,50,000
2.	Major bridges (that is, bridges with linear waterway more than 30 m)	-do-	3,50,000 to 4,50,000
3.	Culverts and slab drains, 6 m	number	6,00,000
4.	Hume pipe drains (1.2 m dia.—one pipe)	number	1,00,000

2 | Valuation

2.1 GENERAL

An engineer or architect frequently has to deal in practice with cases in which he has to find out, not the cost but the value of a property. Examples of such instances are sale and purchase, insurance, taxation, compulsory acquisition of property and so on. Although, many a time the cost of the property is required as an evidence of its value, the cost estimated by the approximate methods is sufficiently accurate and serves the purpose. However, a fair knowledge of the principles underlying the valuation of a property is essential to arrive at a worthwhile figure of the value of a property. The purpose of this chapter is to outline the main principles underlying the valuation of some types of interests which exist in land and buildings.

2.2 DEFINITIONS

Before going into a detailed study, it is considered desirable to commence by defining some of the terms which are extensively used in this chapter, in order that their meaning shall generally be understood.

1. *Value*:   Value is a measure of desirability of ownership of property.
2. *Valuation*:   Valuation is the art of fixing the value of specific properties for specific purposes *or* the art of estimating fair monetary measure of the desirability of ownership of specific property for specific purposes.
3. *Market value*:   Market value may be defined as that amount of money which at any given time can be obtained for the property from persons able and willing to purchase it.
4. *Potential purchaser*:   Potential purchaser is a person who proposes to invest a certain amount of capital in land and building. He may look upon the transaction as an investment capable of yielding an annual return in the form of income or he may be merely interested in commercial, or other personal benefits which he hopes to derive from his occupation of the property.
5. *Book value*:   Book value is defined as the original investment as carried on the company books less any allowance for depreciation entered on the books.

6. *Replacement value*:   Replacement value is that value of a property determined on the basis of what it would cost (at the prevailing price level) to replace the property or its service with at least equally satisfactory and comparable property and service.

7. *Reproduction cost*:   Reproduction cost of a property is the estimated cost of reproducing substantially the identical property at a price level as of the date specified.

8. *Capital value*:   The capital value of a uniform perpetual income is that sum of money whose annual return at the highest rate which it can certainly earn, is equal to the given perpetual income.

9. *Salvage value*:   Salvage value is the net sum (actual or estimated), over and above the cost of removal and sale realised for a property retired from service but sold as it is without being broken into pieces.

10. *Scrap value*:   Scrap value is the net sum (actual or estimated), over and above the cost of removal and sale realised for a property retired from service and scrapped for the value of its materials.

11. *Depreciation*:   Depreciation may be defined as a loss (in value or utility of property), not restored by current maintenance, which is due to one or more of the forces operating to bring about the ultimate retirement of the property. The forces so operating include wear and tear, decay, inadequacy and obsolescence.

12. *Depletion*:   The word 'depletion' is applied to the consumption of the natural resources, which are not replaceable such as reserves of natural gas and oil, coal, timber, ore of all kinds, and so on.

13. *Obsolescence*:   Obsolescence is the loss in value due to the functional inadequacy or undesirability of the property brought on by progress in arts, changes in fashions, new inventions or improvements in designs, and so on.

14. *Sinking fund*:   It is the fund created by allocating a portion of terminable income for reinstatement of capital or repayment of debt.

15. *Gross income*:   The total amount of revenue received from a property during a year is known as gross income.

16. *Outgoings*:   The outgoings in an enterprise may be taken to be the sum of all expenditures.

17. *Net income*:   Net income of the enterprise for the accounting period is the difference between the gross income and outgoings.

18. *Years' purchase*:   Years' purchase represents the capital value of each ₹1 of the income *or* it is that figure which when multiplied by the net income from a property gives the capital value of that property.

19. *Freehold property*:   The term freehold implies a property which the owner holds absolutely and in perpetuity and of which he is either in physical possession or in receipt of rents arising from leases and tenancies which have been created out of the freehold interest.

20. *Leasehold property*:   The term leasehold property implies a property which is held in lease for a term of years, subject to payment of an annual rent and to the observance of covenants contained in the lease.

21. *Lessor and lessee*:   The person who grants the lease is called the 'lessor' and the person who takes it the 'lessee' or 'leaseholder'.

22. *Rent*:   An annual or periodic payment for the use of land or of land and buildings.
23. *Rental value*:   Rental value is the rent which may reasonably be expected to be obtained in the open market.
24. *Ground rent*:   Ground rent represents the rental value of a bare site at the time the lease is granted.
25. *Occupation lease*:   Occupation lease is the lease of land and building for occupation by the lessee.
26. *Rack rent*:   Where the rent reserved under an occupation lease represents the full rental value of land and buildings, or near it, it is called a rack rent.
27. *Head rent*:   Where the rent reserved under an occupation lease is less than the full rental value although obviously not a ground rent, it is called the head rent.
28. *Reversion*:   Reversion is the right at common law retained by a freeholder, who grants leases, to regain physical possession of his property at the expiry of the lease period.

2.3   THE NATURE OF VALUE

The word 'value' is used in many well-established meanings. In one sense, the value of a thing to a person is the importance he attaches to it. For example, a building within an area to be flooded by a reservoir under construction is only of scrap value to anyone so far as its construction materials are concerned. The owner, however, might place a high value on the building.

To the economist 'value' means the exchange worth of a commodity in terms of any other commodity. The value of a piece of cloth, for instance, might be expressed in terms of so many kilos of wheat or jowar. In the modern world, the relative values of commodities and services are measured in terms of money.

2.4   RELATIONSHIP OF COST TO VALUE

The words 'cost' and 'value' need to be distinguished from each other. Price is the amount of money paid to the seller by the purchaser of the property. Cost is price paid plus all other expenses incurred by the purchaser in the acquisition of the property. The cost of the property is not necessarily equal to the value; on the other hand, cost is considered as an evidence of value. While establishing the value of a property it is customary to investigate both original cost and reproduction cost. The cost represents the minimum value to the purchaser, because a shrewd purchaser pays less than maximum value to him to acquire his property.

2.5   POINTS TO BE NOTED REGARDING VALUE

Value is not a fixed and permanent part of the thing to be valued, but depends on outside factors; the chief of such factors are:

1. location of the property,
2. time, and
3. supply and demand conditions.

1. *Location of the property*:   Value of a piece of land or building is dependent to a large extent on the locality in which it exists. A building situated in a good residential area will fetch more price than other similar buildings situated in a bad residential area. The value of a property situated in a business area will be much higher compared to a similar property elsewhere.
2. *Time*:   Money—the medium of exchange—itself has changing value in its purchasing power. The 1934 rupee, the 1986 rupee and the 2014 rupee are widely different in their values. Each year sees some change in monetary values. The changing of value with time comes about not only because of the changing of value of money, but in many cases, due to the rise or fall in the value of property in general or due to supply and demand conditions.
3. *Supply and demand*:   The desire of others to possess a property to be valued, and their ability to pay for it creates '*demand*'. Value depends on the extent to which the '*supply*' of the particular type of property is capable of meeting the 'demand'. The effect of supply and demand on value is that where demand exceeds supply, value will tend to rise and where supply exceeds demand, value will tend to fall.

2.6   SUPPLY AND DEMAND IN RESPECT OF LAND AND BUILDINGS

1. *Land*:   Land is a commodity with the following peculiarities:
    (i)   Total quantity of land can neither be increased nor diminished, at the most more use can be made of it.
    (ii)   Land of a particular type cannot be made available where we please.
    (iii)   Even available land may be further limited by town planning, zoning and similar restrictions.

    On the other hand, supply of land can, in effect, be increased by:
    (a)   more intensive use,
    (b)   conversion from other uses to the required use such as deforestation, reclamation, non-agricultural uses and so on, and
    (c)   providing cheap transport thereby reducing the pressure on a particular locality.

    Generally speaking, as the population in a country increases and its wealth increases, the value of land of all types tends to appreciate.
2. *Buildings*:   An important point in relation to the supply of buildings is that while demand for houses, shops, factories and so on, can ultimately be made by the erection of a sufficient number of buildings of the particular type required, the length of time required to make good deficiencies in supply will often result in high prices being paid over a

considerable period. While, if conditions are reversed and the number of buildings of a particular type should exceed the demand for them, there is no quick or easy method of reducing the supply and a lengthy period of depressed values may be the result.

Factors which affect the demand for a particular class of buildings are as follows:

(i)   A change in the break up of population, for example, breaking up of joint families.

(ii) A change in the economic conditions.

(iii) Social changes such as shortage of domestic servants.

(iv) Movement of population from one area to another owing to change in fashions or facilities.

(v)   Rise and decay of industries.

2.7   FREEHOLD AND LEASEHOLD PROPERTIES

1. *'Freehold'* is the highest form of ownership. The owner of a freehold property—freeholder—usually holds the property without any payment in the form of rent. The freeholder has the right to do what he likes with his property, that is, he may develop it, sell or transfer it or lease it without the consent of any other person. However, the freeholder's power, to do as he likes, with his property is limited by:

   (i)   the law of the land, and

   (ii) by the rights of others.

   A freeholder can be forced to sell his land, by the Acts of Parliament, for various public or semi-public purposes. Similarly, the freeholder is not allowed to develop his property unless the development plan is approved by the proper authorities of the town planning department and local bodies. Further, while developing his property, he must not interfere with the natural rights of others as, for instance, erecting a structure on his land in such way as to seriously diminish the access of light to the windows in the adjoining buildings.

2. *Leasehold property:*   When the freeholder, instead of occupying the property himself, grants the exclusive possession of the premises to another for a certain period, usually in consideration of the payment of rent, he is said to have let the property. Such a letting is known as 'lease'. The person who grants the lease is called 'lessor' and the person who takes it the 'lessee' or 'leaseholder'.

   The main forms of lease are as below:

   (i)   *Building lease:* In this case the lessor grants the lease of the land for the purpose of erecting suitable buildings and the lessee undertakes to pay yearly *'ground rent'*, to erect buildings, maintain them in good condition and pay all outgoings in connection with them throughout the term of lease which is generally 99 or 999 years. On erecting buildings the difference between the net rent and ground rent will constitute the lessee's *'profit rent'* or 'net income'.

   (ii) *Occupation lease:* This term is used when a freeholder grants the lease of the land and building for occupation by the lessee. The period of occupation lease is shorter than the building lease and may vary from 3 years in the case of dwelling houses to 21 years in the case of shops.

In this case, if the rent reserved for land and building represents the full rental value of the property it is called 'rack rent' and if it is less than the full rental value (of course not a ground rent) it is termed as a *head rent*.

Sometimes the lessee agrees to pay 'rising rents'. For example, a house may be leased for 11 years, at a rent of ₹1,20,000 per month for the first five years and ₹1,50,000 per month for the balance of the term.

(iii) *Sublease:* If terms and conditions of the lease permit, a leaseholder may sublease the property, for any less term than he himself holds, at any rent he may able to obtain.

(iv) *Leases for life:* When the period of the lease granted by the freeholder is the duration of the life of the lessee or some other person it is called a lease for life. Such a lease for life is determinable on the death of the party in question.

A lessee is much more restricted in his dealings with the property than is a freeholder. For this and other reasons leaseholds are less attractive to an investor than freeholds.

2.8   SINKING FUND

At the end of the useful life of a property, it is to be replaced or rebuilt. Unless a suitable provision is made during the life period of the property, it would be impossible to rebuild the property with the income it earns in the last year of its service period. A provision may be made from the yearly income of the property by setting aside a specific amount which, accumulating at compound interest, will be equal to the sum required to replace the property at the expiry of its life. The fund so created is known as 'sinking fund'. Such a sinking fund may be created by taking a sinking fund policy with any insurance company. Or the fund may be created by specific periodical payments into a bank.

The equal periodical payment to be made can be easily worked out.

Let n be the life of the property in years,

$$S_n = \text{Sinking fund to be accumulated in } n \text{ years,}$$

$$i = \text{interest rate per period,}$$

$$s = \text{equal-annual-year-end payment.}$$

The accumulation in the sinking fund:

at the end of 1st year $s_1 = s$
at the end of 2nd year $s_2 = s + s(1 + i)$
at the end of 3rd year $s_3 = s + s(1 + i) + s(1 + i)^2$ and so on; at the end of the nth year

$$S_n = s + s(1 + i) + s(1 + i)^2 + s(1 + i)^3 + s(1 + i)^4 + \ldots s(1 + i)^{n-1} \tag{1}$$

Multiplying equation (1) by $(1 + i)$ …

$$S_n (1 + i) = s(1 + i) + s(1 + i)^2 + s(1 + i)^3 + s(1 + i)^4 + s(1 + i)^5 + \ldots + s(1 + i)^n \tag{2}$$

Subtract (2) from (1),

$$S_n - S_n(1 + i) = s - s(1 + i)^n$$

$$S_n \times i = s(1 + i)^n - s = s[(1 + i)^n - 1]$$

or $\qquad s = S_n \dfrac{i}{(1 + i)^n - 1}$ $\hfill (3)$

and $\qquad S_n = s \dfrac{(1 + i)^n - 1}{i}$ $\hfill (4)$

The periodical payments are often small and the book-keeping involved is considerable. For this reason, the interest allowed on the sinking fund so created is considerably low. Therefore, it is not wise to invest the sum in any kind of sinking fund. The best way is to invest the amount in an enterprise capable of earning a high return. Any method that is most suitable may be adopted, but in no case can the provision of sinking fund be neglected.

2.9 DEPRECIATION

The word depreciation is often applied loosely and in several meanings. Currently it is used in the following three distinct meanings:

1. Depreciation is the *loss in value* which is not restored by current maintenance and which is due to several factors such as wear and tear, decay, inadequacy and obsolescence that cause ultimate retirement of the property.
2. Depreciation is the *loss in cost* of a property as entered on the books. When used in this sense such loss in the cost is charged to expenses of operation.
3. In the third sense, depreciation is related to the physical condition of the property. It is defined as the *loss in service capacity* of the property. A property which shows a high degree of wear and tear is highly depreciated and vice versa.

It must be clearly understood that the physical condition of a property is neither depreciation nor the sole measure of depreciation but one evidence upon which either cost depreciation or value depreciation may be based.

2.10 METHODS OF ESTIMATING DEPRECIATION

There are several methods of estimating depreciation such as 'good-as-new', 'direct appraisal', 'arbitrary lump sum', 'sum of the digits', 'declining balance', 'straight line depreciation', 'sinking fund depreciation' and 'present worth depreciation'. There is much controversy over the methods of estimating depreciation because each of the above methods, being based on some assumptions, is not suitable under all circumstances and for all purposes. Many of the above methods have some serious drawbacks and limitations. The detailed discussion of all the above methods is beyond the scope of this book. Only those methods which are commonly used are explained below.

Depreciation is normally required for: (i) cost accounting, and (ii) valuation. The best methods suitable for these purposes are the straight line depreciation method and the sinking fund depreciation method.

1. *Straight Line Depreciation Method*

    Of all the methods mentioned above, the straight line method has found universal acceptance for depreciation cost accounting. This method allocates the depreciable base (original cost – salvage value) of a property unit, uniformly throughout its service life except when the estimate of service life is changed. It is simple and easily understood.

    The basic concept of the straight line method is that, as long as the useful service life and salvage value on which the depreciation allocation is based remain unchanged, the depreciation provision remains the same for each year.

    Let B = Original cost
    V_s = Scrap or salvage value
    B_d = Depreciable base
    n = Life of property in years
    D = annual depreciation allocation

    By definition, the annual allocation is

    $$D = \frac{B - V_s}{n} = \frac{B_d}{n}$$

    and the total depreciation up to age x is $x\,(B_d/n)$.

    In addition to cost accounting the straight line method is also used for valuation. However, the sinking fund method is extensively used in valuation and economy studies.

2. *Sinking Fund Depreciation Method*

    In the sinking fund depreciation method, the total depreciation of a property at any date is assumed to be equal to the accumulation of sinking fund and compound interest thereon up to that date. If s is the equal-annual-year-end payment and i represents the interest rate per period on the sinking fund then:

    Depreciation at the end of the nth year would amount to: $s\left[\dfrac{(1+i)^n - 1}{i}\right]$ (see Eq. 4 in Section 2.8).

2.11   OUTGOINGS

Outgoings are expenses to be incurred to maintain the annual income from a property. Most types of property are subject to outgoings of various kinds. As the outgoings affect the net income from the property and may also influence the figure of years' purchase, by which that income is to be capitalised, great care must be exercised by the valuer while estimating the outgoings. Principal outgoings to which land and buildings may be subjected are mentioned below:

1. *Repairs*:   The expenditure to be incurred on (i) immediate repairs if any, (ii) annual repairs for maintaining the property in good condition and (iii) special repairs including rebuilding or modernisation in the future, must be given due regard. The amount to be spent on repairs is generally 1% to 15% of the gross rent. However, these expenses are not to be included in outgoings, if the property is let and the tenant or lessee is made liable for the repairs.
2. *Sinking fund*:   A portion of the yearly income from the property has to be set aside for reinstatement of capital if the income from the property is not perpetual.
3. *Municipal or land taxes*:   These include property tax, water rate and drainage rate. While estimating these taxes, regard must be had not only to the actual amount paid at the present time but also to the possible changes in rates and assessment in the future, if there is no provision to increase the rent on account of increase in rates.
4. *Income tax*:   The income from a real property like any other source, is subject to income tax. However, all types of income being equally liable to this tax, it is sometimes disregarded as an outgoing.
5. *Insurance*:   This outgoing in the form of premium for fire insurance is commonly borne by the landlord. It is about half per cent of the sum insured. Similarly, if the owner has to pay premium for any other special insurance, the same is also considered as an outgoing.
6. *Management*:   Where the property is large and extensive or where the income from the property is received at irregular intervals (such as a hotel where the rooms are let on daily basis), the investor cannot be expected to be troubled with details of management himself and agency charges on lettings and management have to be allowed for as a separate outgoing.
7. *Loss of rent*:   It is customary to make an allowance for loss of rent on the assumption that the whole of the premises cannot be expected to be fully let at all times. While this assumption may be correct in respect of some types of properties such as hotels and hostels, to make such an allowance under existing conditions of acute housing shortage must be regarded as superfluous.

2.12   PROCESS OF VALUATION

Progress in the engineering valuation can be made only after knowing the purpose of valuation and what property is to be valued. The basis of valuation is controlled largely by the purpose for which the valuation is being made. Although the valuation process for specific property will follow a plan appropriate for that property and the purpose of valuation, the overall steps are common for all properties. These steps are:

1. determination of the purpose of valuation, the parties thereto, the date and place of valuation,
2. preliminary examination of the enterprise and its property,
3. determination of the original cost of all fixed capital physical property excluding land,
4. determination of the depreciation of the property,
5. estimating the reproduction cost of property, excluding land,
6. estimating the fair market value of all land,
7. determination of fair rate of return for the enterprise,
8. estimating market value on the basis of average market prices and its outstanding securities, or on the basis of market values of similar properties,
9. finding the capital value by capitalising the net income from the property, and
10. finding the fair value by giving such weight as may be just and right in each case to each item and to all other factors affecting the value.

2.13   METHODS OF VALUATION

The valuation of properties may be undertaken for any one of a variety of purposes, such as, administration, taxation, sale or transfer, rent fixation, insurance, and so on. The purpose of valuation may determine the degree of detail employed, the items of property included and the evidence of value considered. Depending upon the purpose of valuation, the valuer may need to have recourse to any one or more methods of valuation. The principal methods are:

1. valuation from life—determination of book value,
2. valuation from yield—determination of capital or rental value,
3. direct comparison of capital value,
4. valuation based on cost—reproduction cost,
5. valuation by reference to profits, and
6. the residual or development method.

These methods are discussed in the subsequent sections.

2.14   VALUATION OF PROPERTY FROM LIFE

The management of a property or an industrial enterprise must know the investment in, the value of, and especially the annual consumption of the useful service of the property. For this reason the book accounts of the property are properly maintained. These accounts show in addition to other information, the original cost, estimated life, annual depreciation and book value. These details help to work out the book value of the property at any date.

The book value of a property, by definition, is equal to the original cost less the depreciation of the property up to the date of valuation, plus the cost of land.

Book value = Original cost − Depreciation up to the date of valuation + cost of land

Depreciation up to the date of valuation is given by:

$$\frac{B - V_s}{n} \cdot x$$

as already mentioned in Section 2.10.1.

The following illustrative example will clear these points.

Example 2.1:   Find out the book value, after 40 years, of an asset costing ₹2 lakhs, assuming 100 years as life of the asset and the salvage value of ₹10,000. What would be the book value (after 30 years' life) if the salvage value is nil?

Valuation

(i)   Assuming straight line depreciation

$$\text{Annual depreciation} = \frac{\text{Original cost} - \text{salvage value}}{\text{Life of asset in years}}$$

$$= \frac{2,00,000 - 10,000}{100}$$

$$= ₹1900/\text{-}$$

$$\text{Depreciated cost in 40 years} = 1900 \times 40$$

$$= ₹76,000/\text{-}$$

$$\text{Book value after 40 years} = \text{Original cost} - \text{depreciated cost}$$

$$= 2,00,000 - 76,000$$

$$= ₹1,24,000/\text{-}$$

(ii)   If salvage value is nil

$$\text{Annual rate of depreciation} = \frac{2,00,000}{100}$$

$$= ₹2000/\text{-}$$

$$\text{Depreciation cost in 30 years} = 2000 \times 30$$

$$= ₹60,000/\text{-}$$

$$\text{Book value after 30 years} = 2,00,000 - 60,000$$

$$= ₹1,40,000/\text{-}$$

2.15 VALUATION FROM YIELD

Potential purchaser may be interested in properties either for investment or for occupation. The value of a property as an investment will depend on:

1. income produced or likely to be produced by the property, and
2. the return which the buyer expects from his investment.

Valuation for investment purposes generally includes the following procedure:

1. Ascertain the net income from the property. This is equal to total actual or estimated income minus the outgoing.
2. Estimate the rate of interest which may reasonably be expected from the investment in the property.
3. Fix a sum for the value such that the estimated net income will produce a return at the rate expected.

Example 2.2:   If a certain property is expected to produce a net annual income of ₹50,000 and the buyer expects 10% return on his investment, the estimated value will be worked out thus:

For ₹10 net income the buyer is ready to invest ₹100. Therefore, for ₹50,000 net income he will be ready to invest:

$$50,000 \times \frac{100}{10} = ₹5,00,000$$

It will be seen that capital value = net income × some multiplier (in this case 10). This multiplier is known as *'years purchase'* which shows that the buyer is willing to pay a sum representing so many times the net annual income. It is the present value of one rupee per annum received perpetually.

In the case of a perpetual income the years' purchase (YP) can be found simply by dividing one hundred by the desired rate of interest.

Required rate of interest	YP or present value of one rupee per annum received in perpetuity
6%	$\dfrac{100}{6} = 16.67$
7%	$\dfrac{100}{7} = 14.29$
8%	$\dfrac{100}{8} = 12.50$
10%	$\dfrac{100}{10} = 10.00$

Where the income is receivable for a limited period only the value of years' purchase has to be slightly modified. This is due to the fact that the investor in such cases has to set aside a portion of his income in the form of a sinking fund. If no such provision is made, it is obvious that the investor will lose both the capital and the income from the property at the end of the useful life period. To account for the sinking fund, the years' purchase is suitably reduced. In all such cases,

$$\text{Year's purchase} = \frac{1}{i + s}$$

where i = rate of interest in decimal, s = equal-annual-year-end payment to replace ₹1 at the end of the given period.

$$s = \frac{i}{(1 + i)^n - 1} \text{ (after substituting } S_n = 1 \text{ in Eq. 3 of Section 2.8).}$$

The above equation based on single rate interest is true only when the amount of sinking fund is invested in the same enterprise or elsewhere so that it will earn the same rate of interest that is earned by the capital. Where sinking fund is accumulated at a low rate of interest, a bigger sum must be set aside annually than if the accumulation were at the same rate as interest on capital.

For simplification, it would be proper to designate the rate of interest on sinking fund by i_1 and that on capital by i. The dual rate formula would then be:

$$s = \frac{i_1}{(1 + i_1)^n - 1}; \text{ and YP} = \frac{1}{i + s}$$

It is obvious that any given 'dual rate' figure of years' purchase will be found to be lower than its corresponding 'single rate' figure.

Example 2.3:   Find the value of tenement premises consisting of land and a well-built house, let for ₹20,000 per month inclusive of all taxes. The house is recently built and the rent by comparison with other premises is fair and likely to be maintained.

Valuation:

1.   Gross annual income = ₹20,000 × 12 = ₹2,40,000. Outgoings mentioned in Section 2.11 may be assumed to be 30% of the gross rent.

$$\text{Outgoings} = 2,40,000 \times \frac{30}{100} = ₹72,000$$

$$\text{Net income} = (2,40,000 - 72,000) = ₹1,68,000$$

2.   The net yield (or rate of return) expected from this property is 6%.

3. Years' purchase: Assuming the future life of the building to be 50 years and using a single rate of interest:

$$s = \frac{i_1}{(1+i_1)^n - 1}$$

$$= \frac{0.06}{(1+0.06)^{50} - 1} = 0.003444$$

$$YP = \frac{1}{i+s}$$

$$= \frac{1}{0.06 + 0.003446} = 15.76$$

4. Capital value = net income × YP

$$= ₹1,68,000 \times 15.76$$

$$= ₹26,47,680$$

If the rate of interest on sinking fund is assumed to be 3%,

$$s = \frac{i_1}{(1+i_1)^n - 1}$$

$$= \frac{0.03}{(1+0.03)^{50}} = 0.0069$$

$$\text{and} \quad YP = \frac{1}{i+s}$$

$$= \frac{1}{0.06 + 0.0069} = 14.95$$

It may be noted that the YP in this case is lower than before.

Capital value = 1,68,000 × 14.95 = ₹25,11,600

Deferred income

When the whole or part of the income from a property will not commence until after the lapse of a certain number of years, it is termed 'deferred income'. The method of finding out capital value of such deferred income is explained below with the help of examples.

1. *Perpetual Income*
    Rule: From the figure of years' purchase for perpetuity deduct the years' purchase for the number of years which must elapse before the perpetual income commences.

Example 2.4:   What is the present value of reversion to a net income of ₹2000 per annum receivable in perpetuity only after expiry of 5 years? Assume rate of interest 6%. Amount of rupee one in 5 years at 6% interest is equal to 1.339.

Valuation:

 (i)   YP for perpetuity @ 6% 16.67
(ii)   YP for 5 years @ 6%

$$s = \frac{i}{(1+i)^n - 1} = \frac{0.06}{(1+0.06)^5 - 1} = 0.177$$

$$YP = \frac{1}{i+s} = \frac{1}{0.06 + 0.177} = 4.22$$

Future net income		₹2000/-
YP for perpetuity @ 6%	=	16.67
Less YP for 5 years @ 6%	=	4.22
YP for deferred income		12.45
Value of deferred income	=	2000 × 12:45 = ₹24,900/-

The alternative method that may be followed is, first, to estimate the capital value of the income as though it were commencing at once and then to find the present value of that capital sum by multiplying it by the present value of ₹1 receivable at the end of the term for which the income is deferred.

The above example may be solved by this method:

Future net income	₹2000/-
YP for perpetuity	× 16.67
Value if income were receivable at once	₹33,340/-
Present value of ₹1 in 5 years	

$$@\,6\% = \frac{1}{1.339} = 0.7467$$

(Reciprocal of ₹1 in 5 years @ 6%)

Value of deferred income = 33,340 × 0.7467 = ₹24,900/-

2.   *Income Receivable for a Limited Period*
     Rule: Multiply the years' purchase by the present value of ₹1 receivable at the end of the period which must elapse before the income starts.

Example 2.5:   What is the capital value of an income of ₹5000 receivable for a period of 15 years commencing in 5 years' time? Assume 6% to be the rate at which the sinking fund can be accumulated.

Valuation

Years' purchase for 15 years @ 6% interest rate

$$s = \frac{i}{(1+i)^n - 1} = \frac{0.06}{(1+0.06)^{15} - 1} = 0.04296$$

$$YP = \frac{1}{i+s} = \frac{1}{0.06 + 0.04296} = 9.71$$

Since ₹1 will accumulate to $(1+i)^n$ in n years, the present value of ₹1 due in n years equals $\frac{1}{(1+i)^n}$.

$$\text{The present value of ₹1 in 5 years @ 6\%} = \frac{1}{(1+0.06)^5} = 0.7472$$

Now,

Future net income	₹5000/-
YP for 15 years @ 6%	9.71
Multiplied by PV of ₹1	
in 5 years @ 6%	× 0.7472
YP for 15 years deferred 5	
years @ 6%	= 7.255
Value of deferred income = 5000 × 7.255 =	₹36,275/-

2.16   DIRECT COMPARSION OF CAPITAL VALUES

In order to find the capital value, the net income from the property must be known. But certain circumstances arise where there is little evidence of rental value, for example, a property occupied by the owner himself. Under such circumstances if there is extensive evidence of sale prices of similar properties, the capital value of one property can be found out by direct comparison with the capital value of another similar property. A good knowledge of real estate and sound judgement are necessary to weigh the respective advantages and disadvantages of the two properties if they are not very similar in size and character.

This method is seldom practicable in regard to leasehold properties as it is difficult to find the sale price evidence of another property where the length of the term, the conditions of the lease, and all other circumstances are similar to that of the property under consideration.

2.17   VALUATION BY REFERENCE TO COST

If the net annual income from a property or evidences of sale prices of similar properties are not available, a recourse may be to take cost as evidence of value of the property.

In this method the cost of land is estimated by comparison with similar land in the neighbourhood. Then, the cost of providing the same accommodation in substantially the same form, at the prevailing rate of construction, is found out. This is known as 'reproduction cost (new)' of the structure. This has to be adjusted for depreciated cost during the past life of the structure. Finally, the cost of land and the adjusted reproduction cost are added together to get the value of the property. An example will illustrate the procedure.

Example 2.6:   A freehold residential property, consisting of a house having a plinth area of 300 m^2 constructed 20 years ago on a plot of land measuring 900 m^2, is proposed to be purchased by 'A' for his use. The present rate of construction of a similar building is ₹7500 per m^2 of the plinth area and the cost of land is ₹1000 per m^2.

Advise 'A', as to at what cost he should purchase the property.

Valuation

1.   Cost of land = 900 × 1000 = ₹9,00,000
2.   Reproduction cost of the building (new) 300 × 7500 = ₹22,50,000
3.   Depreciation

Assuming: (i) straight line depreciation, (ii) total life of the building as 80 years, and (iii) the salvage value as 10% of the reproduction cost.

$$\text{Annual depreciation} = \frac{22,50,000 - 2,25,000}{80} \qquad = \quad ₹25,312,50$$

Total depreciation in the past 20 years = 25,312 × 20   =   ₹5,06,250
The present cost of the building = 22,50,000 – 5,06,250   =   ₹17,43,750
Total cost of the property = 17,43,750 + 9,00,000   =   ₹26,43,750

'A' is advised to purchase the property at the maximum cost of ₹26,44,000.

2.18   VALUATION BY REFERENCE TO PROFITS

In the case of certain types of property, capital value is usually estimated by reference to the volume of the trade or business. Examples of such properties are hotels, public places, cinemas, theatres, and so on. In such cases, the method of approach is to estimate the gross earnings and deduct therefrom: (i) working expenses incidental to the earnings, (ii) interest upon the capital invested, and (iii) amount of remuneration to the tenant for his risk and enterprise. The balance left is the net profit which can be paid in rent. The capital value is determined by multiplying this net profit by years' purchase. This method is also known as the 'accounts' method. The value obtained by this method should preferably be checked by any other valuation method.

2.19   RESIDUAL OR DEVELOPMENT METHOD

There are certain types of properties the value of which can be increased by capital expenditure or by change in the use to which the property is put or possibly by a combination of capital expenditure and change of use. The increase in the value results due to sudden rise in the income.

In finding the value of such property, the valuer has to give due regard both to the capital value of such increased income and also to extra expenditure incurred in order to produce it. In principle, the method of approach is to ascertain the present capital value of the estimated future income and to deduct therefrom the cost of the development of the property and an allowance for risks.

2.20   RENT FIXATION

Rent is defined as the annual or periodic payment for the use of land or of land and buildings. From the definition, it follows that the rent is made up of two things:

1.   an annual payment for the use of land, and
2.   an annual payment for the use of capital expended on it in the form of buildings, and so on.

This analysis of rent suggests a possible method of estimating the rent of a property. The rent from a property may be expressed as a percentage of the value of the property, the percentage being the fair annual return expected from the investment in properties of a similar nature. The rent so fixed is justified only if the tenant undertakes to bear the cost of all outgoings except, of course, the income tax. If the owner has to meet expenses on some or all the outgoings, the amount of rent determined as above will have to be increased by an amount equal to the actual outgoings to be paid by the owner. The following example will illustrate these points.

Example 2.7:   A person purchases a land measuring 600 m^2 at the rate of ₹1000/m^2 and constructs a building of plinth area 200 m^2 at the cost of ₹15,00,000. He desires to have 8% return on the cost of the building and 4% return on the land cost. Suggest the suitable rent for the property.

Rent fixation
1.   Cost of land = 600 × 1000 = ₹6,00,000
2.   Amount of return required on the cost of land

$$6,00,000 \times \frac{4}{100} = ₹24,000$$

3.   Amount as return required on the cost of the building

$$15,00,000 \times \frac{6}{100} = ₹90,000$$

Net rent per annum $= ₹1,14,000$

Let this be equal to R.

Outgoings

(i)   Repairs— $\frac{1}{2}$% of the cost of construction of the building $\frac{1}{2} \times \frac{1}{100} \times 15,00,000 = ₹7500$

(ii) Other outgoings assuming

25% of the gross rent $₹0.25\,R$

Outgoings $= 0.25\,R + 7500$

Net income $= R - (0.25\,R + 7500)$

$= 0.75\,R - 7500$

$0.75\,R - 7500 = 1,14,000\ R = ₹1,62,000$

Monthly rent $= \dfrac{1,62,000}{12} = ₹13,500$

2.21   VALUATION FOR MORTGAGE

A mortgage of a freehold or leasehold property is a transaction whereby the owner of a freehold or leasehold property—the mortgagor—grants an interest in his property to another person—the mortgagee—as security for a loan advanced by that person.

The transaction is effected by means of a mortgage deed in which the mortgagor usually agrees to pay interest on the loan at a given rate per cent and also to carry out repairs and to insure the property. The mortgagor retains the right to recover the property on full repayment of debt and interest thereon. This is known as 'equity of redemption'.

In the process of valuation for mortgage purposes, the ordinary principles of valuation apply, but the valuer must bear in mind the following important points:

1.   It may be necessary to realise the security in the future and that unless the sale price then is sufficient to cover the mortgage debt, arrears of interest and costs, the mortgagee will suffer loss. In other words, the valuer, while basing his estimate on the present market value, has to see to it that the value is likely to be maintained in the future and would be readily available.

2.   Due consideration must be given to any factors such as, 'master plan' for the area, which may unfavourably affect the value of the property in the future.

3. The net income from the property must be estimated with the utmost care to make sure that it will adequately cover the interest on the proposed loan.
4. Anything which is purely speculative in nature should be disregarded.
5. In the case of business premises, goodwill should be excluded from the valuation.
6. Anything which can easily be sold or removed by the mortgagor should also be excluded.
7. The price paid for property by the mortgagor may not be the correct basis; on the other hand, it may prove to be misleading. The mortgagor might have paid excessive price for some special reasons or at a time when market values were high.
8. The actual cost of construction of the building is usually best disregarded for mortgage purposes. The owner might have spent large amounts on the gratification of his tastes. Similarly, considerable changes in the building costs may have occurred since the building was constructed.
9. In addition to the valuation of property, the valuer may be required to advise as to the nature of the security advanced and as to the sum which may reasonably be advanced. Generally, the sum of money advanced as to the debt on the security of the property is about 50% to 60% of the value of the property as judged by an expert valuer.

2.22 VALUATION OF LAND

If it is required to find the value of land alone the following methods are used:

1. *Comparative Method*
   In cases where the nearby lands frequently change hands, there may be ample evidence of sale price of land similar to the land of which the value is to be determined. It is then possible to decide fair price for the land by carefully studying the various transactions and sale statistics and giving due weightage to the physical characteristics of the land such as length, width, situation, and so on.
2. *Abstractive Method*
   If no information regarding sale price of the neighbouring land is available, the comparative method mentioned above fails. In such cases the capital value of the neighbouring property consisting of land and building may be found out on the basis of its net income. The reproduction cost (new) of the building at the prevailing rates is calculated and adjusted for the depreciation in the past life. The adjusted reproduction cost of the building is then deducted from the capital value of land and building. The balance left is the value of land. This value of land is divided by the total area in square metres to get the rate per square metre of land. This rate is then used to find the value of the land in question.

2.23 NATURE AND USE OF VALUATION TABLES

It would be clear from the above discussion that the various methods of valuation involve elaborate mathematical calculations in addition to the skill, judgement and practical experience of the valuer. Valuation tables are, therefore, prepared to assist the valuer in his work, to save time and to reduce the chances of errors in mathematical calculations.

The nature of some important valuation tables, the formulae on which they are based, and explanation as to what the figures in them represent are mentioned below:

1. *The Amount of ₹1 Table*
    The figures in this table represent the amount to which ₹1 invested at various rates of compound interest will accumulate over any given number of years.

    Formula
    The amount of ₹1 in n years $= (1 + i)^n$

    where $i =$ rate of interest and $n =$ number of years.
    Use: The table is sometimes useful in calculating the loss of interest involved on a property which, for the time being, is unproductive of income.

2. *Present Value of ₹1 Table*
    The figures in this table are the exact opposite of those in the amount of ₹1 table. They show the amount which invested now at various rates of compound interest will accumulate to ₹1 over any given number of years.

    Formula
    Present value of ₹1 in 'n' years $= 1/(1 + i)^n$
    Use: To calculate the value at the present time of sums receivable in the future, and to make allowances for future expenditure in connection with the property.

3. *Amount of ₹1 per Annum Table*
    The figures in this table represent the amount to which a series of deposits of ₹1 at the end of each year will accumulate in a given period at a given rate of compound interest.

    Formula

    $$\text{Amount A} = \frac{(1 + i)^n - 1}{i}$$

    Use: To calculate the total expense involved over a period of years where annual outgoings are incurred in connection with the property which is for the time being unproductive.

4. *Annual Sinking Fund Table*
    The figures in this table show the sum which must be deposited annually at compound interest in order to produce ₹1 in any given number of years.

    Formula

    $$s = \frac{i}{(1 + i)^n - 1}$$

Use: To know what sum ought to be set aside annually out of income for reinstatement of the capital or payment of debt.

5. *Present Value of ₹1 per Annum Table*

The figures in this table show, at varying rates of interest what sum might reasonably be paid for a series sum of ₹1 receivable at the end of each of a given number of successive years.

Formula

$$\text{Present value V} = \frac{[1 \times 1/(1+i)^n]}{i}$$

Dual rate table.

$$V = \frac{1}{i + i_1 /[(1+i_1)^n - 1]}$$

Use: To find years' purchase for determination of the capital value of an asset.

2.24 VALUATION REPORT SPEICIMEN

REF: 3503/ A (VAL)

DATE: 05/11/2009

FORM: 0-1

(see rules under Wealth Tax Act)

VALUATION REPORT OF IMMOVABLE PROPERTY

PART I—QUESTIONNAIRE.

GENERAL

1. Purpose for which valuation is made : To ascertain the market value
2. Date as on which valuation is made : 01/11/2009
3. Name of the Owner/s : M/s ABC (P) Ltd.
4. If the Property is under joint ownership/co-ownership share of each such owner. Are the shares undivided? : Company ownership
5. Brief description of the property : 'INDU' Commercial and Residential Building
6. Location, street, ward no. : C.S.Nos 2113/k-5, 2113/k-6&2113/k-23,
7. Survey/Plot no. of land Tararani Chowk, Kolhapur : E-Ward, Colony, Old NH-4,
8. Is the property situated in residential commercial/mixed area/industrial area : Mixed zone
9. Classification of locality high class/ middle class/poor class : High class

10. Proximity of civic amenities like schools,   : All amenities are in the close vicinity
    hospitals, offices, markets, cinemas etc
11. Means and proximity to surface   : Main (Old NH-4) road on north of
    communication by which the locality is   Property which is treated as entrance of
    served facilities, Kolhapur Railway Station   Kolhapur having all the communication
    is about 2.0 km away, the main road is   constructed with concrete(four lane)
    now being

LAND

1. Area of land supported by documentary : Total area of land is 3084.0 sq.m
   proof shape, dimensions and   and detailed statement is given on
   physical features   Drawing No. 1 attached herewith.
       The P.R.Cards and C.S.Maps also are
       enclosed.

2. Roads, street or lanes on which the land : OLD(Pune–Bangalore)NH-4 road
   is abutting west and south sides   on north side andcolony roads on
3. Is it freehold or lease-hold land?   : Freehold
4. If lease-hold, the name of lessor/   :
   lessee, nature of lease, date of
   commencement and termination
   renewal of lease
     i) Initial premium   :⎫
     ii) Ground rent payable per annum   : ⎬Not applicable
     iii) Unearned increase payable   :⎭
         to the lessor in the event of sale
         or transfer
5. Is there any restrictive covenant   : No
   in regard to use of land? If so, attach
   a copy of the covenant
6. Are there any agreements of easements? : No
   If so, attach copies
7. Does the land fall in an area included   : The land falls in the D.P. of Kolhapur
   in any Town Planning Plan of the   Municipal Corporation.It is in Mixed
   Government or any statutory body?   Zone (residential andcommercial use).
   If so, give particulars.   The other details are given on site plan
       drawing.
8. Has any contribution been made   : No
   towards development or is any demand
   for such contribution still outstanding?
9. Has the whole or part of the land   : No. The road modifications are
   been notified for acquisition by   already done and the land is finally

Government or any statutory body?   compounded
Give date of the notification

10. Attach a dimensioned site plan   :   Please refer to Drawing No.1

IMPROVEMENTS AND CONSTRUCTION

1. Year of commencement of construction   :   The structure/s standing on the land as
    phase wise. The original building was   on date of valuation are constructed
    and added with first floor works.   for showroom and it was altered, expanded

2. What was the method of construction?   :   The building as existing is well
    By contract/By employing labour   maintained and the average present life
    directly/Both   of the building is @ 30 years

3. For items of work done on contract   :
    produce copies of agreements

4. For items of engaging labourdirectly,   :
    give basic rates of materials and labour
    supported by documentary proof

5. Attach plans and elevations of the   :   Please refer to Drawing No.1 for
    structures standing on the land and   Layout and site plan and Nos 2 to 4 for
    a layout planmain building details.   The Annexure gives all other specification
    details

6. Furnish technical details of the building   :
    on a separate sheet

7. Is the building owner occupied,   :   Owner occupied.
    tenanted/both   (XYZ. commercial and ABC residential)

8. If partly owner occupied, specify   :   Not applicable
    Portion and extent of area under owner
    occupation

9. What is the floor space index   :   F.S.I. Permissible is 1.0 and the land is
    permissible and percentage   capable of using additional T.D.R to
    actually utilised   the extent of 0.4 making total F.S.I
    1.4 as per present D.C. rules of KMC

RENTS

1. Names of tenants/lessees/licensees etc   :
2. Portions in their occupation   :
3. Monthly or annual/rent/   :   The property is in self use of the
    compensation/licence fees, etc paid   owner'scompany and hence this is not
    by each   applicable. The premises are having
    bores, electrical supply and water
4. Gross amount received for the whole   :   storage facilities and all expenses are
    property.   borne by the owner

5. Are any of the occupants related to, or close business associates of the owner? :

6. Is separate amount being recovered for the use of fixtures like fans, geysers, refrigerators, cooking, ranges, built-in wardrobes, etc or for service charges? If so, give details :

7. Give details of water and electricity charges, if any to be borne by the owner :

8. Has the tenant to bear the whole or part of the cost of repairs and maintenance? Give particulars :

9. If a lift is installed, who has to bear the cost of maintenance and operation—owner or tenant? :

10. If a pump is installed who has to bear the cost of maintenance and operation—owner or tenant? : Not applicable as stated earlier

11. Who has to bear the cost of electricity charges for lighting of common space like entrance hall, stairs, passage, compound etc—owner or tenant? :

12. What is amount of property tax? Who is to bear it? Give details with documentary proof : Property Tax to the tune of @ ₹ 21,000/- is paid annually

13. Is the building insured, if so, give the policy No., amount for which it is insured and the annual premium :

14. Is any dispute between landlord and tenant regarding rent pending in a court of law? : Not applicable

15. Has any standard rent been fixed for the premises under any law relating to the control of rent? :

SALES

1. Give instances of sales of immovable property in the locality on a separate sheet, indicating name and address of the property, registration no., sale price and area of land sold details : Please refer to Part II for these

2. Land rate adopted in this valuation   :
3. If sale instances are not available or   :
    not relied upon, the basis of arriving at
    the land rate

PART II—VALUATION

1. <u>INTRODUCTION</u>

    As per the instructions from ---------- of M/s ABC (P) Ltd. the property Owner, we inspected the property at C.S.No-------------------------,
    E-Ward, Tararani chowk, Kolhapur, to ascertain the Market Value of the said property known as 'INDU' building as on 09/09/2009.

2. <u>DESCRIPTION OF PROPERTY</u>

    The property under valuation consists of three developed layout plots which are contiguous and well defined and demarked on site. The structures are erected thereon in phases and as on date of valuation there are the following structures:

    a) Main INDU building with ground floor, first floor andterrace works at second floor level.
    b) Guest house with ground and first floor.
    c) Other constructional units like watch and ward, store shed, garage shed andcompound wall around with entrance gate/s.
    d) Other developmental works such as lawn, pavings, asphalting for the entrance area, grilled security railings for the demarked portions in the premises.
        All these details with area statements are furnished and given on the drawings (four sheets) attached herewith.

        The photographic views are also enclosed.

        The specifications of the various items of the structure are given in the ANNEXURE enclosed herewith. The services, facilities and amenities available to the premises are in the said ANNEXURE. The details of three plots (deemed to be amalgamated for the time being as on date of valuation) as per City Survey office records, that is, City Survey maps are enclosed along with the extract from the Property Registar (P.R.Cards).

3. <u>OWNERSHIP AND USE</u>

    As on date of valuation, the entire property is owned and possessed by the Owner Company. The Company became the owner of the property in 1988. The property was in commercial use and the structure thereon was revamped from time to time.

    The property as on date of valuation is under self use of Company and the ground floor of the main building is used for commercial purpose and the first floor for residential purpose.

    The developed terrace of the building is attached to the residential use having sports facilities and other amenities of the building.

    The ground floor commercial portion is having front lawn, pavings, entrance from the main northern road. The building is having vertical communication by two staircases. The details of this building are shown on the Drawing nos. 2, 3 and 4 attached.

The Guest House structure is with ground floor and first floor units. The first floor unit is connected to the main building at that level by a covered passage.

The property has other development works as shown on the Drawing No 1. attached.

4.   <u>VALUATION APPROACH</u>

The property is having both the components, namely, the LAND and the STRUCTURE as predominant. The purchaser's view points will be positive from immediate occupation purpose for any use and also from further developments for utilising the available FAR. The possible TDR utilisation is also a plus point.

Following other facts in addition to those mentioned herein above are considered for evaluating the MARKET VALUE of this property.

1.   The land has all the benefits of approach, size, shape and it is abutting the main road (Old NH-4). The strata available is murumand most suitable for foundation.
2.   The land has roads on three sides and colony roads (west and south) are at the upper level and may be useful for entry at the upper level for future complex planning and full utilisation of full FSI.
3.   The main structure is well maintained and can be taken for any commercial use.
4.   The ground coverage available can be made use of for further complex development without disturbing the present main building utilisation.
5.   The provisions of Urban Land (C&R) Act has not affected this property and now the said Act is repealed and not applicable to this property.
6.   The Development Control Rules of Corporation and provisions of the Development Plan of Kolhapur are applicable to this property and there is no adverse effect of any of the provisions.

The valuation of this property is worked out by Building and Land Method.

Other approaches such as Market Approach and Income Capitalisation (Rental) Approach are also considered.

The market approach considers only the present main building units but the land component effect on the evaluation of these units becomes difficult.

The building being under self use of the company, the income computations of the premises become assumptive and hence this method also is not feasible.

Hence the valuation by Building and Land Method is worked out. The land rate is adopted by considering all the points and facts given earlier and the recent transactions in the vicinity.

The structure and other constructions are valued by considering the cost of reproduction based on specifications the rates of various materials used and the labour rates applicable, and deducting therefrom the depreciation worked out by considering the age (average) of the building and future life of the same.

The depreciated value of all other works and constructional units is added thereto.

5. <u>VALUATION OF PROPERTY BY LAND AND BUILDING METHOD</u>

I) **<u>Valuation of Land</u>**

a) Total area of land   : 3084.0 sq.m

b) Rate of land on the date of valuation
considering all points and potentials   : ₹ 15,750/- per sq.m

Hence valuation of land only: ₹ 4,85,73,000/- (I)

II) **Valuation of INDU Building**

A) <u>Valuation of structure</u>

a)	Built up area of G.F.	:	624.50 sq.m
	Built up area of F.F.	:	680.40 sq.m
b)	Total built up area	:	1304.90 sq.m
c)	Rate of reproduction as on date of valuation	:	₹ 12,000/- per sq.m
d)	Total undepreciated cost of reproduction	:	₹ 1,56,58,800/-
e)	Life of structure as on date of valuation	:	@ 30 years
f)	Future life of structure	:	@ 50 years
g)	Percentage rate of depreciation	:	Gross 35%
h)	Total amount of depreciation	:	₹ 54,80,580/-
i)	Total depreciated cost of reproduction	:	₹ 1,01,78,220/-
	say	:	₹ 1,01,78,000/- (A)

B) <u>Valuation of other works of structure</u> a)

a)	Valuation of façade (west) on both floors	:	₹ 60,000/-
b)	Verandah works (south) G.F.	:	₹ 50,000/-
c)	Corridor of F.F near north stairs	:	₹ 40,000/-
		:	₹ 1,50,000/- (B)

C) <u>Valuation of works on terrace</u>

a)	Valuation of sports room including pergola and toilet	:	₹ 3,00,000/-
b)	Valuation of terrace finishing with stair mumty, corridor	:	₹ 2,50,000/-
c)	Valuation of O.H. water tank with supporting structure	:	₹ 50,000/-
			₹ 6,00,000/- (C)

Hence valuation of INDU building
adding A, B and C above   : ₹ 1,01,78,000/-
                                              : ₹ 1,50,000/-
                                              : ₹ 6,00,000/-

                                              : ₹ 1,09,28,000/- (II)

III) Valuation of Guest House

a)	Built up area of G.F.	:	78.10 sq.m
	Built up area of F.F.	:	78.10 sq.m
b)	Total built up area	:	156.20 sq.m
c)	Rate of reproduction as on date of valuation	:	₹ 7000/- per sq.m
d)	Total undepreciated cost of reproduction	:	₹ 10,93,400/-
e)	Life of structure as on date of valuation	:	@ 30 years
f)	Future life of structure	:	@ 40years
g)	Percentage rate of depreciation	:	Gross 40%
h)	Total amount of depreciation	:	₹ 4,37,360/-
i)	Total depreciated cost of reproduction	:	₹ 6,56,040/-

Hence valuation of guest house, say   **: ₹ 6,56,000/- (III)**

IV) Valuation of developmental and other works in the premises

a)	Valuation of store shed and garage shed	:	₹ 2,50,000/-
b)	Valuation of lawn, asphalting, paving and PCC pavement in the premises	:	₹ 2,50,000/-
c)	Valuation of compound and internal railings with gates	:	₹ 2,50,000/-
d)	Valuation of W/W bores with pumping arrangements, common urinals andstage/ kitchen garden	:	₹ 1,00,000/-

₹ 8,50,000/- **(IV)**

Hence valuation of the property under
reference adding I, II, III and IV above

I	:	₹ 4,85,73,000/
II	:	₹ 1,09,28,000/-
III	:	₹ 6,56,000/-
IV	:	₹ 8,50,000/-
Total	:	₹ 6,10,07,000/-
Say	:	₹ 6,10,00,000/-

Rupees Six Crore and Ten Lakh only.

6. <u>CONCLUSION</u>

The valuation of the property is worked out by Building and Land Method and considering all the relevant factors discussed herein and it is FAIR, JUST and Reasonable.

Prepared by
Name of Valuer
Govt. Registered Valuer
Registration No. Cat-A/131/1988

PART III—DECLARATION

I hereby declare that

1. The information furnished in Part I is true and correct to the best of my knowledge and belief.
2. I have no direct or indirect interest in the property valued.
3. I have personally inspected the property on 09/09/2009.

Prepared byName of valuer
Govt. Registered Valuer,
Registration No. Cat-A/131/1988

General Principles of Taking out Quantities

3.1 GENERAL

Accuracy of the estimated cost depends to a great extent upon the accuracy with which the quantities are taken out. As such, great care has to be taken, not only in arriving at the correct dimensions from the drawings, but also that no item of work is overlooked. To ensure accuracy, some fundamental principles of taking out quantities must be borne in mind and a definite procedure should be adopted. Some of the principles are outlined in the subsequent articles.

3.2 STUDY OF DRAWINGS

It would be extremely premature if an attempt were to be made at the taking off, immediately on receipt of the drawings. Much preliminary work has to be accomplished before proceeding to take out quantities from the drawings. Many errors can be avoided if the following steps are taken before entertaining any thought of taking off:

1. Look over the drawings and attempt to visualise the work entailed.

   The drawings should not be looked at 'en masse' but an effort should be made to singularise the items of work and then visualise the work under each item. This is one of the best ways of understanding a job and correctly interpreting the architect's drawings.
2. Study the drawings carefully to see that plans, elevations, sections and details, if any, are in agreement with one another.
3. Check the dimensions on the drawings to make sure that each overall dimension agrees with the total room dimensions. If serious errors are discovered, the architect should be informed; but if the discrepancy is due to the result of slight arithmetical error, the drawings should be corrected.

4. If the dimensions do not exist in some places, then it is always better to write them in. These missing dimensions should be worked out from other dimensions, as far as possible, scaling them from the drawing being the last resort.

3.3   VISIT TO THE SITE OF WORK

The estimator should if possible, visit the site of the proposed work. During his visit he may collect much valuable information regarding the removal of debris, buildings to be demolished, access to the site, nature of the surface soil, and so on. Also, he may find that some required information is not given on the drawings and which must be obtained from the architect. All this will help him prepare a realistic estimate of the cost of the work.

3.4   ORDER OF TAKING OFF

The order of taking out quantities largely follows the order of construction. This has the advantage that the estimator mentally erects the structure step by step and is less likely to miss something. In a simple building, the order of taking off would probably be as follows:

1. Excavation,
2. Foundation concrete,
3. Foundation masonry,
4. Damp-proof course,
5. Filling in plinth with hard *murum*,
6. Superstructure—brickwork or stone masonry,
7. Doors, windows and cupboard—frames,
8. Window sills and lintels over openings,
9. Roof,
10. Plastering and pointing,
11. Flooring,
12. Shutters of doors, windows and cupboards,
13. Water supply and sanitary fittings, and
14. Electrical work.

While taking out quantities under each item of work, the estimator should always follow a particular 'system'. In the English System he must decide whether he is going to take on clockwise or anti-clockwise. To illustrate, consider a building simple in layout (see Fig. 3.1).

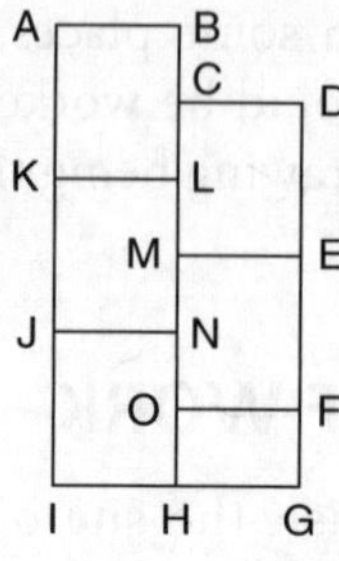

Fig. 3.1

Start must be made to measure from some point on this plan, usually one of the corners, say A. If the estimator proposes to take off clockwise he will continue via B, C, D, E, F, G, H, I, and J back to A, and if anti-clockwise, then via K J I H G F E D C and B to A. While measuring internal work he will measure round the rooms in the same order.

In the PWD system, the estimator groups the walls into two classes, for example, walls running N–S as long walls and others at right angles to these walls as short walls (see Fig. 3.1). He should then proceed from one face back to the other face. For example, in the above figure, he may consider the long walls proceeding in the E–W direction such as walls A K J I, B C L M N O H and D E F G. Then short walls from the front of the building to the rear such as walls I H HG, OF, JN, ME, KL, CD and AB.

3.5 ACCURACY

It is a general principle in taking out quantities that the costlier the item of work the greater should be the accuracy with which the measurements are taken out. As a general rule it may be remembered that dimensions should be measured to the nearest 0.01 metre, areas should be worked out to the nearest 0.01 square metre, and cubic contents should be worked out to the nearest 0.01 cubic metre.

3.6 ADJUSTMENT OF OPENINGS AND VOIDS

A principle to be observed in quantity surveying when measuring areas of excavation or concrete over site, plastering, pointing, and so on, is to measure the full area in the first instance and then deduct the redundant area (voids or openings) afterwards. For example, while calculating the area of excavation of the figure, in Fig. 3.2 overall area would be found out first, thus:

10 m × 6 m = 60 sq m and then the redundant area 4 m × 2 m = 8 sq m would be deducted.

Similarly, while measuring the cubic contents of brickwork or masonry (or area to be plastered or pointed) the walls are considered to be without openings in the first instance and then from the gross quantity of work total volume (or area) of the openings is deducted.

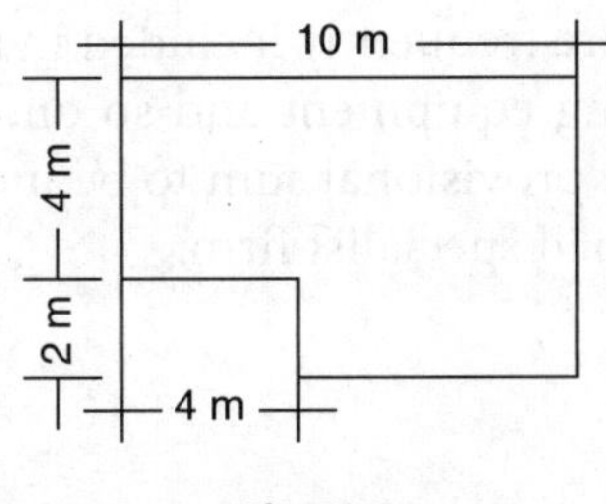

Fig. 3.2

3.7   DESCRIPTIONS

Probably the most difficult aspect of 'taking out' work is to frame adequate, and yet at the same time concise descriptions. It is an art of utmost value and one which beginners should take great pains to master.

The contractor's price build up depends to a great extent upon the description of the item. The estimator, therefore, must always aim at expressing himself clearly, being careful in his choice of words using the various technical terms in the proper sense and using the same wording when referring to the same thing in different parts of the estimate.

In order to save time, the estimator may write short descriptions in the measurement sheets using standard abbreviations, but at the same time, without leaving out essentials. However, descriptions in the abstract (the PWD system) or bill of quantities must be full and proper.

Regarding the order of wording in the description, it should be borne in mind that the reader should not be required to wait almost to the end of the description to determine the subject of the item. The first few words should clearly indicate the nature of the item being described. The following example serves to illustrate this point. The first type of description is quite frequently produced by students.

'Laying 10 cm thick cc (1:5:10) bedding, watering and curing complete, below 4 cm thick Shahabad stone paving laid in cm (1:4) including pointing, cleaning and so on.'

This description would be far better worded as follows:

'Providing and laying 4 cm thick Shahabad stone flooring in cm (1:4) over 10 cm thick cc (1:5:10) bedding including pointing, watering, curing and cleaning.'

3.8   PROVISIONAL QUANTITIES AND SUMS

Work which is not detailed on the drawings and which by its nature cannot be accurately taken off, or which requires site measurements, is described as 'provisional'. The quantity of such an item of work is roughly estimated and marked as 'provisional quantity'. For example, provisional quantities of excavation, concrete and foundation masonry may be included to cover the extra depth of foundation if the results of subsurface investigations are unreliable. Similarly in the case of pile foundation if there is possibility of some of the piles being required to be lengthened in position, provisional quantities for this work are provided separately.

In addition, provisional sums are frequently included in the estimate to cover works such as electric fittings, air-conditioning equipment and so on, which would be carried out by specialist firms. In such cases, the provisional sum to be included in the estimate should be decided by consulting the concerned specialist firm.

3.9   PRIME COST

Drawings received by an estimator from the architect at the time of taking off work, rarely include the details of certain items such as water supply fittings, sanitary fittings, hand railing to concrete staircases and the like which are to be decided by the architect in consultation with his client during the progress of the work. The details of such items are not worked out at the time the estimate is framed but a sum is provided to cover the cost of the same. This sum provided represents the net sum and the contractor is given an opportunity to insert a percentage as profit, charges for packing, carriage and delivery of materials to the site (where these are not included in the prime cost sum) and also the cost of unloading, unpacking, fixing, returning empties and other incidental expenses.

Methods of Taking out Quantities

4.1 GENERAL

The English method of taking out quantities is used in the Military Engineering Services of the Government of India and therefore it is briefly outlined here. The method commonly used in India for taking out quantities is known as the 'Public Works Department Method' (PWD method). The main features of these methods are outlined in this chapter and also in Chapters 16 and 17. The PWD method of taking out quantities is fully explained with the help of examples in Chapters 5 to 15.

4.2 ENGLISH METHOD OF TAKING OUT QUANTITIES

1. *Dimension paper*: The dimensions are taken off from the drawings and recorded on dimension paper. The dimension paper consists of a foolscap sheet divided into two main sections, each section having four columns as shown in Fig. 4.1.

   The columns (not of course normally numbered) are numbered here for identification. The purpose of each column is given below:

   Column 1 is termed 'timesing column', and is used for 'timesing' and 'dotting on'.

   Column 2 is called 'dimension column', in which the measurements are set down as taken.

   Column 3 is known as the 'squaring column' and is used for recording the results obtained by multiplying together the figures in columns 1 and 2.

   Column 4 is known as the 'description column'. It is used for writing the description of the work to which the dimensions apply. The extreme right-hand side of this column is termed 'waste' and used for side notes and calculations.

2. *Entering dimensions*: Since all the dimensions on the architect's drawings are in feet and inches, the measurements are booked in feet and inches. The standard practice is to put a dot separating feet from inches and to avoid the use of 'or' sign. For example 9'6" would be written thus: 9.6.

<table>
<tr><td>1</td><td>2</td><td>3</td><td>4</td><td>1</td><td>2</td><td>3</td><td>4</td></tr>
<tr><td></td><td></td><td></td><td></td><td></td><td></td><td></td><td></td></tr>
</table>

Fig. 4.1   Dimension paper

Note that the dot is not a decimal point.

There are seven different forms of measurement namely,

1. cubes,
2. super,
3. runs,
4. numbers,
5. PC sums,
6. provisional sums, and
7. provisional quantities.

Examples of these items are given below together with the nature and use of the description column.

	69.9		Exc. fdn. tr. n. e. 5′ dp from u/s veg. soil excavn.	A cubic quantity
	3.9			
	4.3			
	69.9		P. & S. to sides of, trs. n. e. 5 total dp from g. 1.	A superficial quantity
	4.3			
	45.0		Levelling. unused rubble walling, 12″ wide to rec. copg., and	A linear quantity
			Copg. of ro. stones to match general walling 12″ wide and av 6″ ht.	

	2 or N̄o. 2 or 2/		C. i. m. h. cover and fr. to BS 497, grade c, 18″ × 18″ weighg. 68 lb and settg. fr. in c m (1.3) and bedding cover in grease.	An enumerated quantity (Cast iron manhole cover)
			Provide the PC sum of ₹2000 for electric light installation executed by specialist firm. Add for profit.	A PC sum
			Allow the provisional sum of ₹1000 for underpinning.	A provisional sum
			ex. fdn. tr. as before described (provisional)	Provisional quantity

The following points should be noted from the above examples.

(i)   A line is always drawn under each measurement.

(ii) It is not necessary to mention the units such as cubes, super, and so on, the number of dimensions in the measurements being an indication of the same.

(iii) The dimensions are set down in the following order:

    (a)   horizontal length,

    (b)   horizontal width, and

    (c)   vertical depth or height.

Although any change in the above order would not affect the calculation of volume or area, tracing of measurements after they are written would be a difficult task.

3.   *'Timesing'*:   When an item which has been booked previously, occurs again or for that matter, occurs a number of times, instead of repeating the dimensions, it is 'timesed', thus:

	4/	18.6 3.3 2.6		exc. fdn. tr. n. e. 5′ dp.	Indicating that the item occurs four times and the cubic measurement is to be multiplied by 4

The number of repetitions are booked in the 'timesing' column and separated from the dimension column by a small diagonal line placed against the first dimension. An item 'timesed' once can be 'timesed' again each multiplier multiplying everything to the right of it, thus:

	3/4/	18.6 3.3 2.6		exc. fdn tre. n. e. 5′ dp.	Indicating that the original measurement is multiplied by $3 \times 4 = 12$

To illustrate further the principle of 'timesing' consider the following example:

There are eight internal doors in a flat. There are four flats in a block. There are 30 blocks on a site. There are five sites.

The superficial item will be booked thus:

| 5/30/4/8/ | 3.0 | | internal door | |
| | 6.6 | | | |

4. *'Dotting on'*:   In repeating a dimension the surveyor may find that instead of multiplying an item which has been already 'timesed' he needs to add to the previous multiplier. For example, having measured four trenches as follows:

4/	18.6		exc. fdn. tr. n. e. 5 ft. dp.	
	3.3			
	2.6			

Suppose he finds 3 more. He could times by multiplying by $1\frac{3}{4}$ thus:

$1\frac{3}{4}$/4/	18.6		exc. fdn. tr. n.e. 5 ft. dp.	
	3.3			
	2.6			

But, to avoid fractions, these 3 trenches can be 'dotted on' or added to the multiple of 4 trenches. The 'dotting on'

3·4/	18.6		exc. fdn. tr. n.e. 5 ft. dp.	
	3.3			
	2.6			

is a simple way of writing $4 + 3$.

'Timesing' and 'dotting on' can be combined thus:

2/3.4/	18.6		exc. fdn. tr. n. e. 5 ft. dp.	Indicating that the cubic measurement is to be multiplied by $2 \times (4 + 3) = 14$.
	3.3			
	2.6			
3·2/4/	18.6		Exc. fdn. tr. n. e. 5 ft. dp.	Indicating that the cubic measurement is to be multiplied by $(2 + 3) \times 4 = 20$.
	3.3			
	2.6			

5. *Booking the dimensions of irregular figures*:   It is often necessary to ,set down the dimensions for triangles, circles and other such figures. The usual method of setting down the dimensions is illustrated in the following examples:

(a)	$\frac{1}{2}/$	3.6 5.0	Area of a triangle $3'6' \times 5'0'$
(b)	$\frac{1}{2}/$	3.6 5.0 2.6	Volume having a triangular base $3'6'' \times 5'0''$
(c)	$\pi r^2 = 3\frac{1}{7}/$	⑥ or 3.0 3.0	Area of a circle of $6'0''$ diameter. or —Ditto—
		or	
	$\pi/4d^2 = \frac{1}{4}/3\frac{1}{7}/$	6.0 6.0	or —Ditto—
(d)		④ or	Area of a semicircle of $4'0''$ diameter.
	$\frac{1}{2}\pi r^2 = \frac{1}{2}/3\frac{1}{7}/$	2.0 2.0	or
		or	—Ditto—
	$\frac{1}{2}\pi/4d^2 = \frac{1}{2}/\frac{1}{4}/3\frac{1}{7}$	4.0 4.0	—Ditto—
(e)		1.6 18.0	Area of a segment having chord height $1'6''$ and chord span $18'0''$.
(f)		4.8 6.0	Volume of a cylinder $4'8''$ diameter and $6'0''$ high.

6.   *Deductions*: When it is necessary to deduct for voids or openings in the main area or volume, a deduction item is entered immediately following the main item, thus:

	76.0 11.0	One-and-a-half bk. wall in Eng. Bond in cm (1:3) Ddt.—do—
4/	3.0 6.6	
	2.6	
2/	6.6	

7. *Corrections*: When a dimension has been written down incorrectly it is not necessary to alter the figure or to erase it. It is either neatly crossed out and the new figure written in, or the word 'nil' is written against it in the squaring column to indicate that it shall be taken as cancelled:

$$
\begin{array}{c}
24.0 \\
8.0
\end{array} \quad \text{nil} \quad \tfrac{1}{2} \text{ bk. wall in cm (1:3).}
$$

8. *Waste*: The dimensions should not be added together mentally or on scrap paper but should be written clearly on the extreme right-hand side of the description column such as:

$$
\begin{array}{c}
1.0 \\
9.9 \\
\underline{24.6} \\
35.3
\end{array}
$$

$$
\text{Less} \quad
\begin{array}{c}
0.9 \\
\underline{3.0} \\
3.9
\end{array}
\qquad
\begin{array}{c}
\underline{3.9} \\
31.6
\end{array}
$$

$$
\begin{array}{c}
31.6 \\
2.6 \\
\underline{3.0}
\end{array}
$$

This not only eliminates the risk of error, because the written dimensions can be checked, but makes clear the process by which the figures were arrived at.

9. *Measurement of length of wall*: For the measurement of foundations and masonry or brickwork the walls are divided into two groups:

  (i)   external walls, and

 (ii)   internal walls, the former being dealt with first.

The main length of all external walls is added together. For example, consider the plan of the residential building shown in Fig. 4.2.

Mean length of the 20 cm external wall will be calculated thus:

Internal length

Front bedroom	=	4.00 m
Internal wall	=	0.20 m
Sanitary block	=	2.80 m
Internal wall	=	0.20 m
Rear bedroom	=	4.00 m
Internal length		11.20 m

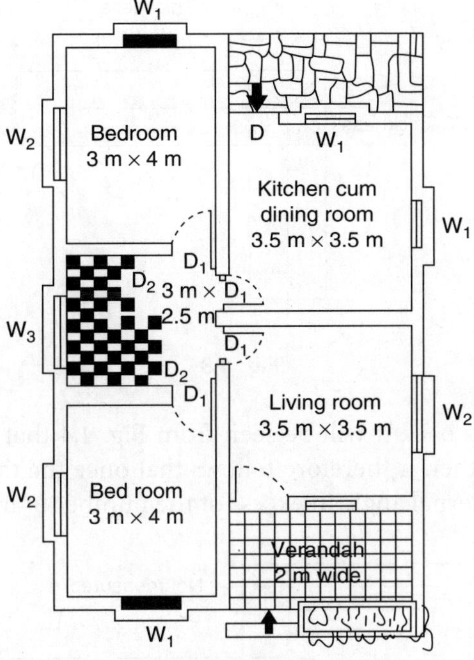

Fig. 4.2

Internal width

Front bed	=	3.00 m
Internal wall	=	0.20 m
Verandah	=	3.50 m
Internal width	=	6.70 m

Mean length

Twice internal length = 2/11.20	=	22.40 m
Twice internal width = 2/6.70 m	=	13.40 m
Four times width of external wall = 4/0.20	=	0.80 m
		36.60 m

It will be observed that the mean length is the length along the centre line of the wall, to calculate which it is necessary to add once the thickness of the wall for each corner (Fig. 4.3).

If outside measurements are shown on the drawing, the whole process can be reversed and once the thickness of the wall for each corner would be deducted from the sum total of external length.

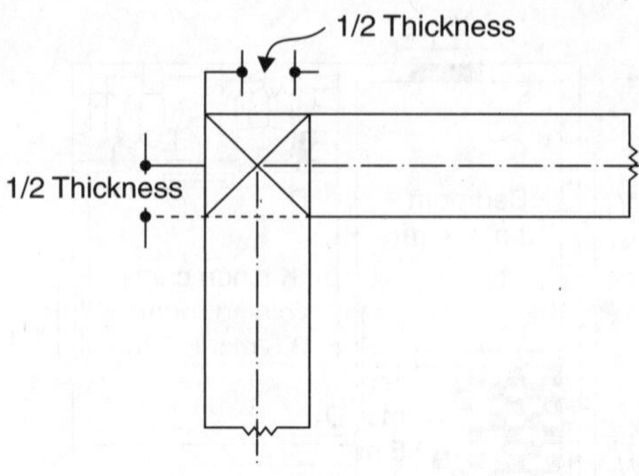

Fig. 4.3

Where the wall breaks back it will be seen from Fig. 4.4 that the external and internal angles balance each other. It therefore follows that once the thickness of the wall has to be added for every external angle in excess of the number of internal angles.

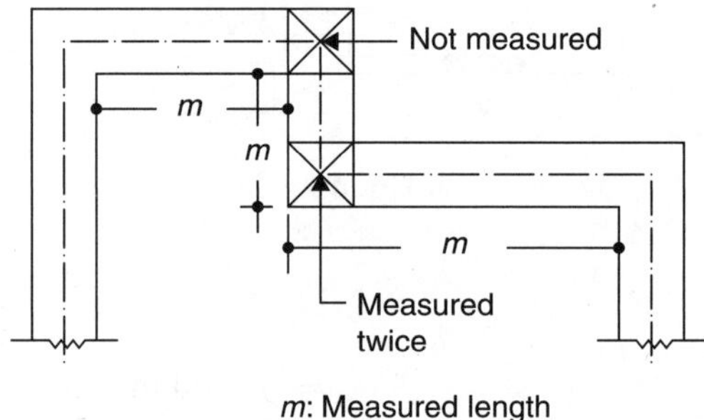

m: Measured length

Fig. 4.4

It is assumed that the external walls have a uniform width all round. In case the width varies each width will be dealt with separately.

Internal walls are collected in groups, width and depth in each group being uniform. The length is measured between the internal faces of the external walls at appropriate levels as shown in Fig. 4.5.

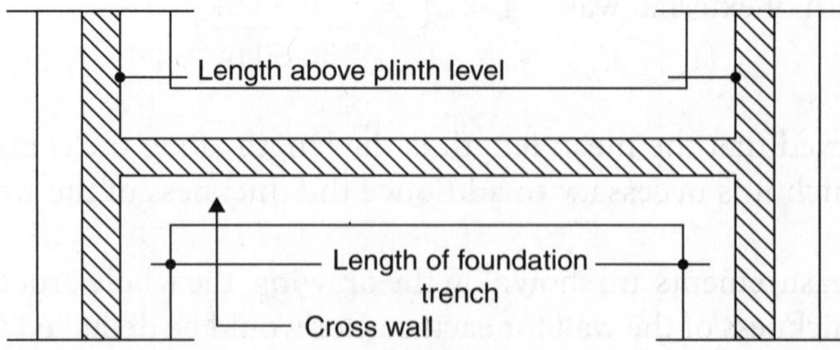

Fig. 4.5

10. *Working up*: Taking out quantities as explained above is the first step in the preparation of an estimate. The second and third steps, namely, squaring out dimensions, and abstracting and preparation of bill of quantities, are discussed in Chapters 16 and 17 respectively.

4.3   MES METHOD

The English method of taking out quantities is used with most of its essential features by the Military Engineering Services Department of the Government of India and is known as the 'MES Method'. However, with the introduction of metric units, the dimensions are in metres and the decimal system has replaced the duodecimal system.

4.4   PWD METHOD

1. *Measurement Sheet*

    The dimensions are taken out from the drawings and entered on a foolscap paper ruled as shown below:

Item No.	Particulars	No.	Length	Breadth	Depth of height	Quantity	Total

Fig. 4.6

The serial number of an item of work is entered in the first column, and its description is written in the second column. The figure showing number of times the measurements occur on the drawing is set down in the third column. This column is, thus, similar in use to the 'timesing' column on the dimension paper as used in the English Method. It saves the labour of repeating the dimensions. The dimensions are entered in the next three columns marked length, breadth and depth or height. All the dimensions are in metres. The results obtained by multiplying together the figures in the 3rd, 4th, 5th and 6th columns are recorded in the column marked quantity. From the gross quantity of work of an item, the deductions, if any, are made and the net result is entered in the last column marked 'total'.

2. *Measurement of Length of Walls*

Every intending quantity surveyor is advised to prepare the foundation plan showing centre-to-centre dimensions, at least in the beginning of his career. After sufficient practice this step would appear to be superfluous and can be dispensed with.

Centre-to-centre dimensions are worked out by adding half the width of each cross wall (as shown on plan) to the inside dimensions of a room. Figure 4.7 shows a plan of a building, of which the foundation plan with the centre-to-centre dimensions between adjacent walls is shown in Fig. 4.8.

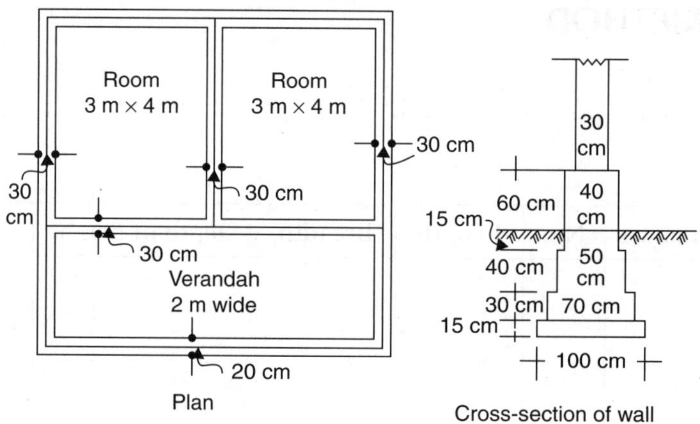

Fig. 4.7

After the foundation plan is prepared, the walls are divided into two groups. Walls in any one direction are called 'long walls' and the remaining walls at right angles to long walls are termed as 'short walls' or cross walls.

For convenience and to avoid the possibility of errors creeping in at a later stage, this grouping of walls is not to be changed even though such a change may seem convenient at times. The long wall is measured outside to outside of the cross walls at its extreme edges. The length of a short wall is always measured between inside faces of the two consecutive long walls. In other words, the common portion at the junction of walls is always measured in long walls and it is not considered at all in the measurement of short walls. This is shown in Fig. 4.8.

If the above-mentioned points are strictly followed, the calculation of lengths of long and short walls at various levels is a mechanical procedure and the following rules can be blindly applied:

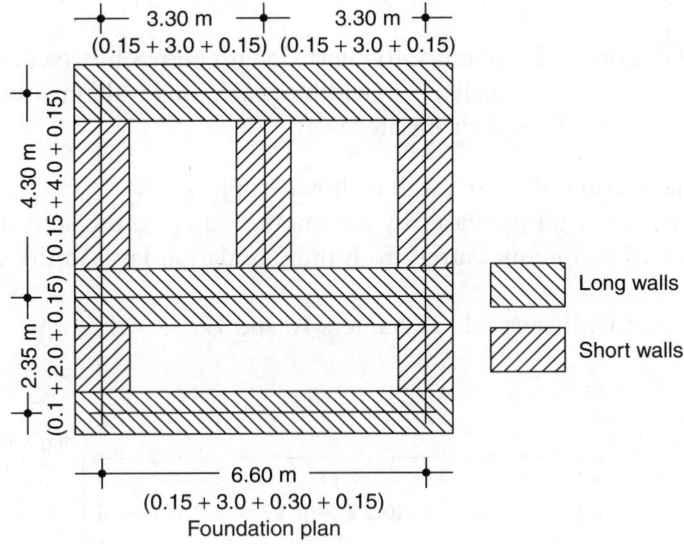

Fig. 4.8

Rule 1

Length of a long wall = (Centre-to-centre length between the short walls at its extreme edges) + $\frac{1}{2}$ (width of the short walls at the extreme edges of the long wall at the appropriate level)

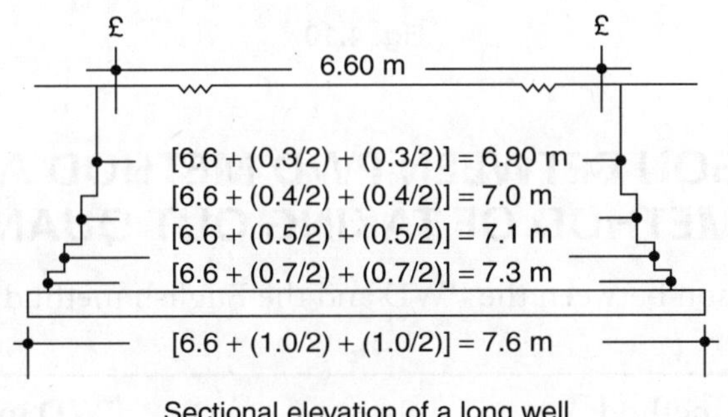

Fig. 4.9

It must be remembered that the length of a 'long wall' does not remain constant but varies at different levels as shown in Fig. 4.9.

It is obvious from the above figure that the length of a long wall decreases by an amount equal to the sum of offsets on either side while measured from the foundation level to the superstructure in successive stages.

Rule 2

> Length of a short wall = (Centre-to-centre length between two consecutive long walls) − ½ (width of each long wall on either side at the appropriate level)

The longitudinal section of a cross wall is shown in Fig. 4.10.

The length of a cross wall increases by an amount equal to the sum of the two offsets (as shown in Fig. 4.10), while measured from the foundation level to the superstructure in successive stages.

This method is further discussed in Chapters 16 and 17.

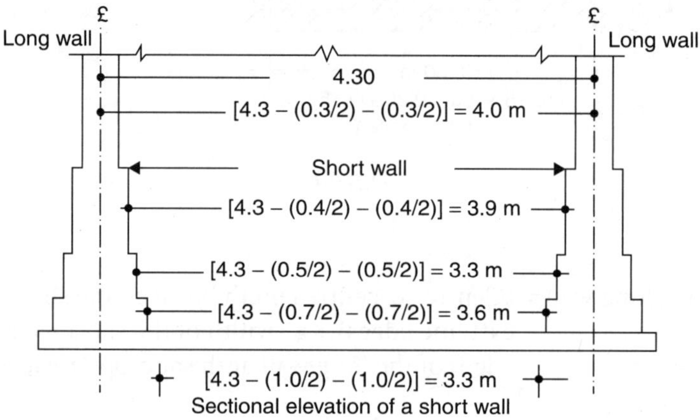

Fig. 4.10

4.5   COMPARISON BETWEEN PWD METHOD AND THE ENGLISH METHOD OF TAKING OUT QUANTITIES

The points of comparison between the PWD and the English method are listed below:

English method	PWD method
1. The items are grouped under trades	1. Items are in order of construction.
2. Splitting of details into separate items.	2. Averaging by rates.
3. Forms of entering:	3. Forms of entering:
(a) taking off in feet and inches separated by dot	(a) measurements are recorded in metres
(b) centre line method	(b) division into rectangles

(*continued*)

Continued

	English method		PWD method
(c)	results are in duodecimals	(c)	results are in decimals
(d)	timesing	(d)	total number is written
(e)	measurements are written one below the other	(e)	measurements are written on horizontal lines
(f)	waste collection of different parts of dimensions	(f)	no waste or collection
4.	Abstract for collection and classification of quantities by trades and units	4.	Nothing to correspond with
5.	Bill of quantities	5.	Abstract
6.	Estimate price shown	6.	Total cost + 5% contingencies = Estimated cost

Excavation and Foundation

5.1 GENERAL

Before beginning the work of taking out quantities, the drawings must be carefully examined to ascertain that the natural ground levels are sufficiently shown for calculating the average depth of excavation. Similarly, the nature of the ground, for example, paved, grass-covered, and so on, the type of soil in the sub-strata, for example, soil, *murum*, rock and so on, and the position of the ground water table must be known. These details are generally shown on the drawings in the form of particulars relating to trial pits excavated on the site. If necessary, the site may be visited to ascertain these and any other details which are likely to affect the cost of the work.

5.2 CLASSIFICATION OF EXCAVATION

In the Standard Method of Measurement of Building Works (IS: 1200–1965), the excavation is classified into the following heads, the measurements of which are to be taken out separately.

1. *Ordinary soil:*   Comprises any material which yields to ordinary digging equipment like pickaxe, *phawara* and shovel. For example, organic soil, sand, soft *murum* and so on.
2. *Hard soil:*   Comprises any material which requires the close application of picks and scarifiers to loosen. For example, stiff clay, hard *murum*, macadam surfaces and so on.
3. *Ordinary rock:*   Comprises hard materials that can be broken up with crowbars or picks. For example, soft rock, stone masonry, unreinforced cement concrete and so on.
4. *Hard rock:*   Comprises any material that requires blasting. For example, rock, reinforced cement concrete and so on.
5. *Hard rock:*   Comprises any material that requires blasting but which is to be excavated with the help of chiselling and wedging when blasting is prohibited for some reason.

5.3 SURFACE DRESSING

Where new buildings are to be constructed over natural ground, one of the first operations performed on a building site is to remove the surface soil containing vegetation. It is a superficial item. The area is measured in square metres to the extreme external edges of the foundation and the average depth, not exceeding 15 cm, is included in the description together with the method of disposal on site. Any further excavation for foundation trenches, basement and so on, is measured from the underside of such surface excavation.

Where the site is covered by some existing road or pavement, an item must be provided for breaking of such pavement and so on, and the surface excavation deducted for that area.

5.4 SURFACE EXCAVATION

Where the excavation exceeds 1.5 m in width and 10 m^2 on plan but where the depth of excavation does not exceed 30 cm it is measured as surface excavation and expressed in units of square metre. Like surface dressing this must also be measured over an area of the whole building including the projection of concrete foundation beyond the external wall.

5.5 EXCAVATION OVER AREAS

The term excavation over areas is used where the depth of surface excavation exceeds 30 cm. The excavation over areas is measured in cubic metres, split into different stages each of 1.5 m depth, that is, up to 1.5 m depth, exceeding 1.5 m depth but not exceeding 3 m depth, and so on, each stage being measured in a separate item. The measurement of area is similar to that defined in surface excavation except where the method of wall construction requires workmen to operate from the outside, an allowance of 60 cm working space from the external face of the wall is made.

To find the volume of excavation, the area and average depth of excavation must be known. The area of the excavation to be found is straightforward enough, but the real difficulty is encountered when obtaining the average depth particularly when the surface is uneven. The methods commonly used to find out the average depth are explained below with the help of examples.

Consider a square plot of land with ground levels marked thereon at each corner.

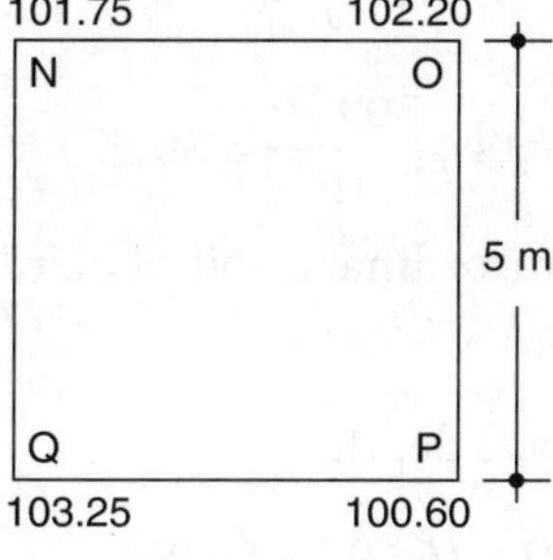

Fig. 5.1

To reduce the level to a datum of RL 100.00 m the average depth of excavation will be:

at	N	1.75 m
at	O	2.20 m
at	P	0.60 m
at	Q	3.25 m
Total		7.80 m

divided by 4 because there are four levels $= \dfrac{7.80}{4} = 1.95$ m.

The volume of excavation will be recorded in the measurement sheet as shown below:

Item No.	Description	No.	L	B	D	Quantity	Total quantity
1.	Excavate over an area to reduce the level to RL 100.00 m and cart away the excavated material up to a lead of 50 m	1	5.00	5.00	1.50	37.50	37.50 m^3
2.	-do-	1	5.00	5.00	0.45	11.25	11.25 m^3

This example of a square plot is only basic and the reader should not think that in order to find out volume of excavation, area can be indiscriminately multiplied by the average depth of the number of levels given.

The correct method to work out the volume of excavation in cases of advanced examples is to divide the area into suitable parts and to work out the volume of each part by using appropriate levels. This is illustrated in the following example (see Fig. 5.2).

Method 1   The obvious approach to this problem would be to find out the average ground level, thus:

$$(51.75 + 50.60 + 50.50 + 51.85 + 51.25 + 50.85 + 50.65 + 50.50 + 50.45$$
$$+ 51.30) = 509.70$$

divided by 10, number of levels, that is, $\dfrac{509.70}{10} = 50.97$ m.

If the type of soil changes from ordinary soil to hard soil at RL 50.00 the quantity of excavation in ordinary soil will be:

Area of excavation × Average depth

that is, $[(20 \times 10) - (8 \times 4)] \times (50.97 - 50.00)$

$$= 168 \times 0.97 = 162.96 \text{ m}^3$$

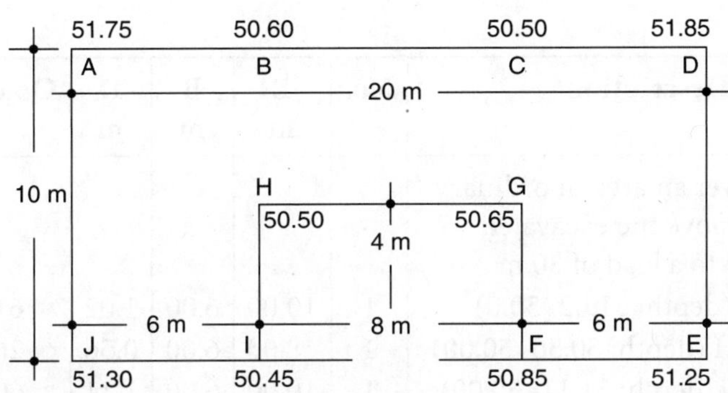

Fig. 5.2

Method 2   In this method the figure is divided into three suitable parts (see Fig. 5.2) and the average ground level and ultimately the average depth of excavation is found out separately.

(a) Area ABIJ

$$\text{Average ground level} = \frac{51.75 + 50.60 + 50.45 + 51.30}{4}$$

$$= \frac{204.10}{4} = 51.025 \text{ m}$$

$$= 51.02 \text{ m}$$

(b) Area BCGH

$$\text{Average ground level} = \frac{50.60 + 50.50 + 50.65 + 50.50}{4}$$

$$= \frac{202.25}{4} = 50.56 \text{ m}$$

(c) Area CDEF

$$\text{Average ground level} = \frac{50.50 + 51.85 + 51.25 + 50.85}{4}$$

$$= \frac{204.45}{4} = 51.11 \text{ m}$$

The volume of excavation in ordinary soil will be recorded in the measurement sheet as shown below:

Item No.	Description	No.	L m	B m	D m	Quantity	Total quantity
1.	Excavate over an area in ordinary soil and remove the excavated material up to a lead of 50 m						
	Area ABIJ (depth: 51.02–50.0)	1	10.00	6.00	1.02	61.20	
	Area BCGH (depth: 50.56–50.00)	1	8.00	6.00	0.56	26.88	
	Area CDEF (depth: 51.11–50.00)	1	10.00	6.00	1.11	60.60	
						148.68	148.68 m^3

The volume obtained by the second method is more accurate than that obtained by the first method.

The second method explained above is useful for small areas such as excavation for basement and so on. On large sites for example, excavation for a wasteweir channel in the case of a dam, the levels are taken at more frequent and regular intervals. A drawing containing these levels would be similar to that shown in Fig. 5.3.

In this case, the conventional method of finding out volume of excavation for each square would be a laborious one. A simple method involving less of manual effort and error is explained below:

If it is proposed to take out quantities of each square separately, it will be observed that some levels which are common to more than one square will be repeated twice (for example,

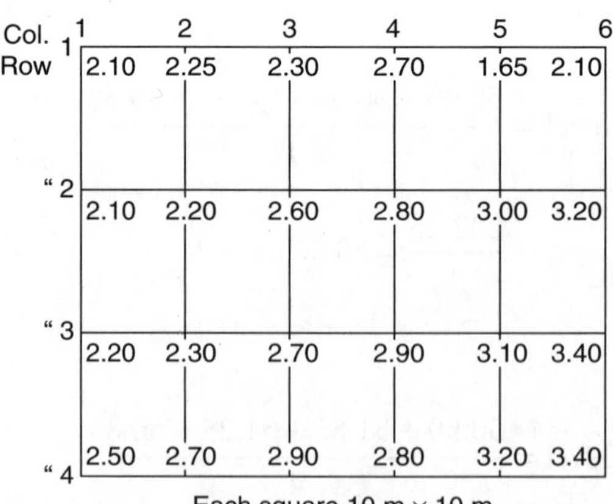

Fig. 5.3

the intermediate levels of second and third row). Then it is possible to simplify the work of averaging by considering each level the number of times it would have been considered on the basis of dealing with each square separately. This is illustrated below:

Column	1	2	3
Row 1	2.10	2×2.25	2×2.30
" 2	2×2.10	4×2.20	4×2.60
" 3	2×2.20	4×2.30	4×2.70
" 4	2.50	2×2.70	2×2.90
Totals	13.20	27.90	31.60

Column	4	5	6
Row 1	2×2.70	2×1.65	2.10
" 2	4×2.80	4×3.00	2×3.20
" 3	4×2.90	4×3.10	2×3.40
" 4	2×2.80	2×3.20	3.40
Totals	33.80	34.10	18.70

Total of Column	1	13.20
"	2	27.90
"	3	31.60
"	4	33.80
"	5	34.10
"	6	18.70
Grand totals		159.30

This grand total has to be divided by the total number of levels considered in the sum, which are equal to:

(Number of levels per square) × (Number of squares)

That is, $4 \times 15 = 60$

Average level of the grid, therefore, is equal to $\dfrac{159.30}{60} = 2.655$ m.

In order to find out the volume of excavation necessary to reduce levels down to a datum RL 0.00, the measurements will be recorded thus:

Item No.	Description	No.	L m	B m	D m	Quantity	Total quantity
1.	Excavation over area to reduce level and removing the excavated material, lift 1.5 m and lead 50 m	15	10.00	10.00	1.50	2250.0	2250.0 m^3
2.	-do- (Except that lift to be 3 m)	15	10.00	10.00	1.16	1740.0	1740.0 m^3

5.6 FOUNDATION TRENCHES

Excavation in trenches (width not exceeding 1.5 m) is to be measured separately from excavation over areas. The unit of measurement is cubic metre. The measurements are to be grouped in separate items for successive stages of 1.5 m depth. For example, excavation of a trench 2 m deep will be measured in two items, that is, up to 1.5 m deep in the first item and exceeding 1.5 m depth but not exceeding 3 m depth in the second item.

In order to find out the depth of excavation, two sets of levels must be known. These are:

1. bottom of foundation, and
2. natural ground level.

The best way is to mark these levels on the foundation plan. If the ground level is uneven, one way is to find out the average ground level for the complete structure and knowing the bottom of foundation levels, work out the average depth of excavation for each trench. If the bottom of the foundation is in a level plane, the average depth of excavation will be the same for all trenches. The second and more accurate way is to find out the average ground level and depth of excavation for each trench separately.

Description of the item of excavation should include the following operations if they are not proposed to be given separately:

1. Removal and disposal of excavated material, specifying lead in units of 50 m and lift in units of 1.5 m.
2. Returning, filling and ramming of selected material in layers not exceeding 20 cm in thickness including watering and levelling.

5.7   TRENCHES FOR PIPES, CABLES AND SO ON

Excavation in trenches for pipes, cables and so on, is classified into the following three groups:

1.   pipes, cables, not exceeding 75 mm in diameter,
2.   pipes, cables, between 75 mm and 300 mm in diameter, and
3.   pipes, cables, more than 300 mm in diameter.

The unit of measurement is running metre. The description is to include returning, filling, ramming and removal of surplus soil, in addition to concrete foundations, if any.

5.8   DISPOSAL OF EXCAVATED MATERIAL

The term 'excavate and get out' ordinarily includes removing the material at least one metre clear of the edge of excavation. The subsequent disposal of the material may be accounted for in any one of the two ways mentioned below:

1.   The description of the item of excavation may include the way of disposal such as 'return, fill and ram' or 'cart away'. In the latter case, the distance of removal—lead—is to be stated. Then, while deciding the unit rate of excavation this additional work is to be duly accounted for.

   This method of accounting for the disposal is suitable when the excess material can be suitably disposed of within a reasonable lead, say, 50 m to 100 m. If the distance of removal exceeds this limit the expenses for removal are considerable and it is better to account for the same separately as mentioned below.
2.   An item may be included in the bill of quantities to account for loading, unloading and means of removal, specifying the distance through which the material is to be conveyed in units mentioned below:
   (i)   50 m up to one kilometre, and
   (ii)   100 m for distances exceeding one kilometre. The quantity of material to be disposed of is expressed in the unit of cubic metre. For taking out measurements, any one of the following methods would be suitable, depending upon the circumstances.
      (a)   The excavated material may not be suitable for refilling, in which case, the quantity to be disposed of would be the total quantity that is excavated.
      (b)   Part of the material may only be needed for filling in the sides of masonry. In such cases, the remaining quantity of material that is to be disposed of can be found out in a simple way as explained below.

         From the total quantity of excavation in trenches (not including top 15 cm surface dressing) deduct the quantity of foundation concrete and masonry up to 15 cm below ground level (see Fig. 5.4). The advantage of this method is that all the different quantities involved, namely, excavation in trenches, foundation concrete and foundation masonry, are worked out separately under respective headings and no measurements need be taken out specially for this purpose.

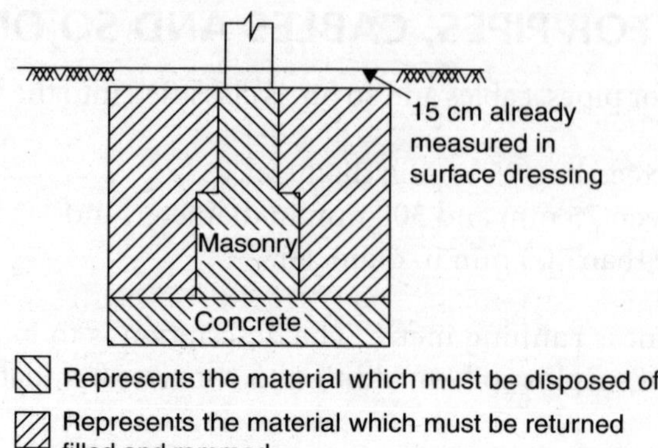

Fig. 5.4

(c) In the case of basement excavation, it is convenient to account for the whole of the excavated material as 'cart away' in the first instance and later on to adjust the returning, filling and ramming of soil in the working space.

5.9   PLANKING AND STRUTTING

Planking and strutting to retain the sides of excavation is not to be described in the item of excavation but is to be measured in square metres of the face area supported.

Where the depth of excavation to be supported exceeds 1.5 m, the measurements are taken out in two or three items as below:

1.   depth less than 1.5 m,
2.   depth exceeding 1.5 m but not exceeding 3 m, and
3.   depth more than 3 m.

Similarly, planking and strutting in respect of: (i) trenches, (ii) areas, (iii) shafts, wells, manholes, and the like, and (iv) situations where planking and strutting are required to be left permanently in position, are each to be separately measured and so described.

5.10   CONCRETE IN FOUNDATIONS

Concrete in foundations is measured in cubic metres and the mix is stated in the description. The length and width measured for excavation of trenches will usually be found to serve for the measurements of the foundation concrete. The thickness of concrete is generally shown on the drawings. Where the concrete foundation is stepped it is easier to measure it as if it were at a uniform level and then add extra concrete for laps, the planking and strutting to the face of the steps being negligible.

Example 5.1   Foundation for a small building (Fig. 5.5)

Item No.	Description	No.	Length	Breadth	Depth or height	Quantity	Total quantity
1.	Surface dressing 15 cm thick and removing the material 50 m away (11.40 + 0.80) = 12.20 (6.90 + 0.80) = 7.70	1	12.20	7.70		93.94	
	For front steps and flower bed	1	3.90	0.50		1.95	
						95.89	95.89 m^2
2.	Excavation in trenches in ordinary soil including the disposal of excavated material—lift 1.5 m and lead 50 m *Long walls*:						
	Bed, bath and WC: (11.40 + 0.80)	2	12.20	0.80	0.50	9.76	
	Living room and kitchen: (9.65 + 0.80)	1	10.45	0.80	0.50	4.18	
	Short walls: (3.20–0.80)	4	2.40	0.80	0.50	3.84	
	Verandah, living and kitchen: (3.70 − 0.80)	4	2.90	0.80	0.50	4.64	
						22.42	22.45 m^2
3.	Excavation in fdn. trenches in hard soil including disposal of excavated material—lift 1.5 m and lead 50 m, returning, filling and ramming the selected material around foundation *Long walls*						
	Bed, bath and WC	2	12.20	0.80	0.30	5.86	
	Living room and kitchen	1	10.45	0.80	0.30	2.51	
	Short walls						
	Bed, bath and WC	4	2.40	0.80	0.30	2.30	
	Verandah, living room and kitchen	4	2.90	0.80	0.30	2.75	
						13.42	
	Alternately	Qty.	of item	2 ×	$\frac{0.30}{0.50}$	13.45	13.45 m^3

(continued)

Example 5.1   Continued

Item No.	Description	No.	Length	Breadth	Depth or height	Quantity	Total quantity
4.	Providing planking and strutting up to 1.5 m deep. External walls—outer faces (including 15 cm depth of surface dressing) $(11.40 + 0.80 + 3.20 + 0.80 + 1.75 + 3.70 + 9.65 + 0.80 + 0.90 + 0.80 = 39.80)$ Internal faces (measured from under side of surface dressing) Length: Beds $4 \times (4.20 - 0.80) = 13.60$ $2 \times (3.0 - 0.80) = 4.40$ Bath and WC $6 \times (3.20 - 0.80) = 14.40$ Verandah $2 \times (2.0 - 0.80) = 2.40$ Living $2 \times (3.70 - 0.80) = 3.80$ Kitchen $2 \times (3.95 - 0.85) = 6.20$ and short walls of ver. living and kit. $6 \times (3.70 - 0.80) = \underline{17.30}$	1	39.80		0.65	25.87	
	Grand total 62.20	1	62.20		0.50	31.10	
						56.97	56.97 m²
5.	Providing and laying cement concrete M75 in foundation including ramming and curing						
	Long walls	2	12.20	0.80	0.15	2.93	
		1	10.45	0.80	0.15	1.25	
	Short walls	4	2.40	0.80	0.15	1.15	
		4	2.90	0.80	0.15	1.39	
						6.72	
	Alternatively		$\left(\dfrac{0.15}{0.30} \times \text{Qty. of Item 3}\right)$			6.72	6.72 m³

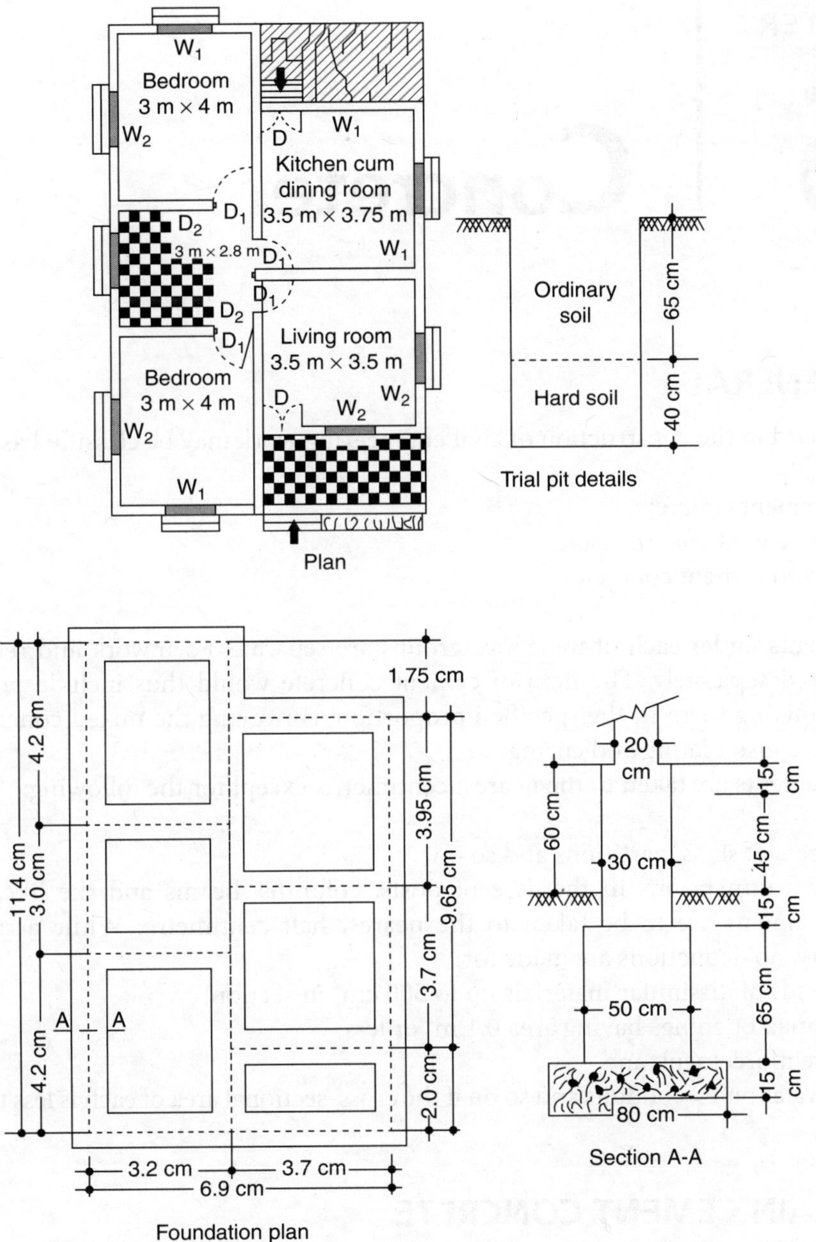

Fig. 5.5   Example 5.1 (Foundation for a small building)

CHAPTER 6 | Concrete

6.1 GENERAL

Concrete used in the construction of civil engineering work may be classified as follows:

1. plain cement concrete,
2. precast cement concrete, and
3. reinforced cement concrete.

Measurements under each of the above groups are separate. Formwork and reinforcement are measured separately. The item of cement concrete would thus include providing the materials, mixing them in the specified proportion, conveying the mixed concrete, placing in position, consolidating and curing.

Measurements are taken to the nearest centimetre, except for the following:

1. Thickness of slabs, partitions and so on,
2. Sectional dimensions in the case of posts, columns, beams and the like, where the measurements are to be taken to the nearest half centimetre. While arriving at the quantity, no deductions are made for:
   (i) ends of dissimilar materials up to 500 cm^2 in section,
   (ii) small openings having area 0.1 m^2 or less,
   (iii) reinforcement, and
   (iv) water pipes, conduits and so on if the cross-sectional area of each is less than 25 cm^2.

6.2 PLAIN CEMENT CONCRETE

Plain cement concrete is grouped into the following sections and measurements under each section are kept separate.

1. *Foundations*: The method of measurement of plain cement concrete in foundation is already explained in Chapter 5 and also illustrated in Example 5.1.
2. *Mass concrete*: Mass concrete is measured in cubic metres.

3. *Walls*: Walls including attached buttresses, pilasters and their caps and bases are measured in cubic metres. While finding out the cubic contents, the deductions are made for openings and dissimilar materials excepting the cases mentioned in Section 6.1.
4. *Independent piers and columns*: These are measured separately and expressed in units of cubic metres.
5. *Damp-proof course*: Damp-proof course is fully described and given in square metres stating the thickness. No separate measurements are taken out for (i) levelling and preparing brickwork or masonry, (ii) formwork, and (iii) fair finish to edges.

6.3 PRECAST CEMENT CONCRETE

Precast cement concrete may be used in the construction of works in the form of:

1. solid articles (reinforced or otherwise),
2. light weight partitions,
3. precast concrete block construction,
4. jaalis, and
5. piles.

1. *Solid articles*: The item includes all moulds, finished faces, and hoisting and setting in position mortar of specified proportion. The unit of measurement is cubic metre. Reinforcement is, however, measured separately and given in the units of quintal.
2. *Lightweight partitions*: These are fully described and measured in square metres stating the thickness.
3. *Precast block construction*: The precast block construction is measured in accordance with the rules for measurement of brickwork (Chapter 7). However, the following points may be noted here:
    (i) The work includes all moulds, finished faces and hoisting and setting in position.
    (ii) Hollow block construction should be measured separately and no deductions should be made for the hollows.
    (iii) Any special finish to block construction, such as imitation of quarry faced natural stone, is to be fully described and given in square metres.
4. *Jaalis*: Precast concrete jaalis are fully described and given in square metres. The reinforcement is included in the item.
5. *Piles*: Concrete piles are classified according to section and length. Any requisite moulds, necessary strappings, bolts and lifting holes are included in the item. The unit of measurement is cubic metre.

Heads and shoes of steel or iron are enumerated. The weight of each head or shoe is stated in the description.

Cutting or breaking of heads of piles are enumerated.

Steel reinforcement may be fully described and included with the item.

6.4 REINFORCED CEMENT CONCRETE

Reinforced cement concrete is generally measured in three parts:

1. concrete,
2. formwork, and
3. steel reinforcement.

Measurements of only one item of the above should be taken out at a time.

6.4.1 Concrete

A general examination of drawings will indicate that concrete is used in different parts of a structure at different levels. Although the unit of measurement is cubic metre in each case it would be incorrect to measure all concrete work in one item. It is obvious that concrete of a given mix provided in the upper floor slab in a multi-storeyed building will cost more than concrete of the same mix provided in a ground floor slab, due to greater handling costs. Secondly, the cost per unit quantity of concrete will be influenced by the volume of concrete to be laid and the ease with which concrete can be placed and compacted in position.

For the above reasons, measurements of concrete in different parts have to be so grouped that the same unit rate is applicable to each part in a group of items. Classification of concrete based on this principle is given in Clause 4.7 (a) of SMM of BWIS 1200-1965. It is suggested therein that the measurements under each group should be recorded under separate items. For example, in a small building the following items may have to be included in the estimate:

1. foundations, bases for columns, and so on,
2. walls,
3. columns,
4. lintels, beams, and
5. stairs.

At the junction of two members in a monolithic structure, the common portion is measured with only one item. The different cases are given below and illustrated in Fig. 6.1 to 6.4.

Common portion between	Measured with
1. Beam and column	Column
2. Slab and rectangular beam	Slab
3. Slab and T or L beam	T or L beam
4. Chajja and lintel	Lintel

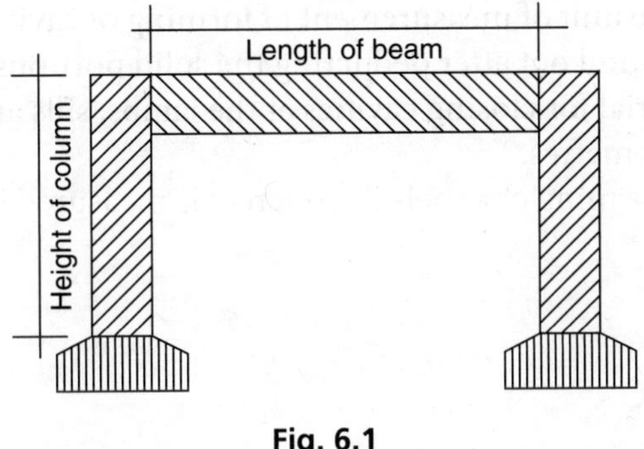

Fig. 6.1

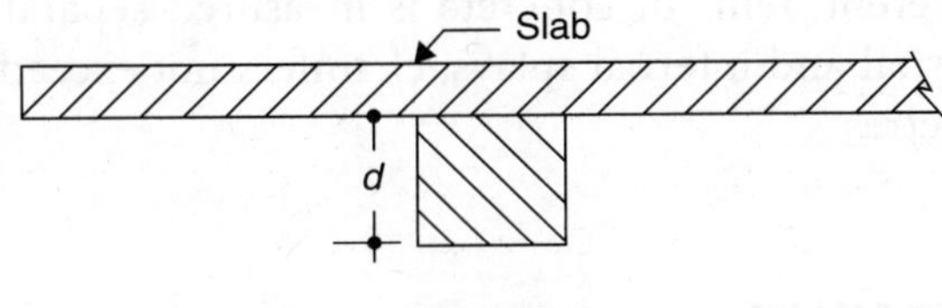

Fig. 6.2

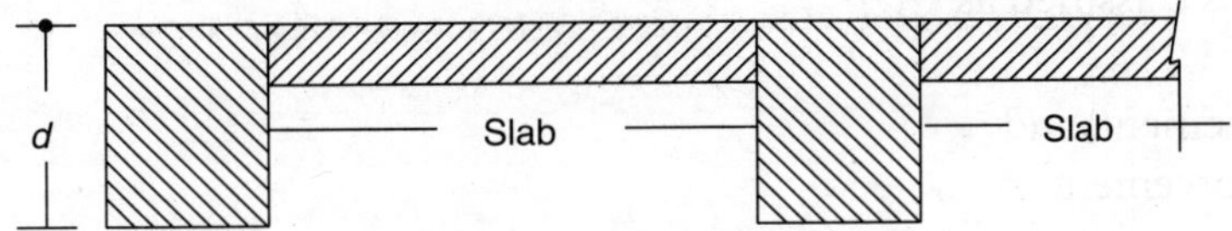

Fig. 6.3

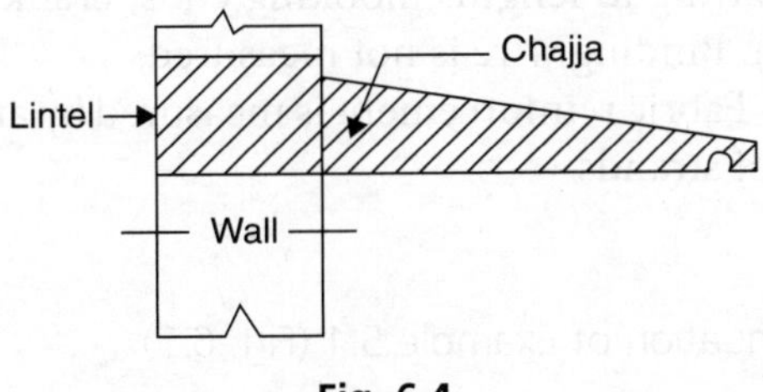

Fig. 6.4

6.5 MISCELLANEOUS ITEMS

The work under each of the following items is kept separate and measured in the unit specified below:

1. *Expansion joints in concrete*:   These are measured in running metres stating the depth and the width of the joint. The unit rate includes the cost of formwork and labour necessary to form the joint.

2. *Cavity walls:*   (i) The unit of measurement of forming of cavity is square metre. The net area of the cavity is found out after deducting the solid portions of wall and all openings. (ii) Labour and material for closing cavities at the jambs, sills and heads of openings, are measured in running metres.
3. *Surface channels:*   Forming channels in concrete is measured in running metres stating the inner girth.

6.6 FORMWORK

The actual surface of the formwork in contact with the concrete is measured in square metres. Formwork used for different items of concrete is measured separately. Fillets required to produce throating, external and internal splays, chamfers not exceeding 20 cm in girth are measured in running metres.

6.7 REINFORCEMENT

Reinforcement is classified as

1. bar reinforcement, and
2. fabric reinforcement.

1. *Bar reinforcement:*   Bar reinforcement is measured by weight in quintals, and the description includes cutting to lengths, hooking ends, cranking or bending, conveying and placing in position. Binding wire is not measured.
2. *Fabric reinforcement:*   Fabric reinforcement is measured by area in square metres stating the mesh and the size of strands.

Example 6.1 (A)   In continuation of Example 5.1 (Fig. 6.5)

DATA

1. Schedule of doors and windows: Openings

Type of door or window	Size in cm	Size of opening in cm	Number of openings
D	107.5×247.5	110×250	2
D_1	97.5×197.5	100×200	4
D_2	77.5×187.5	80×190	2
W_1	87.5×157.5	90×160	4
W_2	107.5×157.5	110×160	4
W_3	67.5×97.5	70×100	2

2. *Design of Lintels*
    (i)   Over doors and windows except door D_2: Size 20 cm × 15 cm
          Main bars: 4 nos, 10 mm dia., 2 bent up
          Anchor bars: 2 nos, 10 mm dia.
          Stirrups: 6 mm dia., 15 cm c/c
    (ii) Over door D_2 : Size 10 cm × 15 cm
          Main bars: 3 nos, 6 mm dia., one bent up
          Anchor bars: 2 nos, 6 mm dia.
          Stirrups: 6 mm dia., 30 cm. c/c

3. *Design of Beam*
    Size: 20 cm × 30 cm
    Main bars: 4 nos, 12 mm dia., 2 bent up
    Anchor bars: 2 nos, 10 mm dia.
    Stirrups: 6 mm dia., 30 cm c/c

4. *Design of Slab:* Thickness 12 cm
    Over verandah
    Main bars: 10 mm dia. @ 15 cm c/c
    Secondary bars: 6 mm dia. @ 20 cm c/c
    Bedroom and kitchen: Two-way slab thickness 12 cm
    Along shorter span: 12 mm dia. @ 15 cm c/c
    Along longer span: 10 mm dia. @ 20 cm c/c
    Living room and sanitary block: Two-way slab, 12 cm thick
    12 mm dia. bars @ 15 cm c/c both ways

5. *Design of Chajja*
    Average thickness: 10 cm
    Main bars: 10 mm dia. 15 cm c/c
    Secondary bars: 6 mm dia. 20 cm c/c
    (The arrangement of the reinforcement is shown in Fig. 6.5.)

Example 6.1 (A)

Item No.	Description	No.	Length in m	Breadth in m	Depth or height in m	Quantity	Total
1.	Formwork to lintels and beams. For openings of:						
	Door: D. Soffit	2	1.10	0.20		0.44	
	Sides	4	1.40		0.15	0.84	
	Edges	4	0.20		0.15	0.12	

(continued)

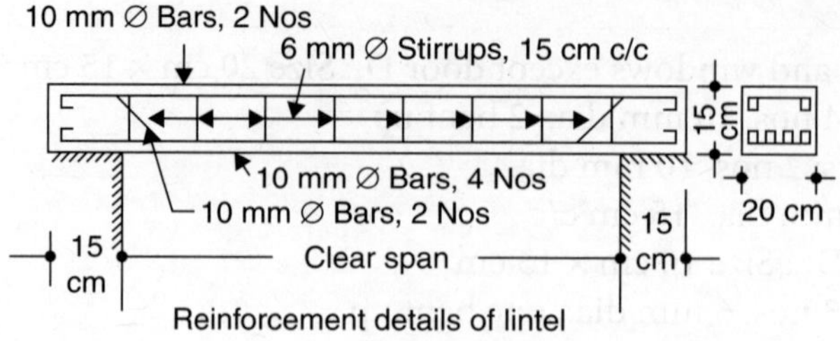

Reinforcement details of lintel

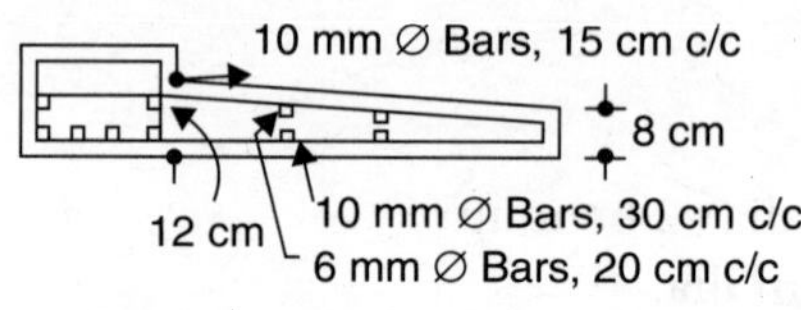

Reinforcement details of *chajja*

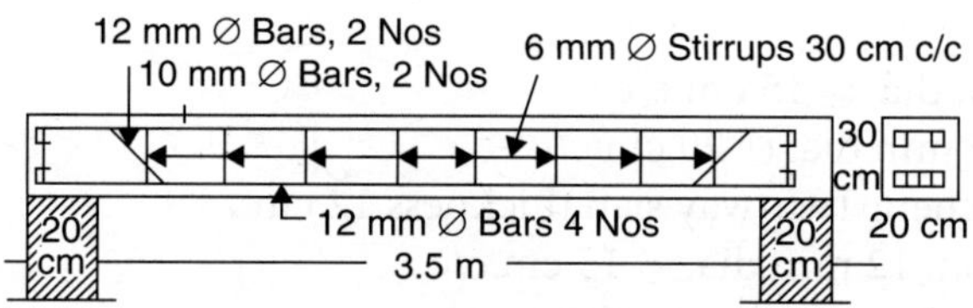

Reinforcement details of beam

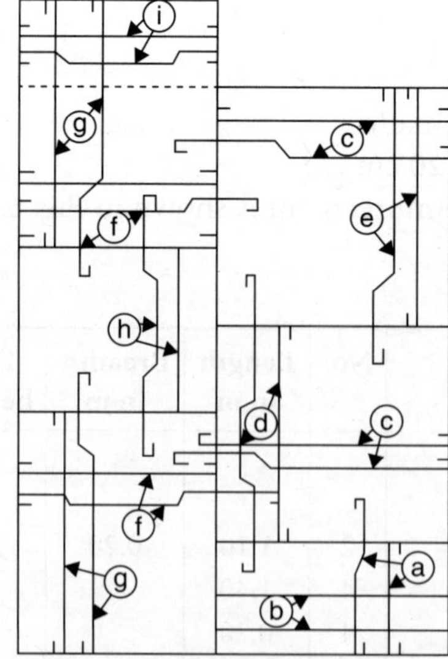

Arrangement of reinforcement in slab

Fig. 6.5   Example 6.1(A)

***Example 6.1* (A)**   Continued

Item No.	Description	No.	Length in m	Breadth in m	Depth or height in m	Quantity	Total
	Door: D_1 Soffits	4	1.00	0.20		0.80	
	Sides	8	1.30		0.15	1.56	
	Edges	8	0.20		0.15	0.24	
	Door: D_2 Soffits	2	0.80	0.10		0.16	
	Sides	4	1.10		0.15	0.66	
	Edges	4	0.20		0.15	0.12	
	Windows: W_1 Soffits	4	0.90	0.20		0.72	
	Sides	8	1.20		0.15	1.44	
	Edges	8	0.20		0.15	0.24	
	Windows: W_2 Soffits	4	1.10	0.20		0.88	
	Sides	8	1.40		0.15	1.68	
	Edges	8	0.20		0.15	0.24	
	Windows: W_3 Soffits $(0.70 + 0.10 + 0.70) = 1.50$ m	1	1.50	0.20		0.30	
	Sides	2	1.80		0.15	0.54	
	Edges	2	0.20		0.15	0.06	
	Verandah beam						
	Bottom	1	3.50	0.20		0.70	
	Sides	2	3.90		0.30	1.17	
	Edges	2	0.20		0.30	0.12	
						13.03 m²	13.03 m²
2.	Formwork to *chajja*						
	Door: D Bottom	1	1.40	0.60		0.84	
	Sides	1	1.40		0.10	0.14	
		2	0.60		0.10	0.12	
	Window: W_1 Bottom	4	1.20	0.60		2.88	
	Sides	4	1.20		0.10	0.48	
		8	0.60		0.10	0.48	
	Window: W_2 Bottom	3	1.40	0.60		2.52	
	Sides	3	1.40		0.10	0.42	
		6	0.60		0.10	0.36	
	Window: W_3 Bottom	1	1.80	0.60		1.08	
	Sides	1	1.80		0.10	0.18	
		2	0.60		0.10	0.12	
						9.62 m²	9.62 m²

(*continued*)

Example 6.1 (A)   Continued

Item No.	Description	No.	Length in m	Breadth in m	Depth or height in m	Quantity	Total
3.	Providing and laying cement concrete M 150 in lintels and beams including compacting and curing						
	Lintel over						
	Door D	2	1.40	0.20	0.15	0.084	
	Door D_1	4	1.30	0.20	0.15	0.156	
	Door D_2	2	1.10	0.10	0.15	0.033	
	Window W_1	4	1.20	0.20	0.15	0.144	
	Window W_2	4	1.40	0.20	0.15	0.168	
	Window W_3 (Continuous lintel)	1	1.80	0.20	0.15	0.054	
	Verandah beam	1	3.90	0.20	0.30	0.234	
						0.873	0.87 m³
4.	Providing and laying c.c. M 150 in *chajja* including compacting and curing						
	Over Door D	1	1.40	0.60	0.10	0.084	
	Window W_1	4	1.20	0.60	0.10	0.288	
	Window W_2	3	1.40	0.60	0.10	0.252	
	Window W_3	1	1.80	0.60	0.10	0.108	
						0.732 m³	0.732 m³
5.	Formwork to roof slab over:						
	Verandah	1	3.50	1.80		6.30	
	Living room	1	3.50	3.50		12.25	
	Kitchen	1	3.50	3.75		13.13	
	Bedrooms	2	3.00	4.00		24.00	
	Sanitary block	1	3.00	2.80		8.40	
	Formwork to sides of slab						
	$L = 11.40 + 0.20 = 11.60$ m	1	11.60		0.12	1.39	
	$L = 3.20 + 0.20 = 3.40$ m	1	3.40		0.12	0.41	
	$L = 3.70 + 0.10 - 0.10$ $= 3.70$ m	1	3.70		0.12	0.44	
	$L = 9.65 + 0.20 = 9.85$ m	1	9.85		0.12	1.18	
	$L = 6.90 + 0.20 = 7.10$ m	1	7.10		0.12	0.85	
						68.35 m²	68.35 m²

(*continued*)

Example 6.1 (A)   Continued

Item No.	Description	No.	Length in m	Breadth in m	Depth or height in m	Quantity	Total
6.	Providing and laying cement concrete M 150 in slab including compacting and curing Length $(11.40 + 0.20) = 11.60$ m Width $(6.90 + 0.20) = 7.101$ m	1	11.60	7.10	0.12	9.88	
	Deduct portion of redundant area and	1	3.70	1.75	0.12	0.77	
	L beam in Verandah: Length $(3.70 + 0.20) = 3.90$	1	3.90	0.20	0.12	0.09	
	Total					0.86 9.02 m^3	9.02 m^3
7.	Providing, cutting, bending, hooking, conveying, placing and binding in position m s reinforcement		As per bar bending schedule attached			12.33 q	12.33 q

BAR BENDING SCHEDULE

Shape of bar	No.	Length in m			Total length in m		
		6 mm	10 mm	12 mm	6 mm	10 mm	12 mm
Lintels over door D and window W_2							
Bottom straight	12		1.53			18.36	
Bottom bent up	12		1.63			19.56	
Anchor bars	12		1.53			18.36	
Stirrups no. $= \dfrac{1.35}{0.30} + 1 = 5$							
Total no. $= 6 \times 5 = 30$	30	0.68			20.40		
Lintel over door D_1							
Bottom straight	8		1.43			11.44	
Bottom bent up	8		1.53			12.24	
Anchor bars	8		1.43			11.44	
				C.O	20.40	91.40	

(continued)

BAR BENDING SCHEDULE   Continued

Shape of bar	No.	Length in m			Total length in m		
		6 mm	10 mm	12 mm	6 mm	10 mm	12 mm
Stirrups no. $= \dfrac{1.25}{0.30} + 1 = 5$				B.F	20.40	91.40	
Total no. $= 4 \times 5$	20	0.68			13.60		
Lintel over door D_2							
Bottom straight	4	1.16			4.64		
Bottom bent up	2	1.26			2.52		
Anchor bars	4	1.16			4.64		
Stirrups nos $= \dfrac{1.05}{0.30} + 1 = 5$							
Total no. $2 \times 5 = 10$	10	0.40			4.00		
Lintel over window W_1							
Bottom straight	8		1.33			10.64	
Bottom bent up	8		1.43			11.44	
Anchor bars	8		1.33			10.64	
Stirrups no: $= \dfrac{1.15}{0.30} + 1 = 4$							
Total no. $= 4 \times 4 = 16$	16	0.68			10.88		
Lintel over window W_3							
Bottom straight	2		1.93			3.86	
Bottom bent up	2		2.03			4.06	
Stirrups nos $= \dfrac{1.75}{0.30} + 1 = 6$	6	0.68			4.08		
Verandah beam							
Bottom straight	2			4.02			8.04
Bottom bent up	2			4.27			8.54
Anchor bars	2		3.98			7.96	
Stirrups no. $= \dfrac{3.80}{0.30} + 1 = 13$	13	0.91			11.83		
Reinforcement in chajja							
Over door D:1 no. &							
Window W_2: 3 nos							
Total $= 4$							
Nos per *chajja* $= \dfrac{1.30}{0.30} + 1 = 5$							
				C.O	76.59	140.00	16.58

(*continued*)

BAR BENDING SCHEDULE   Continued

Shape of bar	No.	Length in m			Total length in m		
		6 mm	**10 mm**	**12 mm**	**6 mm**	**10 mm**	**12 mm**
				B.F	76.59	140.00	16.58
Total no = $4 \times 5 = 20$	20		1.03			20.60	
	20		1.81			36.20	
Secondary bars Nos per *chajja* = 9 Total no = $4 \times 9 = 36$	36	1.35			48.60		
Over window W_1 Main bars (alternate) Nos per *chajja* $= \dfrac{1.10}{0.15} + 1 = 9$	18		1.03			18.54	
(Total no. $4 \times 9 = 36$)	18		1.81			32.58	
Secondary bars Total no = $4 \times 9 = 36$	36	1.10			39.60		
Over window W_3 (Shape as above) Numbers $= \dfrac{1.70}{0.15} + 1 = 13$	7		1.03				7.21	
	6		1.81			10.86	
Secondary bars	9	1.70			15.30		
Reinforcement in slab Verandah slab Main bars straight (a) Nos $= \dfrac{3.80}{0.30} + 1 = 14$	14		2.48			34.72	
Main bars bent up (a)	14		3.36			47.04	
Secondary bars Nos $= \dfrac{2.30}{0.20} + 1 = 13$	13	3.80			49.40		
Living Room Main bars: straight (c) and (d) length $= 3.80 + 0.22 = 4.02$ m Nos $= \dfrac{3.80}{0.30} + 1 = 14$ both ways: 2×14	28			4.02			112.56
				C.O	229.49	347.75	129.14

(continued)

BAR BENDING SCHEDULE   Continued

Shape of bar	No.	Length in m			Total length in m		
		6 mm	10 mm	12 mm	6 mm	10 mm	12 mm
				B.F	229.49	347.75	129.14
Main bars: bent up (c) length $0.70 + 3.80 + 0.22 + 0.08$ $= 4.80$ m $\text{Nos} = \dfrac{3.80}{0.30} + 1 = 14$	14			4.80			67.20
Main bars: bent up (d) length $0.50 + 3.80 + 0.85 + 0.22$ $= 5.45$ m	14			5.45			76.30
Slab over kitchen Along 3.5 m span Main bars: straight (c) Length $= 3.80 + 0.22 = 4.02$ m $\text{Nos} = \dfrac{4.05}{0.30} + 1 = 14$	14			4.02			62.40 56.27
Main bars: bent up (c) length $= 0.70 + 3.80 + 0.08 + 0.22$ $= 4.80$ m	14			4.80			67.20
Along 3.75 m span Main bars: straight (e) Length $= 4.05 + 0.18$ $= 4.23$ m $\text{Nos} = \dfrac{3.80}{0.40} + 1 = 11$	11		4.23			46.53	
Alternate bars: bent up (e) Length $= 0.80 + 4.05 + 0.18 + 0.08$ $= 5.11$ m	11		5.11			56.21	
Slab over bedrooms 2 nos *Along 3 m span* Main bars: straight (f) Length $= 3.30 + 0.22 = 3.52$ m $\text{Nos} = \dfrac{4.30}{0.30} + 1 = 16$	32			3.52			112.64
Main bars: bent up (f) Length $= 3.30 + 0.80 + 0.22 + 0.08 = 4.40$ m							
				C.O	229.49	450.49	571.15

(*continued*)

BAR BENDING SCHEDULE   Continued

Shape of bar	No.	Length in m			Total length in m		
		6 mm	10 mm	12 mm	6 mm	10 mm	12 mm
				B.F	229.49	450.49	571.15
Front bed $Nos = \dfrac{4.30}{0.30} + 1 = 16$ $Rear\ bedroom = \dfrac{(4.30 - 1.75)}{0.30} + 1 = 9$ Total no. $16 + 9 = 25$	25			4.40			110.00
Main bars: bent up (i) $Nos = \dfrac{1.75}{0.30} + 1 = 7$ Length $= 3.30 + 0.08 + 0.22 = 3.60$ m	7			3.60			25.20
Along 4 m span Main bars: straight (g) Length $= 4.30 + 0.18 = 4.48$ m $Nos\ \dfrac{3.30}{0.40} + 1 = 10$	10		4.48			44.80	
Main bars: bent up (g) Length $4.30 + 0.66 + 0.08 + 0.18$ $= 5.22$ m	10		5.22			52.20	
Over sanitary block Along 2.8 m span Main bars: straight (h) Length $= 3.10 + 0.22 = 3.32$ m $Nos = \dfrac{3.30}{0.30} = 12$	12		3.32			39.84	
Main bars: bent up (h) Length $= 0.90 + 3.10 + 0.90$ $+ 0.22 + 0.08 = 5.20$ m	12			5.20			62.40
Along 3 m span Main bars: straight (f) Length $= 3.30 + 0.22 = 3.52$ m $Nos = \dfrac{3.10}{0.30} + 1 = 12$	12			3.52			42.24
				C.O	229.49	587.33	810.99

(*continued*)

BAR BENDING SCHEDULE   Continued

Shape of bar	No.	Length in m			Total length in m		
		6 mm	10 mm	12 mm	6 mm	10 mm	12 mm
				B.F	229.49	587.33	810.99
Main bars: bent up (f) Length = 3.30 + 0.80 + 0.08 + 0.22 = 4.40 Nos = $\dfrac{3.10}{0.30}$ + 1 = 12	12			4.40			52.80
Extra bars: secondary reinforcement at the top for bent up bars Front to rear (11.40 + 0.20 − 0.10) Length = 11.50 m 2 bars at each end	4		11.50			46.00	
(9.65 + 0.20 − 0.10) = length = 9.75 m At one end: 2 nos	2		9.75			11.50	

From left to right over beds and sanitary block
Length
   = 3.20 + 0.20 − 0.10
   = 3.30 m at 4 nos
in each room
	12		3.30			39.60	

Over verandah living room and kitchen (3.70 + 0.20 − 0.10) = 3.80 m at 4 in each room	12		3.80			45.60	
Add for laps length = 45 × 0.10 = 4.5 m	6		4.5			27.00	

	6 mm	10 mm	12 mm
TOTAL LENGTH	229.49	757.03	863.79
Weight in kg. per m run	0.22	0.62	0.89
Total weight	50.49	469.36	713.14

Grand total weight

6 mm dia.: 50.49

10 mm dia.: 469.36

12 mm dia: 713.24

            1233.09 kg

        Say, 12.33 quintals

Example 6.1 (B)   Column and column footing (Fig. 6.6)

DATA

RCC Column	Size: 0 cm × 40 cm
	Height: 5 m
Reinforcement	Main: 8 nos, 20 mm dia.
	Stirrups: 6 mm dia., double stirrups
	@ 3 cm c/c
RCC Column Footing	Size: 2.0 m × 2.0 m

Depth at the face of column: 0.60 m
Depth at edges: 0.30 m
Reinforcement: 10 mm dia. bars @ 7 cm c/c both ways
Arrangement of the reinforcement is shown in Fig. 6.6

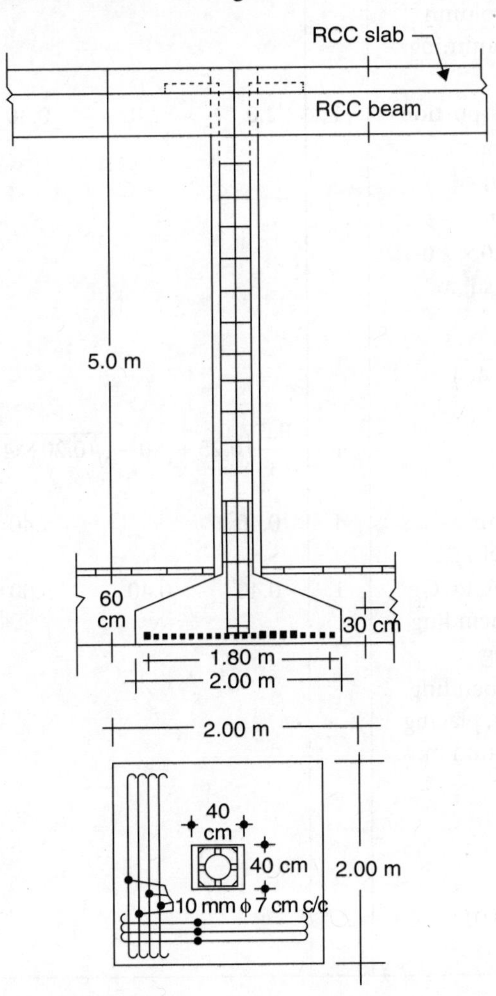

Fig. 6.6   Reinforcement details of column and column footing

Column and Column Footing
MEASUREMENT SHEET

Item No.	Description	No.	Length	Breadth	Depth or height	Quantity	Total
1.	Formwork to column footing						
	Sides	4	2.0		0.30	2.40	
	Inclined sides	4	$\dfrac{2+0.50}{2}\times\sqrt{(0.3)^2+(0.75)^2}$			$\dfrac{4.00}{6.40}$ m^2	6.40 m^2
2.	Providing and laying in position plain cement concrete M150 in column footing including ramming and curing						
	Bottom 30 cm thick portion	1	2.0	2.0	0.30	1.20	
	Above 30 cm portion						
	Area at top $= 0.5\times0.5$ $\qquad\quad = 0.25$ m^2						
	Area at bottom $= 2.0\times2.0$ $\qquad\qquad\quad = 4$ sq. m						
	$\therefore$ Volume $=\dfrac{d}{3}\left(A_1+A_2+\sqrt{A_1A_2}\right)$						
		1	$\dfrac{0.3}{3}\left(0.25+4.0+\sqrt{(0.20\times4)}\right)$			$\dfrac{0.525}{1.725 \text{ m}^3}$	1.73 cm
3.	Formwork to column Height $= (5.00-0.60)$	4	0.40		4.40	7.04 m^2	7.04 m^2
4.	Providing and laying PCC M 150 in column including ramming and curing	1	0.40	0.40	5.00	0.80 cu. m	0.80 cu. m
5.	Providing, cutting, bending, hooking, conveying, placing and binding in position m. s. reinforcement						
	Column footing						
	10 mm dia. bars						
	Length $= 1.80+2\times9\times0.01$ $\qquad\; = 1.98$ m						

(continued)

***Example 6.1* (A) MEASUREMENT SHEET**   Continued

Item No.	Description	No.	Length	Breadth	Depth or height	Quantity	Total
	$\text{Nos} = \dfrac{1.80}{0.07} + 1 = 27$						
	in two layers 2×27	54	1.98	@0.62	kg/m	66.42 kg	
	Column bars	8	5.45	@2.47	kg/m	107.69	
	20mm dia						
	$0.50 + 5.60 + 0.50$						
	$-$(cover at bottom and top						
	$= 0.05 + 2 \times 0.01 + 0.08$)						
	$= 6.60 - 0.15 = 5.45$						
	Lap length	8	0.75	@2.47	kg/m	14.82	
	Stirrups in height						
	$4.40 + 0.60 - 0.07$ m						
	$\text{No.} = \dfrac{(5.00 - 0.07)}{0.30}$						
	$\times (2) + 1 = 35$ nos						
	Length of one stirrup	35	1.31	@0.22	kg/m	10.09	
	$4 \times 0.30 + 0.11$						
	for hook $= 1.20 + 0.11$						
	$= 1.31$ m						
	Total weight					199.02 kg	
	Add 5% waste					9.95 kg	
	Grand total					208.97 kg	
						Say 209 kg	2.09 q

Stone Masonry and Brickwork

7.1 GENERAL

It is obvious that the unit rate of stone masonry or brickwork will increase with the increase in the height of the structure because of increased handling charges. The stone masonry or brickwork is therefore classified into the following categories, the measurements under each category being kept in separate items.

1. Stone masonry or brickwork from foundation up to plinth level.
2. From plinth level up to first floor level.
3. From first floor level to second floor level, and so on for each subsequent floor.

The parapet wall is included with the wall of the storey immediately below it.

Further classification of masonry or brickwork and the mode of measurement are explained in the subsequent sections.

7.2 STONE MASONRY

Stone masonry is primarily classified into two main sections:

1. stone work, and
2. stone walling.

Measurements under each of the above sections are kept separate.

7.3 STONEWORK

Under this category is grouped the stone masonry in sills steps, columns, caps, copings, lintels, and the like. Stonework is measured in cubic metres. The length, breadth and height of the smallest rectangular dressed block from which the finished dressed stone can be worked are considered for finding out the cubic contents. For example, measurements of a dressed circular stone column shown in Fig. 7.1 will be:

Length: 30 cm; Breadth: 30 cm; Height: 300 cm.

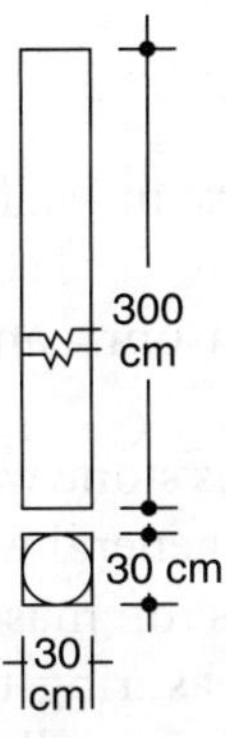

Fig. 7.1

The item includes the labour involved for dressing on rectangular faces, beds and joints. All other labour is considered 'extra-over' and measured separately in square metres as facework. The facework in the following cases is measured in separate items:

1. dressing hard stones like granite, trap and so on,
2. dressing executed on stones already built in position,
3. circular work having width more than 7.5 cm, and
4. circular work as in domes and so on.

The cost of stonework per unit volume will depend upon the size of the stones. The stonework in general is kept separate according to the following sizes or, alternatively, volumes.

Length of each stone	(or volume of each stone)
(1) Up to 1m	Up to 60 dm^3
(2) More than 1 m but less than 2 m	Exceeding 60 dm^2 but less than 180 dm^3
(3) More than 2 m but less than 4 m	Exceeding 180 dm^3 but less than 360 dm^3
(4) Exceeding 4 m in length	Exceeding 360 dm^3

String courses and cornices which are grouped under stonework are measured in running metres. The additional labour for dressing is however measured in square metres as usual.

7.4 STONE WALLING

Stone masonry in footings, walls (including battered or tapered) and in chimney breasts, chimney stack, smoke or air flues is grouped with the 'general walling'.

Stone walling is classified into the following three categories:

1. general item of walling,
2. facework, and
3. angles in facings.

1. *General walling*: Stone masonry work in walls is further classified as below and kept separate:

(i)   uncoursed or random rubble masonry, and
(ii) coursed rubble masonry.

Under each of the above classifications stone walling curved on plan up to a mean radius of 6 m will be kept separate from the general walling.

In calculating the cubic contents of masonry (or brickwork) in superstructure, beginners may have difficulty in measuring the external walls, the tops of which are sloping and also the gable portions of the walls.

The height of the external wall above the plinth normally shown on the drawing is the height of the lower external face as shown in Fig. 7.2.

To get the correct volume, the mean height along the centre of the wall should be considered (see Fig. 7.3.).

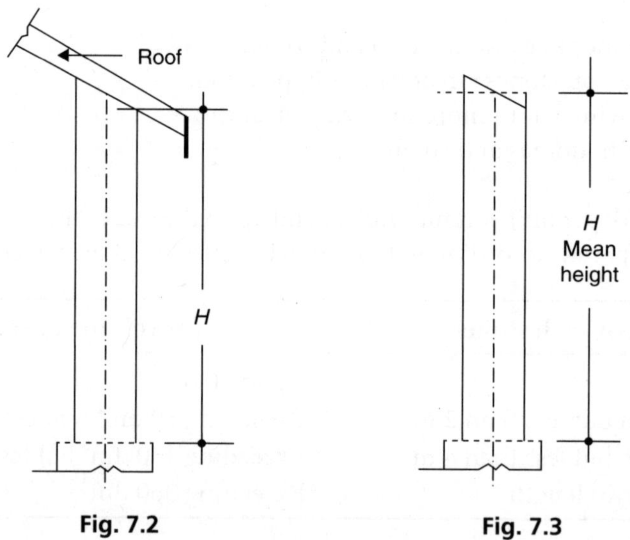

Fig. 7.2   **Fig. 7.3**

The higher inside face compensates for the lower outside face as shown by the dotted line in Fig. 7.3.

In the case of the gable ends, the wall should be divided into two parts (Fig. 7.4). The lower part is measured along with the external walls as usual and the second triangular portion should then be entered as:

$$\frac{1}{2} \times \text{Base} \times \text{Width} \times \text{Height of triangle}$$

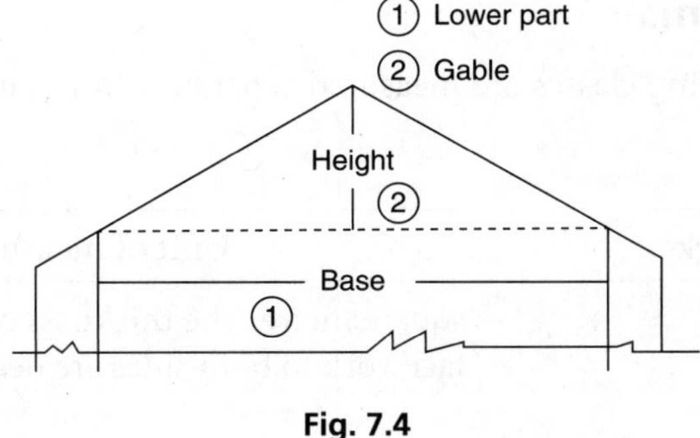

Fig. 7.4

From the gross volume of stone walling all the openings are deducted except for the following:

   (i) opening with area not exceeding 0.1 m^2,
  (ii) wall plates, bed plates and bearing of slabs, and the like not exceeding 10 cm in thickness and the bearing not extending over the full thickness of the wall,
 (iii) ends of dissimilar materials such as beams, lintels, trusses, joists, purlins, rafters and so on, if their cross-sectional area is less than 500 sq cm. For example, measurements for deductions of the window opening shown in Fig. 7.5 would be recorded as below:

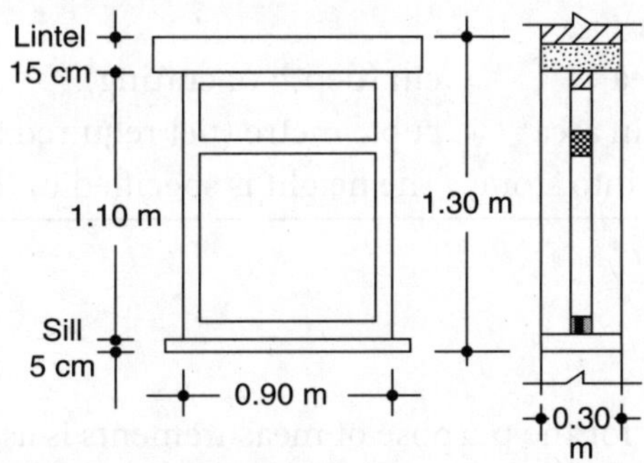

Fig. 7.5

No.		Length		Breadth		Height
1	×	0.90 m	×	0.30	×	1.30

It would be noted that the ends of lintel and sill are not deducted.

2. *Facework*: Additional work involved in dressing the stones is described as an extra-over walling and measured separately in square metres. As the unit rate of dressing will depend upon the type of dressing, measurements of each type of dressing are kept separate.

3. *Angles in facing*: External angles in facework are measured separately in running metres. The size of the quoin should be specified in the description of the item.

Miscellaneous Items

Works under the following classes are measured separately in the units mentioned against each of them.

Item of work	Unit of measurement
1. Stone-nogging	square metre. The thickness of wall and the facework to both sides are described.
2. Wall steining	
(a) generally	cubic metre. Mean radius is specified.
(b) sinking of steining	running metre.
3. Arches	
(a) generally	cubic metre including centering for spans up to 6 m.
(b) centering	square metre for spans exceeding 6 m.
(c) facings	square metre (if different from general facework).
4. Under-pinning	
(a) general	cubic metre.
(b) extra labour and material to the existing structure	square metre (length $\times$ width of top course).
5. Cuttings and openings	
(a) up to 0.1 m^2 in area	cm (depth of cutting).
(b) exceeding 0.1 m^2 in area	cubic metre (net required area $\times$ thickness).
6. Figures and letters cut into stone	The height is specified in the description.

7.5 BRICKWORK

Classification of brickwork for the purpose of measurements is as below:

1. *General brickwork*:   This includes footings, walls, brickwork in chimney stacks, smoke or air flues, pilasters, plain copings and sills.
   The following work will be measured separately:
   (i) fire brickwork,
   (ii) independent chimney shafts, and
   (iii) brickwork curved on plan to a mean radius not exceeding 6 m.
   The battered or tapered walls are included with general brickwork but the battered or tapered surface is to be measured as 'extra-over' in square metres.
2. *Brickwork* in corbels, string courses, projecting pilasters and friezes is measured in running metres.
3. *Half-brick thick walls*: Walls 10 cm or less in thickness are measured in square metres. Reinforcement, if any, is measured separately in quintals.

4. *Pillars*: If the ratio of sides of brickwork is equal to or less than 3:1, the work is called pillars and measured separately in cubic metres. Where the sizes and shapes of pillars are different, an additional item enumerating the pillars will be included in the estimate.

5. *Brickwork in backing to masonry*: Where the wall has stone facing and brickwork in the backing, the brickwork is measured separately in cubic metres. Average thickness of brickwork is considered for finding the volume of brickwork.

6. *Honeycomb brickwork*: Honeycomb brickwork is measured in square metres. The gross area is found out neglecting the honeycomb openings.

7. *Brick-nogging*: If the thickness is 10 cm or less brickwork is measured in square metres. If the thickness is more it is included with general brickwork neglecting the volume occupied by timber framing.

   Timber framing is measured separately in cubic metres.

8. *Brickwork in arches*: Brickwork in arches is kept separate from general brickwork and measured in cubic metres. Centering is measured separately in square metres, only if the span is more than 6 m.

The general procedure for calculating the net cubic contents is similar to that mentioned in stonework (Section 7.4).

Example 7.1   MASONRY AND BRICKWORK: In continuation of Example 5.1 (See Fig. 7.6)

MEASUREMENT SHEET

Item No.	Description	No.	Length	Breadth	Depth or height	Quantity	Total
1.	Providing uncoursed rubble masonry in c.m. (1 : 6) in foundation and plinth including curing **Long walls** Beds, bath and WC (11.40 + 0.50) = 11.90 m	2	11.90	0.50	0.65	7.74	
	Living room and kitchen (9.65 + 0.50) = 10.15 m	1	10.15	0.50	0.65	3.30	
	Short walls Beds, bath and WC (3.20 − 0.50) = 2.70 m	4	2.70	0.50	0.65	3.51	
	Verandah, living and kitchen (3.70 − 0.50) = 3.20 m	4	3.20	0.50	0.65	4.16	

(continued)

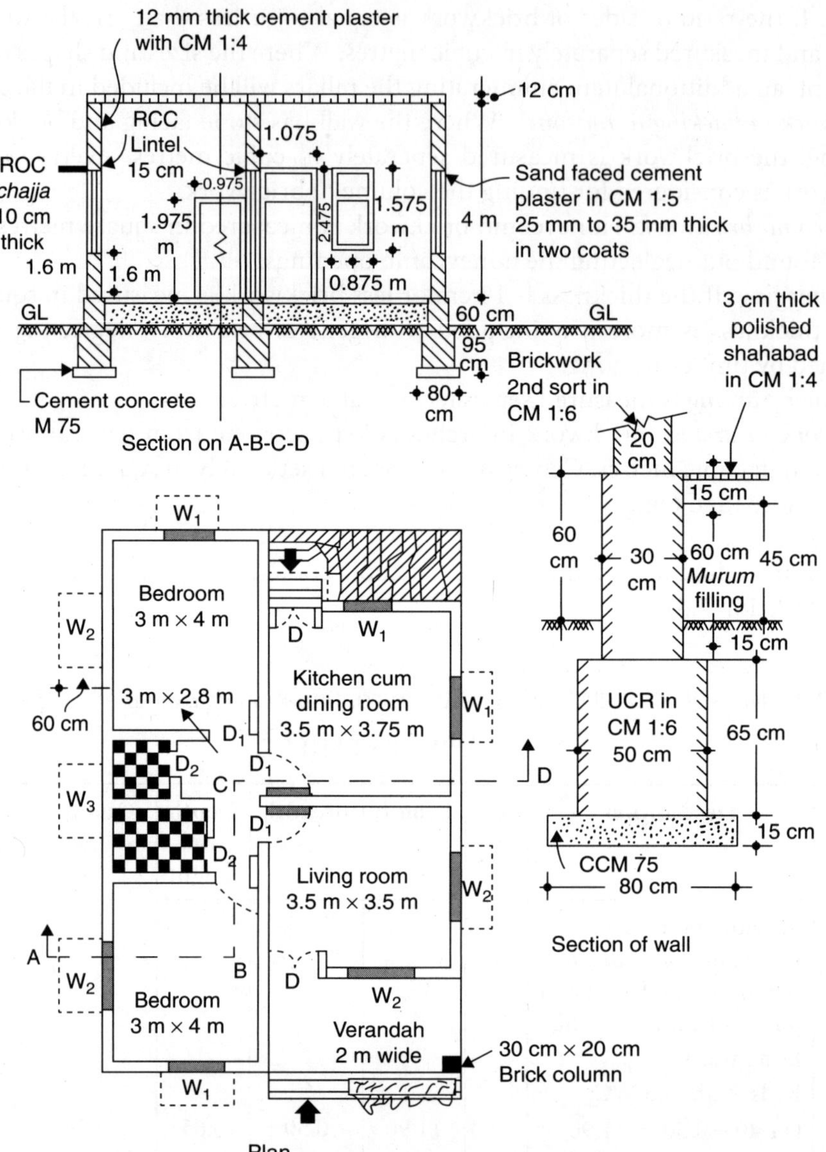

Fig. 7.6   Example 7.1: Masonry and brickwork

Example 7.1   Continued

Item No.	Description	No.	Length	Breadth	Depth or height	Quantity	Total
	Plinth Course						
	Long walls						
	Beds, bath and WC						
	$(11.40 + 0.30) = 11.70$ m	2	11.70	0.30	0.75	5.27	
	Living room and kitchen						
	$(9.65 + 0.30) = 9.95$ m	1	9.95	0.30	0.75	2.24	
	Short walls						
	Bed, bath and WC						
	$(3.20 - 0.30) = 2.90$	4	2.90	0.30	0.75	2.61	
	Verandah, living and kitchen						
	$(3.70 - 0.30) = 3.40$ m	4	3.40	0.30	0.75	3.06	
						31.89 cu. m	31.89 cu. m
2.	External angles in facings						
	$H = (0.65 + 0.75)$	5	1.40			7.00 m	7.00 m
	$\quad = 1.40$ m						
3.	Providing brickwork II sort in cm (1 : 6) proportion, in superstructure including scaffolding and curing						
	Long walls						
	Beds, bath and WC						
	$(11.40 + 0.20) = 11.60$ m	2	11.60	0.20	4.00	18.56	
	Living room and kitchen						
	$(3.70 + 3.95 + 0.20)$	1	7.85	0.20	4.00	6.28	
	$= 7.85$ m						
	Short walls						
	Bed, bath and WC						
	$(3.30 - 0.20) = 3.00$ m	4	3.00	0.20	4.00	9.60	
	Living room and kitchen						
	$(3.70 - 0.20) = 3.50$ m	3	3.50	0.20	4.00	8.40	
						42.84	
	Deductions						
	Door openings (includings lintels)						

(*continued*)

Example 7.1 Continued

Item No.	Description	No.	Length	Breadth	Depth or height	Quantity	Total
	Doors: D	2	1.10	0.20	2.65	1.17	
	D_1	4	1.00	0.20	2.15	1.72	
	Windows W_1	4	0.90	0.20	1.75	1.26	
	W_2	4	1.10	0.20	1.75	1.54	
	W_3						
	$L = (0.70 + 0.10 + 0.70)$						
	$= 1.50$ m	1	1.50	0.20	1.15	0.35	
						6.04	
						36.80 cu. m	36.80 cu. m
4.	Providing brickwork II sort in cm $(1:6)$ prop. in columns in superstructure including scaffolding and curing Verandah						
	Height $= 4.0$ m $+ 12$ cm thickness of slab -0.30 m depth of beam $= 3.82$ m	1	0.30	0.20	3.82	0.23 cu. m	0.23 cu. m
5.	Providing 10 cm thick brickwork in cm $(1:4)$ including m. s. reinforcement of 6 mm dia. 2 bars at every third course.						
	Partition wall between WC and bath						
	Length $(1.80 + 0.10$ $= 1.9$ m)	1	1.90		4.00	7.60	
	Front wall of bath	1	1.50		4.00	6.00	
	Front wail of WC	1	1.20		4.00	4.80	
						18.40	
	Deductions						
	Door D_2	2	0.80		2.05	3.28	
						15.12 sq. m	15.12 sq. m

Doors and Windows

8.1 GENERAL

The current practice, in India, is to measure the doors and windows by the metre superficial, the width and height of the opening being considered to find out the area of doors and windows. The estimator, while analysing the price per unit area has to calculate the woodwork in cubic metres including allowances for joints and waste, cost of fixtures and the cost of painting. Thus, although the work of the taker off is simplified, much is left for the estimator in deciding a unit price. Secondly, frames of doors and windows are fixed at an early stage in the construction of a work (along with the masonry or brickwork) whereas the shutters are fixed when the work is nearly completed. Thus, a contractor has to be paid for at part rate for the cost of frames, the full rate being paid only on completion of the whole item. These drawbacks are eliminated by breaking up the work into different items. According to the procedure laid down in the Standard Method of Measurement of Building Works (IS: 1200–1965), the woodwork and joinery are measured by splitting the work into the following items:

1. frames of doors, windows, cupboards and so on,
2. doors and window leaves,
3. fitting to frames and fixing or hanging with hinges and so on—labour only,
4. fixtures and fastenings (ironmongery),
5. painting.

Steel doors and windows are measured separately.

8.2 FRAMES OF DOORS AND WINDOWS

The frames of doors and windows are measured in cubic metres. For working out the cubic contents, the sectional area of least square or rectangle from which the mouldings and so on may be cut, is considered. To the sight length, the lengths of tenons, horns and others are added.

Portions of frames, segmental or circular in shape, are measured separately.

8.3   DOORS AND WINDOW LEAVES

The wood work and joinery of doors and window leaves is measured as area superficial and expressed in the unit of square metre. It is obvious that unit price will vary with the type of joinery and the thickness of the shutter. One item should, therefore, include shutters having only one type of joinery and uniform thickness. Thus, if in a building, the shutters of doors or windows are of three types, namely, fully panelled, partly panelled and partly glazed, framed and louvered, then there would be three separate items covering the shutters of each type.

A schedule of doors and windows should be prepared, if it is not included on the drawings. The schedule will include the doors and windows classified according to the type of joinery. Total number of doors and windows of each type, on each floor should also be stated in the schedule. If the schedule is properly prepared, it would not be necessary for the taker off to refer to the drawing while taking out the quantities of doors and windows.

It must be noted by the beginner that unless the type of joinery or thickness of the shutter changes, a separate item is unwarranted for a mere change in the sizes of doors and windows or thickness of walls in which they are fitted.

8.4   FIXING OR HANGING OF LEAVES TO FRAMES

The labour involved in fitting to frames and fixing or hanging the shutters with hinges and so on is measured separately. The unit of measurement in both cases is the area of shutters in sq. m.

8.5   IRONMONGERY

The various articles of ironmongery are enumerated items and only such articles which are of the same material, type and size are included in one item.

8.6   PAINTING

Painting of woodwork in doors, windows and cupboards is measured by the metre superficial. The preparatory work such as priming is included in the item and special colour or finishes are stated in the description of the item.

For simplifying the measurement, the areas of the uneven surfaces such as panelled, framed and braced and so on, are converted into equivalent plain area by multiplying the plain area including frame by a multiplying factor.

The multiplying factors for some types of doors and windows are given in Table 8.1.

Table 8.1 Equivalent plain areas of uneven surfaces

S. No.	Type of doors and windows	Multiplying factor for each side
1.	Fully panelled	$1\frac{1}{8}$
2.	Fully glazed	$\frac{1}{8}$
3.	Partly panelled and partly glazed	1
4.	Flush	1
5.	Ledged and battened or ledged, battened and braced	$1\frac{1}{8}$
6.	Fully venetianed or louvered	$1\frac{1}{2}$
7.	Trellis (no deductions for openings)	2
8.	Steel roller shutter	$1\frac{1}{4}$
9.	Collapsible gates and expanded metal (painting all exposed faces)	$\frac{1}{2}$

Painting to steel doors and windows is kept separate.

8.7 STEEL DOORS AND WINDOWS

Steel doors and windows are classified as below:

1. *Collapsible gates:*   These are measured by metre superficial or alternatively by weight in quintals.
2. *Rolling shutters:*   Rolling shutters are measured by metre superficial of the opening covered.
3. *Steel doors and windows:*   These are measured by metre superficial of the area covered. The sizes of the various members are specified in the description of the item.

The above items include fixing, hanging sand fastenings and these need not be measured separately. Painting is measured as described in Section 8.6.

8.8 GRILLS

Openings of windows are sometimes provided with grills, bars or expanded metal work. In such cases grills and bars are measured separately by quintal weight or in sq. metres. Expanded metal work is measured in sq. metres. Net area of expanded metal work will be worked out after deducting openings exceeding 0.2 m^2, if any.

***Example* 8.1**   In continuation of Example 7.1 (Fig. 8.1)

DATA
Schedule of doors and windows

Type	Particulars	Size	Number
D	Fully panelled double shutter with ventilator	107.5×247.5	2
D_1	Flush door, single shutter	97.5×197.5	4
D_2	Ledged, braced and framed door, single shutter	77.5×187.5	2
W_1	Fully panelled with ventilator fully glazed, double shutter	87.5×157.5	4
W_2	-do-	107.5×157.5	4
W_3	Louvered	67.5×97.5	2

***Example* 8.1**   (Continued)

MEASUREMENT SHEET

Item No.	Description	No.	Length	Breadth	Depth or height	Quantity	Total
1.	Providing and fixing country teak wood frames 70 mm × 95 mm size scantlings for second class doors and windows *Door D type* Vertical posts Length 247.5 + 12 cm for shoe = 259.50 cm	4	2.60	0.07	0.095	0.069	
	Head Length = 107.5 cm + 30 cm for horns = 137.5 cm	2	1.38	0.07	0.095	0.068	
	Transom and sill Length = 107.5 say 1.08 m	4	1.08	0.07	0.095	0.029	
	Door D_1 type Vertical posts Length = 97.5 cm + 12 cm	8	2.10	0.07	0.095	0.112	
	Head length = 97.5 cm + 30 cm for horns = 127.5 cm	4	1.28	0.07	0.095	0.034	

(*continued*)

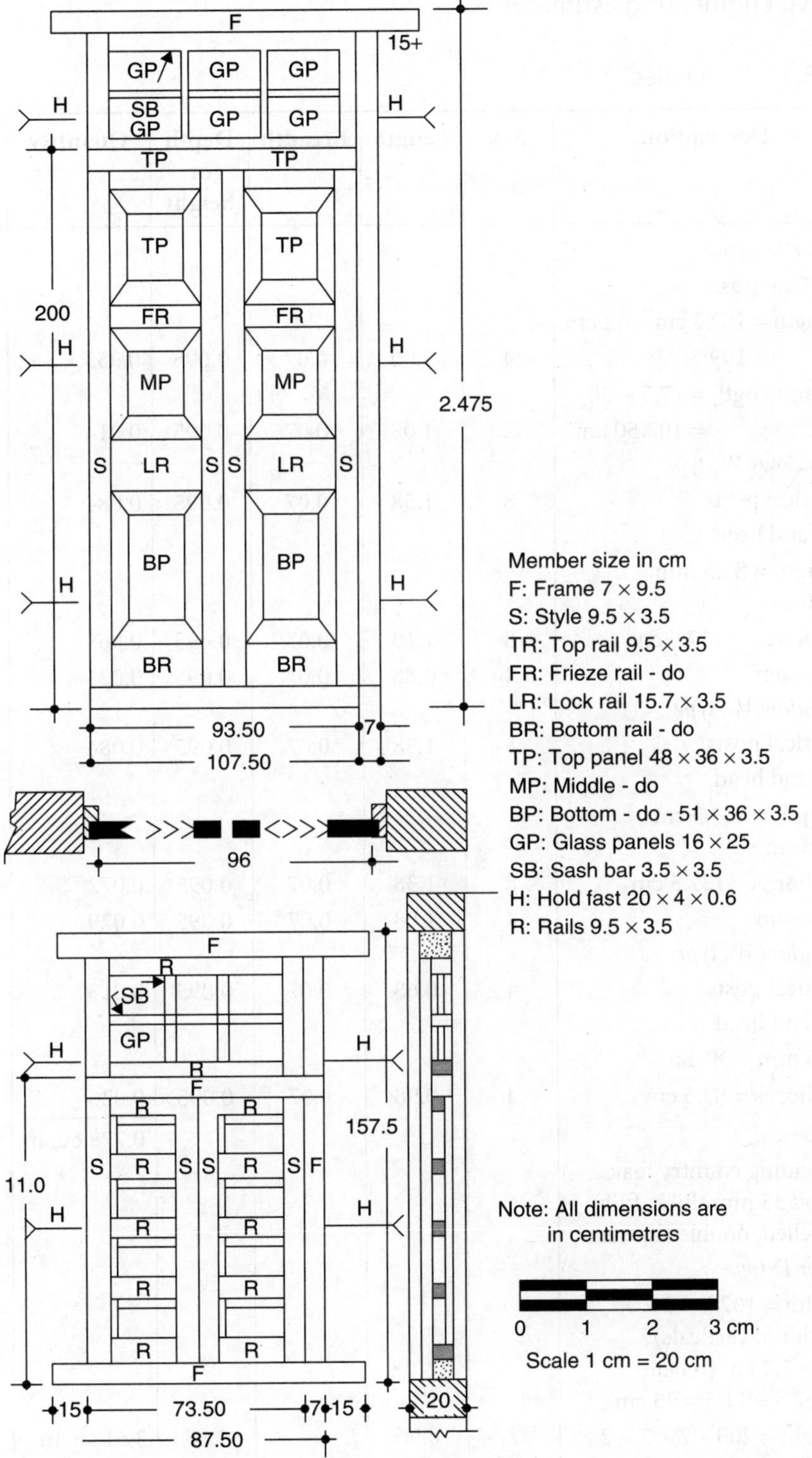

Fig. 8.1 Example 8.1: Doors and windows

***Example* 8.1**   (Continued)

Item No.	Description	No.	Length	Breadth	Depth or height	Quantity	Total
	Door D_2 type						
	Vertical posts						
	Length = 187.5 cm + 12 cm						
	= 199.5	4	2.00	0.07	0.095	0.053	
	Head Length = 77.5 + 30						
	= 107.50 cm	2	1.08	0.07	0.095	0.014	
	Windows W_1 type						
	Vertical posts	8	1.58	0.07	0.095	0.084	
	Sill and head						
	Length = 87.5 mm						
	+ 30 cm						
	for horns = 117.5 cm	8	1.18	0.07	0.095	0.063	
	Transom	4	0.88	0.07	0.095	0.023	
	Window W_2 type						
	Vertical posts	8	1.58	0.07	0.095	0.084	
	Sill and head						
	Length = 107.5 cm						
	+ 30 cm						
	for horns = 137.5 cm	8	1.38	0.07	0.095	0.072	
	Transom	4	1.08	0.07	0.095	0.029	
	Window W_3 type						
	Vertical posts	4	0.68	0.07	0.095	0.018	
	Sill and head						
	67.5 mm + 30 cm						
	for horns = 97.5 cm	4	0.98	0.07	0.095	0.026	
						0.778 cu. m	0.78 cu. m
2.	Providing country teak wood 35 mm thick, fully panelled, double shutters						
	Door D type						
	Width = 107.5 − 2 × 7						
	(width of verticals)						
	+ 2 × 1.2 cm (rebate)						
	= 107.5 − 11.6 = 96 cm						
	Height = 200 − 2 × 7 + 2 × 1.2 (for rebate) = 188.5 cm	2	0.96		1.88	3.61 sq. m	3.61 sq. m

(*continued*)

***Example* 8.1**   (Continued)

Item No.	Description	No.	Length	Breadth	Depth or height	Quantity	Total
3.	Providing solid core flush door in single leaf 35 mm thick Width = $97.5 - 2 \times 7 + 2 \times 1.2$ for rebate = 86 cm Height = $197.5 - 7.00 - 0.5 + 1.2 = 191.2$ cm	4	0.86		1.91	6.56 sq. m	6.56 sq. m
4.	Providing ledged, braced and framed door shutters Width = $77.5 - (2 \times 7) + (2 \times 1.2) = 66$ cm Height = $187.5 - 7 - 0.5 + 1.2 = 181$ cm	2	0.66		1.81	2.38 sq. m	2.38 sq. m
5.	Providing country teak wood 35 cm thick fully panelled shutters for windows W_1 *and* W_2 *type* Width of W_1 type shutter $= 87.5 - (2 \times 7) + (2 \times 1.2) = 76$ cm Height = $110.00 - (2 \times 7) + (2 \times 1.2) = 98$ cm	4	0.76		0.98	2.98	
	Width of W_2 type shutter $107.5 - (2 \times 7) + (2 \times 1.2) = 96$ cm Height = 0.98 m as above	4	0.96		0.98	3.84	
						6.82 sq. m	6.82 sq. m
6.	Providing 6 mm thick glass louvered window shutters Width = $67.5 - (2 \times 7) + (2 \times 1.2) = 56$ cm Height = $97.5 - (2 \times 7) + (2 \times 1.2) = 86$ cm	2	0.56		0.86	0.96 sq. m	0.96 sq. m
7.	Providing fully glazed ventilators						

(*continued*)

***Example* 8.1** (Continued)

Item No.	Description	No.	Length	Breadth	Depth or height	Quantity	Total
	Over door D						
	Width = 96 cm as above	2	0.96		0.43	0.83	
	Height = $47.5 - 7 +$ $(2 \times 1.2) = 43$ cm						
	Over windows W_1						
	Height = $47.5 - 7 +$ $(2 \times 1.2) = 45$ cm	4	0.76		0.43	1.31	
	Over window W_2	4	0.96		0.43	1.65	
						4.75 sq. m	4.75 sq. m
8.	Hanging the shutters of doors and windows with hinges to frames						
	Door D	2	0.96		1.88	3.61	
	Door D_1	4	0.86		1.91	6.56	
	Door D_2	2	0.66		1.81	2.38	
	Window W_1	4	0.76		0.98	2.98	
	Window W_2	4	0.96		0.98	3.84	
	Ventilators over						
	Door D	2	0.96		0.43	0.83	
	Window W_1	4	0.76		0.43	1.31	
	Window W_2	4	0.96		0.43	1.65	
						23.16 sq. m	23.16 sq. m
9.	Fixing the shutters to frames		0.56		0.86	0.96 sq. m	0.96 sq. m
10.	Providing and fixing holdfasts 20 cm × 4 cm × 4 mm to frames with screws						
	Doors						
	D 2 × 6 = 12	12				12	
	D_1 4 × 6 = 24	24				24	
	Windows						
	W_1 4 × 4 = 16	16				16	
	W_2 4 × 4 = 16	16				16	
	W_3 2 × 4 = 8	8				8	
						76 nos	76 nos

(*continued*)

***Example* 8.1**   (Continued)

Item No.	Description	No.	Length	Breadth	Depth or height	Quantity	Total
11.	Providing and fixing with screws 10 cm size brass butt hinges Doors						
	$D = 2 \times 2 \times 3 = 12$ nos	12				12	
	$D_1 = 4 \times 3 = 12$ nos	12				12	
	$D_2 = 2 \times 3 = 6$ nos	6				6	
	Window						
	$W_1 = 4 \times 2 \times 3 = 24$ nos	24				24	
	$W_2 = 4 \times 2 \times 3 = 24$ nos	24				24	
						78 nos	78 nos
12.	Providing and fixing with screws brass tower bolts: 20 cm size						
	Doors D $2 \times 2 = 4$	4				4	
	$D_1 = 4 \times 1 = 4$	4				4	
	$D_2 = 2 \times 1 = 2$	2				2	
	Window						
	$W_1 = 4 \times 2 \times 2 = 16$	16				16	
	$W_2 = 4 \times 2 \times 2 = 16$	16				16	
						42 nos	42 nos
13.	Providing and fixing brass handles						
	Total no @ 2 nos per shutter except W_3	26				26 nos	26 nos
14.	Providing and fixing in position brass aldrop						
	To doors @ one per door	8				8 nos	8 nos
15.	Providing and fixing in position stoppers to doors	10				10 nos	10 nos
16.	Providing and applying two coats of oil paint to woodwork						

(*continued*)

***Example* 8.1**   (Continued)

Item No.	Description	No.	Length	Breadth	Depth or height	Quantity	Total
	Door D both faces	$1\frac{1}{8} \times 2$	1.08		2.08	5.06	
	Ventilator " "	$1\frac{1}{2} \times 2$	1.08		0.48	1.44	
	Door D_1 " "	8	0.98		1.98	15.52	
	D_2 " "	$1\frac{1}{8} \times 8$	0.88		1.10	8.71	
	Window W_1 both faces	$1\frac{1}{8} \times 8$	1.08		1.10	8.71	
	Ventilator over W_1 both faces	$\frac{1}{2} \times 8$	0.88		0.48	1.69	
	W_2	$1\frac{1}{8} \times 8$	1.08		1.10	10.69	
	Ventilator over W_2	$\frac{1}{2} \times 8$	1.08		0.48	2.07	
	W_3 to frame Girth = 7 + 7 + 9.5 cm $\quad$ = 23.5 cm $\quad$ = say 24 cm						
	Verticals	2	0.24		0.98	0.47	
	Top and bottom	2	0.24		0.68	0.33	
						54.69 sq. m	54.69 sq. m

9 | Roofs

9.1 GENERAL

The measurement of roof is divided into two main sections:

1. Woodwork
   (i) trusses,
   (ii) purlins, common rafters, ridge plate, wall plate, eaves board and barge board,
   (iii) ceiling, and
   (iv) painting.
2. Coverings
   (i) main roof,
       (a) battens and tiles,
       (b) steel sheeting,
       (c) asbestos cement sheeting,
       (d) terraced roof,
   (ii) ridge, valleys and hips, and
   (iii) rain water goods.

Before taking out quantities it is advisable to prepare the plan of roof, if it is not provided. One such plan is shown in Fig. 9.1.

9.2 WOODWORK

Woodwork and joinery in roof are measured in cubic metres except the battens and ceiling which are measured in square metres.

1. *Woodwork in roof trusses* is kept separate. While measuring the length of a member in a roof truss, lengths of tenon and scarf are added to the sight lengths.
2. Purlins, common rafters, ridge plate, wall plate and hip and valley rafters are all measured in one item. The cross-sectional dimensions of these members are invariably shown on the drawings. The lengths of ridge, wall plate and purlins can easily be found out from the roof plan. The calculation of number and lengths of common rafters, hips, valleys and roof slopes is explained below.

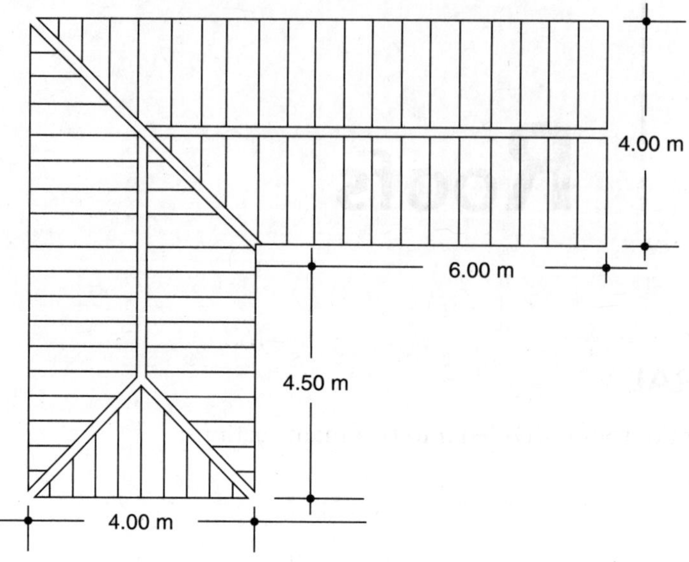

Fig. 9.1

(i)   *Common Rafters*

The number of rafters are calculated by taking the total length between end rafters, dividing it by centre-to-centre spacing of rafters, and adding one to the result because the division by the centre-to-centre spacing gives the number of spaces between rafters. The number of rafters may be measured in the same way even in the case of a hipped end also. The lengths of jack rafters on the hipped end are equal to the lengths (shown dotted in Fig. 9.2) which would have been there if the roof were gabled. However, if the central rafter on the hipped end is entirely extra it should be added to calculate the total number of rafters. The total number of rafters to one slope of the first section of the roof shown in Fig. 9.l would be:

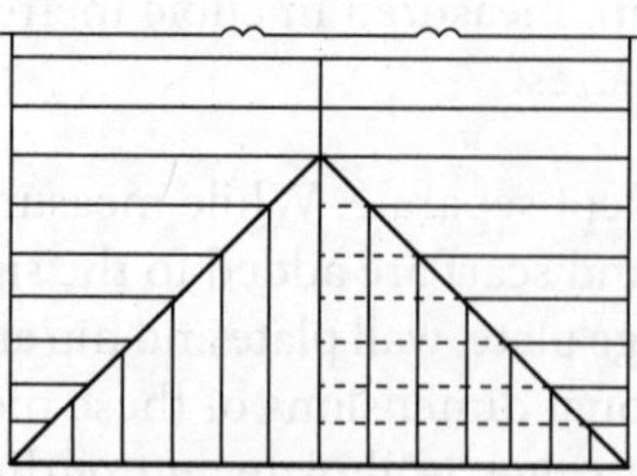

Fig. 9.2

Length of roof = 4.5 metres or 450 cm

Centre-to-centre distance between rafters = 45 cm

$$\text{The total number of rafters} = \frac{450}{45} + 1 = 11$$

There are two roof slopes. The total number of rafters would be: $2 \times 11 = 22$

Add one for central rafter on hipped end   1

Total rafters   23

Similar rafters on the second portion of the roof will be worked out by dividing 10 m by 45 cm and adding one extra rafter: Thus, $1000 \div 45 + 1 = 24$ rafters. There are two roof slopes. The total number of rafters will be: $2 \times 24 = 48$. The length of a common rafter is found by multiplying the natural secant of the angle of pitch by half the total span of the roof. The natural secant of more usual pitches of roof are as follows:

Pitch of roof	26°30′	30°	33°30′	35°
Natural secant	1.118	1.155	1.199	1.221

For example, the length of the common rafter shown in Fig. 9.3 will be calculated thus:

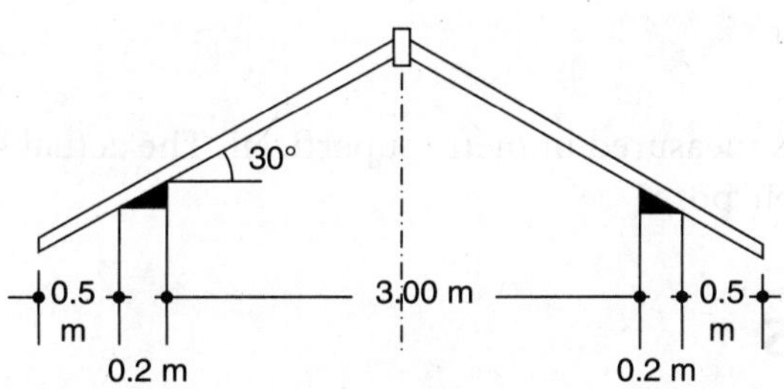

Fig. 9.3

Half total span of roof = Half clear span + thickness of wall
+ projection beyond the face of wall

= 1.50 + 0.20 + 0.50 = 2.2 m

Length of rafter = $2.2 \times \sec 30°$

= $2.2 \times 1.155 = 2.54$ m

(ii) *Hips and Valleys*

The length of a hip or valley rafter has to be calculated from a triangle. Its length is the hypotenuse of a triangle with the base equal to the horizontal length of the hip or valley on plan and height equal to the height of the roof. The best method is to plot this triangle on the roof plan and scale the required length of the hypotenuse as shown in Fig. 9.4.

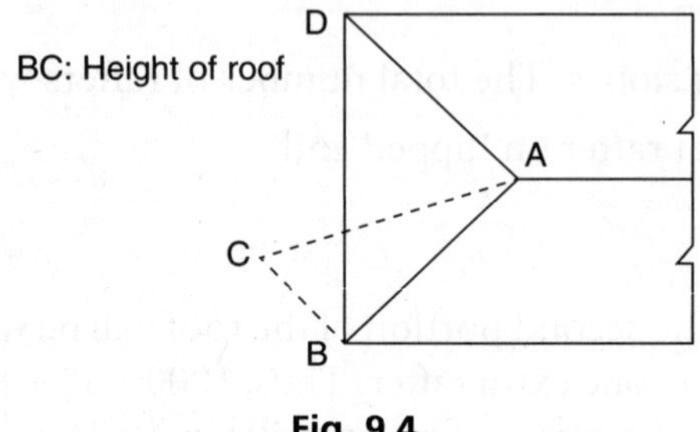

Fig. 9.4

The length AC represents the length of the hip.

9.3   CEILINGS

The supporting woodwork required for the ceiling is measured in cubic metres in the general item of woodwork in a roof. The boarding in ceiling is given in metre superficial. The actual area is measured from the plan. No deductions are made for openings not exceeding 0.4 m^2 each.

9.4   PAINTING

Painting to woodwork is measured in metre superficial. The actual surface area painted has to be measured for the purpose.

9.5   COVERINGS

The covering of a roof slope is a superficial item. Different types of coverings are measured separately. Flat area is measured and no allowance for laps and so on, is made.

Openings up to 0.4 m^2 are not deducted.

In the case of tiled roof, battens are included in the item and no separate measurement of battens is necessary.

The main item of covering is followed by the lineal roofing items measured in running metres. These items may be:

1.   ridges, hips and valleys,
2.   eaves tiles bedded in mortar on walls, and
3.   screwing eaves tiles to battens.

These items are then followed by enumerated items such as glass tiles, ventilating tiles, and so on.

9.6 RAIN WATER GOODS

Gutters and pipes are classified according to the material and diameter and measured in running metres. All bends, elbows, offsets, drop ends and stopped ends are measured as 'extra-over' the length. Lengths of these fittings are not deducted.

Painting of rainwater pipes, gutters and so on is measured in running metres.

Example 9.1 A pitched roof (Fig. 9.5)

MEASUREMENT SHEET

Item No.	Description	No.	Length in m	Breadth in m	Depth or height in m	Quantity	Total
1.	Providing country teak woodwork in roof trusses including joinery, iron fixtures and erecting the trusses in position. Quantity of wood for four trusses Principals nos. = $4 \times 2 = 8$ Length = 2.5 sec 30°						
	$\qquad$ = 2.885 m	8	2.89	0.08	0.13	0.240	
	Ties: nos $4 \times 1 = 4$	4	5.60	0.08	0.13	0.233	
	Struts nos = $4 \times 2 = 8$						
	Length assumed equal to $\frac{1}{2}$ length of principals	8	1.44	0.08	0.08	0.074	
	King posts: Nos $4 \times 1 = 4$ Length = 2.5 tan 30°						
	$\qquad$ = 1.44 m	4	1.44	0.08	0.18	0.083	
						0.63 m^3	0.63 m^3
2.	Providing and fixing in position country teak woodwork in roof Ridge 8.40 + 3.30						
	= 11.70 m	1	11.70	0.08	0.15	0.170	
		1	2.50	0.08	0.15		
	Purlins Purlins over walls	1	14.50	0.08	0.13		
		1	8.10	0.08	0.13		
		1	5.00	0.08	0.13		

(continued)

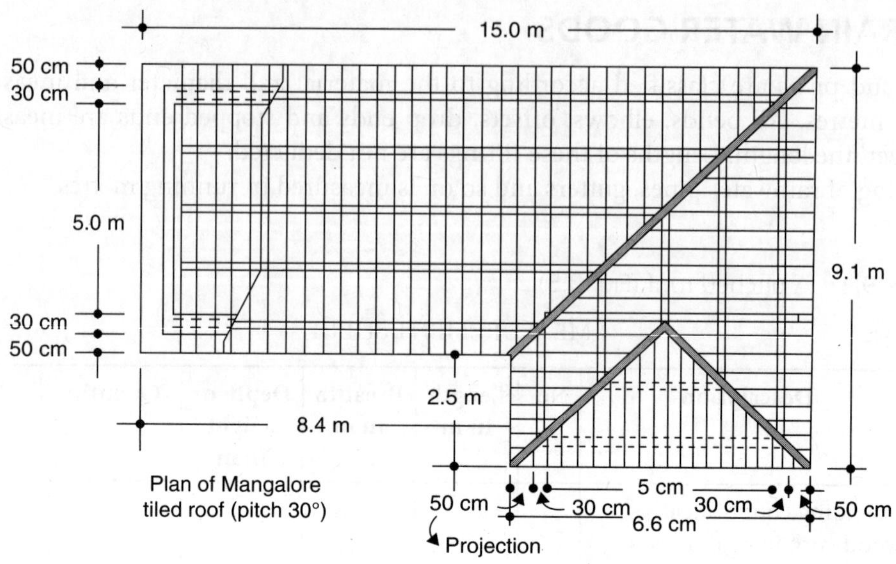

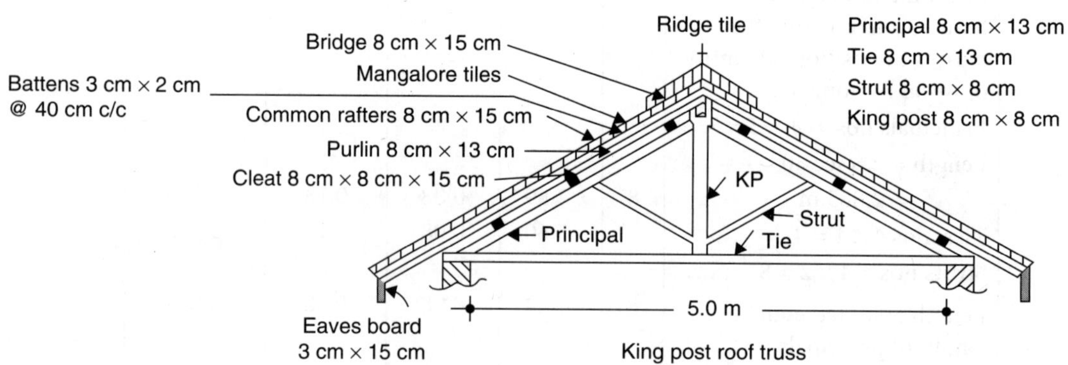

Fig. 9.5   Example 9.1: A pitched roof

***Example* 9.1**   (Continued)

Item No.	Description	No.	Length in m	Breadth in m	Depth or height in m	Quantity	Total
		1	2.50	0.08	0.13		
		1	8.90	0.08	0.13		
	Intermediate	1	13.25	0.08	0.13		
		1	5.60	0.08	0.13		
		2	2.50	0.08	0.13		
		1	10.30	0.08	0.13	0.735	

(continued)

***Example* 9.1**   (Continued)

Item No.	Description	No.	Length in m	Breadth in m	Depth or height in m	Quantity	Total
	Cleats: Nos. $4 \times 4 = 16$	16	0.15	0.08	0.08	0.015	
	Common rafters						
	Nos. $= \frac{15.0}{0.40} = 38$						
	$38 + 1 = 39$						
	Deduct 2 for hip and						
	valley $= 39 - 2 = 37$						
	Over two slopes						
	$2 \times 37 = 74$						
	Length $= 3.30 \times \sec 30°$						
	$= 3.81$ m	74	3.81	0.05	0.08		
	Over front hipped ends						
	Nos. $\frac{2.5}{0.40} = 7$						
	Two slopes $7 \times 2 = 14$						
	Add l for extreme apex						
	Total $= 15$	15	3.81	0.05	0.08	1.356	
	Hip and valley rafters						
	Length in horizontal plane						
	$= \sqrt{3.30^2 + 3.30^2}$						
	$= 4.667$ m						
	True length $=$						
	$\sqrt{(4.66)^2 + (3.3 \tan 30)^2}$						
	$\sqrt{(4.66)^2 + (1.92)^2}$						
	$= 5.04$ m	4	5.04	0.13	0.18	0.470	
	Barge boards	2	3.81	0.03	0.15 ⎤		
	Eaves	2	15.00	0.03	0.15 ⎟	0.221	
		1	9.10	0.03	0.15 ⎟		
		1	2.50	0.03	0.15 ⎦		
						2.967 m^3	2.967 m^3
3.	Providing and applying 3 coats of oil painting to woodwork in roof. Trusses girth $= 2 \times 0.13$ $+ 2 \times 0.08 = 0.42$						

(*continued*)

***Example* 9.1**   (Continued)

Item No.	Description	No.	Length in m	Breadth in m	Depth or height in m	Quantity	Total
	Principals	8	2.89	0.42		9.710	
	Ties	4	5.60	0.42		9.408	
	Struts	8	1.44	0.32		3.686	
	King posts	4	1.04	0.32		1.331	
	(Top and bottom splayed portion)	8	0.20	0.52		0.832	
	Ridge girth $= 2 \times 0.08 + 2 \times 0.15 = 46$ m	1	11.70	0.46		5.382	
		1	2.50	0.46		1.150	
	Purlins girth $= 2 \times 0.08 + 2 \times 0.13 = 0.42$ m	1	14.50	0.42			
		1	8.10	0.42			
		1	5.00	0.42			
		1	2.50	0.42			
		1	8.90	0.42			
		1	13.25	0.42			
		1	5.60	0.42			
		2	2.50	0.42			
		1	10.30	0.42		36.750	
	Cleats Girth $= 4 \times 0.08 = 0.32$	16	0.15	0.32	0.768		
	Common rafters girth $= 2 \times 0.05 + 2 \times 0.08 = 0.26$ m	74	3.81	0.26			
		15	3.81	0.26		88.163	
	Hip and valley rafters girth $= 2 \times 0.13 + 2 \times 0.18 = 0.62$ m	4	5.04	0.62		12.500	
	Eaves: girth $= 2 \times 0.03 + 2 \times 0.15 = 0.36$ m						
		2	15.00	0.36			
		1	9.10	0.36			
		1	2.50	0.36			
	Barge boards	2	3.87	0.36		17.719	
						187.399 m^2	187.40 m^2

(*continued*)

***Example* 9.1**   (Continued)

Item No.	Description	No.	Length in m	Breadth in m	Depth or height in m	Quantity	Total
4.	Providing Mangalore tiled roof with battens 4.5 cm × 2.5 cm at 35 cm c/c Length = length of rafter + 10 cm eaves projection = 3.81 + 0.10 = 3.91 m	2	15.00	3.91		117.30	
		2	2.50	3.91		19.45	
						136.75 m^2	136.75 m^2
5.	Providing and laying in position Mangalore ridge tiles	1	11.70	11.70			
		1	2.50	2.50			
	Over lips	3	5.04	15.12			
						29.32 m	29.32 m
6.	Screwing eaves tiles	2	15.00			30.00	
		1	9.10			9.10	
		1	2.50			2.50	
						41.60 m	41.60 m
7.	Providing valley gutter	1	5.04			5.04 m	5.04 m

10 | **F**loors and **P**avings

10.1 GENERAL

Floors and pavings are grouped primarily into two main types:

1. Ground floor consisting of concrete bed supported on a bed of hardcore or *murum* filling.
2. Upper floors consisting of:
   (i) RCC slabs,
   (ii) timber floors,
   (iii) jack arch floors, and
   (iv) composite floors.

Each of the above types subdivides itself naturally into construction and finishing, each part being further subdivided according to the type of finishing.

10.2 GROUND FLOOR CONSTRUCTION

Unless accompanied by a basement floor, a ground floor usually rests on the filling of *murum* or well-packed rubble or both. Measurements of the complete flooring then divide into two or three of the following items.

1. Hand-packed rubble filling measured in cubic metres.
2. *Murum* filling including watering and ramming also measured in cubic metres.
3. Sand bedding to paving measured in cubic metres.
4. Concrete bedding up to 20 cm measured in sq. metres.
5. Flooring such as Shahabad stone, tiles, Indian patent stone and so on, measured as metre superficial.

1. *Measurements of Filling in Plinth*
   Out of the three dimensions required for calculating the volume of filling, the depth is usually shown on drawings. It must be remembered that the remaining two dimensions,

namely, length and breadth will not be those of the room itself, but the superficial area which lies on the inside of the plinth masonry as shown in Fig. 10.1.

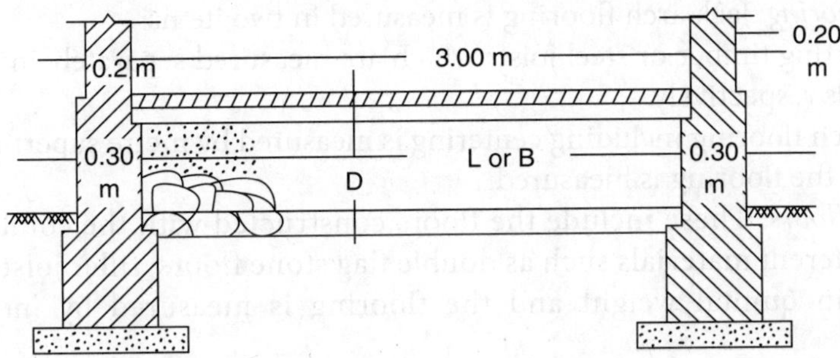

Fig. 10.1

The length and breadth of the filling will be worked out by deducting two offsets at plinth level. For example, if the room dimensions in the above figure are 4 m × 3 m, the length and breadth will be:

$$4.00 - (0.05 + 0.05) = 3.90 \text{ m}$$

$$\text{and} \quad 3.00 - (0.05 + 0.05) = 2.90 \text{ m}$$

2. *Floor Finish*

   The area of the flooring will be worked out by multiplying the length by the breadth of the room between the wall plaster or skirtings or dados if any. Flooring provided over the wall thickness in the openings of doors should be added to the room area.

   Different types of floorings should be kept separate. For example, glazed tiles provided in WC and bathrooms are measured separately from ordinary tiles in other rooms.

3. *Skirting and Dado*

   The skirting and dado area is calculated and kept separate. If the height of skirting is 30 cm or less it is measured in running metres.

10.3   UPPER FLOORS

Construction: The construction of upper floors is measured under the different items as briefly outlined below.

1. *RCC slab:* RCC slab is measured as indicated in Chapter 6.
2. *Timber floors:* The construction of a timber floor divides itself into the following two parts:
   (i) *Supporting members* comprising joists, binders and girders as required. These are measured in cubic metres. Calculation of the volume is simple enough and needs no elaboration.

(ii) *Floor boarding* is measured in square metres. Finished thickness of the boarding is specified in the description of the item.

(iii) *Ceiling* if provided is measured as indicated in Section 9.3.

3. *Jack arch flooring:* Jack arch flooring is measured in two items:

  (i) Supporting timber or steel joists which are measured separately in cubic metres or quintals respectively.

  (ii) Jack arch flooring including centering is measured by metre superficial. Flat overall area of the flooring is measured.

4. *Composite floors:* These include the floors constructed with the combination of two or more different materials such as double flag stone floors; filler joists are measured separately in quintal weight and the flooring is measured by metre superficial as usual.

Example 10.1   Floors and pavings: In continuation of Example 9.1 (Fig. 10.2)

MEASUREMENT SHEET

Item	Description	No.	Length	Breadth	Depth or height	Quantity	Total
1.	Filling in plinth with hard *murum* in layers including watering and ramming						
	Bedrooms						
	$3.20 - 0.30 = 2.90$ m						
	$4.20 - 0.30 = 3.90$ m	2	2.90	3.90	0.60	13.57	
	WC and bathroom						
	$3.00 - 0.30 = 2.70$						
	$3.20 - 0.30 = 2.90$	1	2.90	2.70	0.60	4.70	
	Kitchen						
	$3.70 - 0.30 = 3.40$						
	$3.95 - 0.30 = 3.65$	1	3.40	3.65	0.60	7.45	
	Living room						
	$3.70 - 0.30 = 3.40$	1	3.40	3.40	0.60	6.94	
	Verandah						
	$3.70 - 0.30 = 3.40$						
	$2.00 - 0.30 = 1.70$	1	3.40	1.70	0.60	3.47	
						36.13 cu. m	36.13 cu. m

(continued)

***Example* 10.1**   (Continued)

Item	Description	No.	Length	Breadth	Depth or height	Quantity	Total
2.	Providing and laying 12 cm thick cement concrete (1:5:10) in flooring including ramming and curing						
	Bedrooms	2	2.90	3.90		22.62	
	WC and bath	1	2.90	2.70		7.83	
	Kitchen	1	3.40	3.65		12.41	
	Living room	1	3.40	3.40		11.56	
	Verandah	1	3.40	1.70		5.78	
						60.20 sq. m	60.20 sq. m
3.	Providing and laying 3 cm thick polished Shababad stone on a bed of 1 : 4 cement mortar						
	Bedrooms	2	3.00	4.00		24.00	
	WC and bath	1	3.00	2.80		8.40	
	Kitchen	1	3.50	3.75		13.13	
	Living room	1	3.50	3.50		12.25	
	Verandah	1	3.50	2.00		7.00	
	Add for door sills						
	Door D	2	1.10	0.20		0.44	
	Door D1	4	1.00	0.20		0.80	
						66.02	
	Deduct: For portion below 10 cm thick partition walls						
	WC and bath	1	1.85	0.10		0.19	
	Bathroom front wall	1	1.50	0.10		0.15	
	WC front wall	1	1.20	0.10		0.12	
						0.46	
						65.56 sq. m	65.56 sq. m

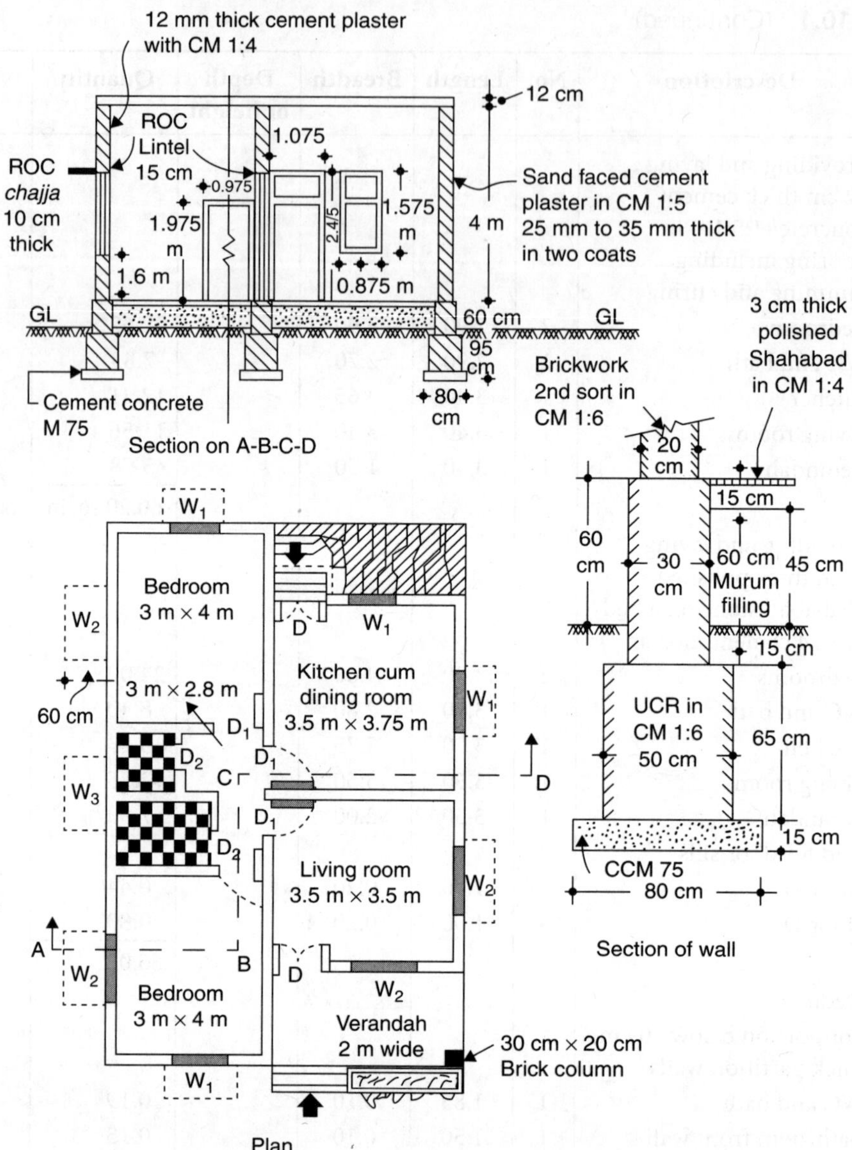

Fig. 10.2   Example 10.1: Floors and pavings

11 | Plastering and Pointing

11.1 GENERAL

Plastering may be classified as below:

1. Type of mortar:
   (i) cement plaster, and
   (ii) lime plaster.
2. Type of surface to be plastered:
   (i) stone masonry,
   (ii) brickwork, and
   (iii) plastering on old walls.
3. Exterior plastering:
   (i) up to 10 m height from ground level, and
   (ii) height above 10 m from ground level is measured in separate items for each storey above.

   Exterior plastering needs to be kept separate as the type of finish is invariably different from the one specified for internal plastering.
4. Internal plastering:
   (i) walls,
   (ii) roofs,
   (iii) ceilings, and
   (iv) plastering bands 30 cm or below.

Pointing: Pointing can similarly be classified on the basis of type of mortar such as (i) cement pointing, (ii) lime pointing, and also on the basis of various types of pointing, for example, struck, keyed, flush, tuck and so on.

White washing, colour washing and distempering are measured similar to plastering and the same measurements hold good for this work also.

11.2   WALL PLASTERING AND POINTING

The unit of measurement of plastering or pointing is square metre. Plastering to walls is measured neglecting the openings in the first instance. Deductions are made later on as explained below.

1.   Area reduced by dissimilar material such as ends of joists, beams and trusses, and openings not exceeding 0.5 m^2 each are not deducted.
2.   For openings such as doors and windows exceeding 0.5 m^2 but not exceeding 3 m^2, deductions are made for one face only. If two faces of walls have plastering of different types or one face has pointing and the other plastering, deductions are to be made from that side on which the width of reveal is less. This is illustrated in Fig. 11.1.

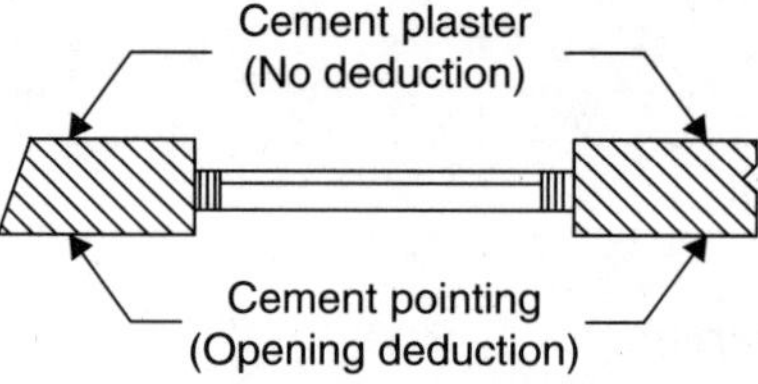

Fig. 11.1

It may be noted that plastering or pointing on one face is not deducted because extra work in jambs, reveals, soffits and sills is not measured.

3.   In the case of openings of area above 3m^2 deductions are made on both faces and actual measurements of plastering or pointing to jambs, reveals, soffits and sills are added.

If there are any projections in the walls such as pillasters, plastering on the sides of such pillasters and so on, is measured and included in the item.

11.3   PLASTERING TO CEILING

Ceilings are measured between walls and partitions. The dimensions before plastering are taken. If ceilings are at a height greater than 5 m the same are measured separately.

If beams are projecting then plastering to the sides of beams will be added to the flat area of the room measured.

Soffits of stairs are included in the item of ceiling and measured accordingly.

11.4   BANDS OF PLASTERING

Plastering bands 30 cm or below are measured in running metres. The height is stated in the description of the item.

11.5   WHITE WASHING, COLOUR WASHING AND DISTEMPERING

White washing, colour washing and distempering are measured separately in square metres. Deductions are made for openings as per item of plastering.

Priming and alkali neutralising treatments are measured separately.

***Example* 11.1**   Plastering and pointing: In continuation of Example 10.1 (Fig. 11.2)

MEASUREMENT SHEET

Item No.	Description	No.	Length	Breadth	Depth or height	Quantity	Total
1.	Providing cement plaster 12 mm thick, single coat in cement mortar 1:4 proportion with *neeru* finish including scaffolding and curing (internal)						
	Bedrooms	4	4.00		4.00	64.00	
		4	3.00		4.00	48.00	
	Bathroom	2	1.75		4.00	14.00	
		2	1.50		4.00	12.00	
	WC	2	1.50		4.00	12.00	
		2	1.20		4.00	9.60	
	Lobby	2	2.80		4.00	22.40	
		1	1.15		4.00	4.60	
		1	1.40		4.00	5.60	
		1	0.25		4.00	1.00	
	Kitchen	2	3.75		4.00	30.00	
		2	3.50		4.00	28.00	
	Living room	4	3.50		4.00	56.00	
	Verandah	1	3.50		4.00	14.00	
		1	2.00		4.00	8.00	
						329.20 sq. m	
	Deductions. Openings of:						
	Door D	2	1.10		2.50	5.50	
	Door D$_1$	4	1.00		2.00	8.00	
	Door D$_2$	2	0.80		1.90	3.04	

(*continued*)

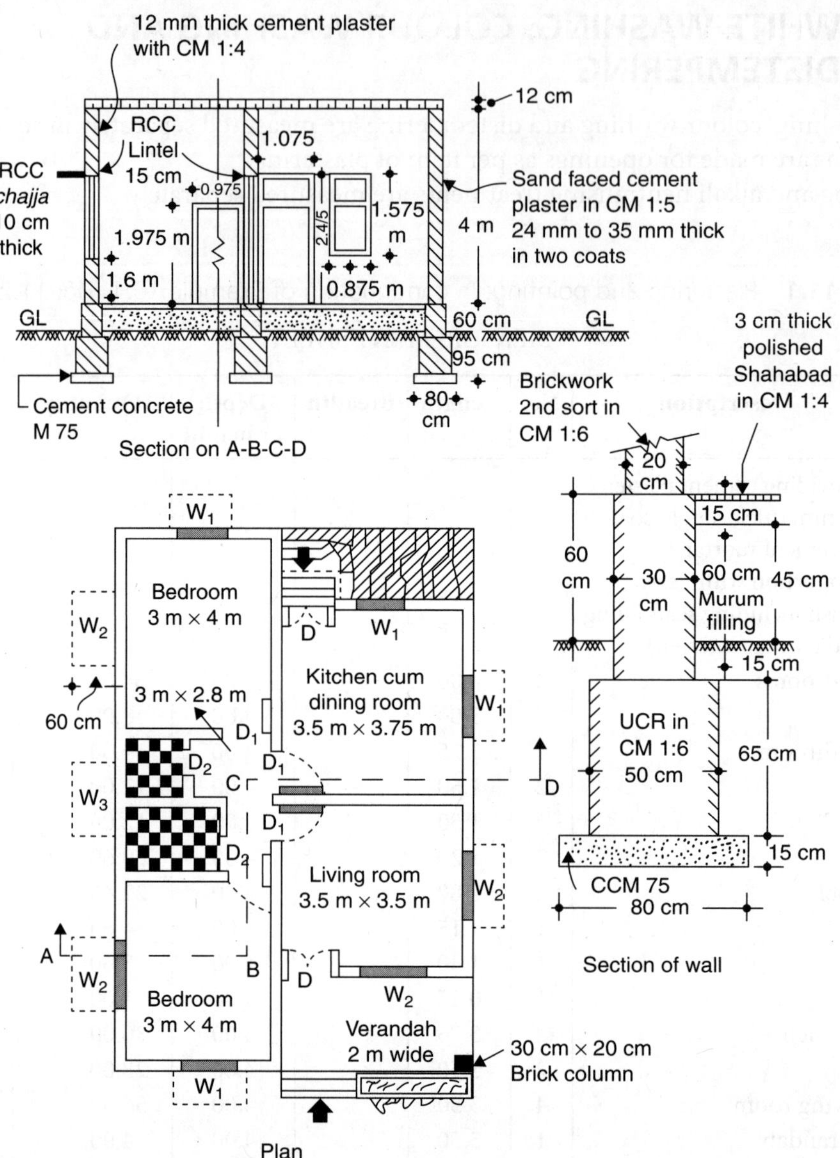

Fig. 11.2   Example 11.1: Plastering and pointing

***Example* 11.1**   (Continued)

Item No.	Description	No.	Length	Breadth	Depth or height	Quantity	Total
	Windows W_1	4	0.90		1.60	5.76	
	" W_2	4	1.10		1.60	7.04	
	" W_3	2	0.70		1.00	1.40	
						30.74	
						298.46 sq. m	298.46 sq. m
2.	Providing sand faced plaster in cm (1 : 5) 25 mm to 35 mm thick in two coats (outside) Height = 0.15 m Below ground 0.60 m Plinth course 0.05 m " offset 4.00 m Superstructure 0.12 m Slab thickness 4.92 m Total Side wall of beds						
	Length = (11.40 + 0.20)	1	11.60		4.92	57.07	
	Front and rear wall of Beds (3.20 + 0.20)	2	3.40		4.92	33.46	
	Kitchen rear wall	1	3.70		4.92	18.20	
	Side wall of kit. and Living room (3.70 + 3.95 + 0.20 = 7.85 m)	1	7.85		4.92	38.62	
	Pillar	2	0.20 ⎫		3.82	3.82	
		2	0.30 ⎭				
						151.17 sq. m	151.17 sq. m
3.	Providing and applying colour wash in two coats						
	Quantity of item No. 1 + 2 = 298.46 + 151.17 = 439.63 sq. m					449.63 sq. m	449.63 sq. m

12 | Stairs

12.1 GENERAL

Stairs are classified according to materials used, as given below:

1. timber stairs,
2. RCC stairs, and
3. steel stairs.

The method of measurement of each of the above types is discussed in the subsequent sections.

12.2 TIMBER STAIRS

Timber stairs are measured in any one of the following two ways.

1. *Area of tread*: If the detailed drawings are ready at the time the estimate is being prepared, the stair may be measured as metre superficial in one item. The area to be measured will be found by multiplying area of one tread by the total number of treads. Description should be carefully worded and should include references to relevant drawings. The landing is measured as per boarding in timber floors and is included in that item.
2. If the detailed drawings are not ready, measurements are divided into the following subsections and each subsection is measured separately:
    (i) treads and risers,
    (ii) landings and bearers,
    (iii) wall strings,
    (iv) outer strings,
    (v) cut strings,
    (vi) hand rails,
    (vii) balusters, and
    (viii) newels.

12.2.1   Treads and Risers

The unit of measurement of treads and risers is square metre. The area is obtained by multiplying the length of the tread by the exposed width of tread (going + projection of nosing) plus the rise from step to step. This is illustrated in Fig. 12.1.

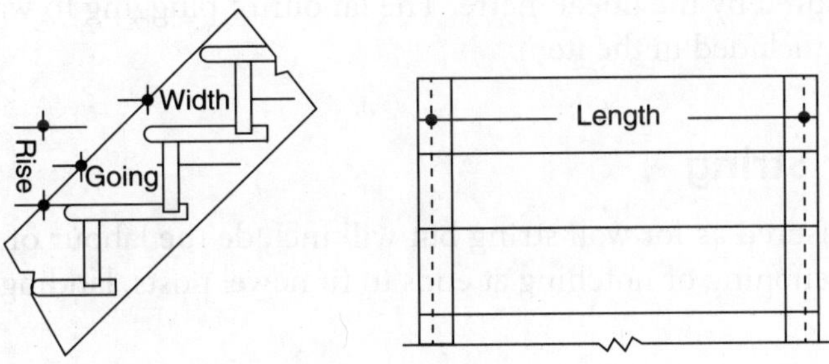

Fig. 12.1

While measuring the length, the portions housed into strings are included (see Fig. 12.1).

In the case of winders and kites, the area of treads should be measured net. To the length, projection of each nosing should be added. For example, consider a typical plan shown in Fig. 12.2.

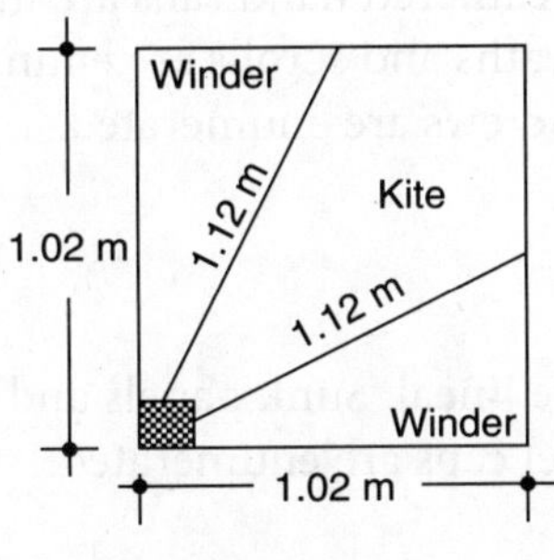

Fig. 12.2

The superficial square is 1.02 m plus two 4 cm nosings = $1.02 + (2 \times 0.04) = 1.10$ m

Area of treads = 1.10×1.10 $\qquad = 1.21$ m^2

Area of risers

Two risers 1.02 m long = $2 \times 1.02 \times 0.15$ $\qquad = 0.306$ m^2

Two risers 1.12 m long = $2 \times 1.12 \times 0.15$ $\qquad = 0.336$ m^2

$\qquad\qquad\qquad$ Total area $\quad = 0.642$ m^2

12.2.2 Landing and Bearers

These are measured as for boarded floors.

12.2.3 Wall String

Wall string is measured by the lineal metre. The labour of plugging to wall and housing for treads and risers is included in the item.

12.2.4 Outer String

Outer string is measured as for wall string but will include the labour of housing for treads and risers and for tenoning or notching at ends to fit newel posts, landing, and so on.

12.2.5 Cut String

Cut string is measured similar to outer string but is held to include labour of cutting of upper edge for treads.

12.2.6 Hand Rail

1. Hand rails are measured by the metre run. Length is measured along top centre line. Circular, level, ramped and weathered hand rails are measured in separate items.
2. Quadrants, short lengths, wreaths and scrolls are enumerated.
3. Housings, mitres, joints and screws are enumerated.

12.2.7 Newels

Newels are measured by the metre lineal. Sunk panels and mouldings planted on or housed into newels are enumerated. Newel caps are enumerated.

12.3 RCC STAIRS

RCC stairs may be divided as below:

1. steps,
2. landing (to be included in floor slabs), and
3. parapet.

Each of the above sections will be subdivided into: (i) concrete, (ii) formwork, and (iii) reinforcement.

Mode of measurement will be as described in Chapter 6.

12.4   STEEL STAIRS

1. Spiral stairs are enumerated. The following details are included in the description of the
    item:
      (i) overall diameter,
     (ii) total number of treads, and
    (iii) height above ground level.
    This item is held to include the cost of:
      (i) treads, risers and sleeves,
     (ii) balusters, and
    (iii) hand rail.
2. Exit landings are described and enumerated.
3. Central shaft, base plate and other attachments are measured in quintals.

Example 12.1   A timber stair

Brief Specifications

Treads and risers: 3 cm treads with rounded nosings and small scotia mould tongued to soffit of
tread and 2.5 cm risers tongued in top and bottom and with treads and risers housed and wedged to
wall string and with returned mitred end to cut outer string and carried on 5 cm × 10 cm carriage.

Wall string: 3 cm × 25 cm moulded wall string plugged to brickwork.

Outer string: 3 cm × 25 cm cut and mitred string housed to newels.

Landing: 3 cm cross-tongued landing on 5 cm × 10 cm bearers provided at 30 cm centre to centre.

Newels: 12.5 cm × 12.5 cm newels to each flight.

Hand rail: 8 cm × 10 cm teak wood handrail, rounded and hollowed on both sides and housed to newels.

Balusters: 2.5 cm × 2.5 cm square balusters housed both ends. Two balusters to each tread.

Painting: Two coats of polishing to all teak wood joinery.

Example 12.1   A timber stair: Refer to Fig. 12.3 for a brief specification

MEASUREMENT SHEET

Item No.	Description	No.	Length in m	Breadth in m	Depth or height in m	Quantity	Total
1.	Providing and constructing teak wood stair as per Fig. 12.3 and the detailed specifications attached Total number of treads = 20 Length of each tread = 1.0 + 0:04 for housing = 1.04 m						

(continued)

Example **12.1** (Continued)

Item No.	Description	No.	Length in m	Breadth in m	Depth or height in m	Quantity	Total
	Width = 0.26 + 0.04 for nosing = 0.30	20	1.04	0.30		6.24 sq. m	6.24 sq. m
	(Note: If the detailed drawings are not ready the measurements are taken out as below. See Section 12.2)						
2.	Providing 3 cm teak wood treads with rounded nosing and small scotia mould tongued to soffit of tread and 2.5 cm risers tongued in top and bottom including housing and wedging						
	Treads	20	1.04	0.30		6.24	
	Risers	23	1.04		0.15	3.59	
						9.83	9.83 sq. m
3.	Extra for curtailed end of steps	1				1.00	1.00 no.
4.	Providing and fixing in position teak wood work in landing						
	Pitching pieces at quarter landing						
	Length = 1.00 + 0.30 for bearing = 1.30 m	2	1.30	0.10	0.15	0.039	
	Beam at first floor landing:						
	Length = 2.90 + 0.30 = 3.20 m	1	3.20	0.12	0.25	0.096	
	Bearers 30 cm c/c						
	No. $= \dfrac{1.0}{0.30} = 4$	2 × 4	1.30	0.05	0.10	0.052	
						0.187	0.19 cu. m
	(Note: In a detailed estimate of a building, measurements of this item are included in the item of woodwork in general.)						
5.	Providing and fixing 3 cm thick teak wood boarding for landing						
	Width = 1.00 m						
	Length = 1.00 + 0.04 for nosing = 1.04 m						

(*continued*)

***Example* 12.1** (Continued)

Item No.	Description	No.	Length in m	Breadth in m	Depth or height in m	Quantity	Total
	There are two quarter space landing	2	1.04	1.00		2.08	2.08 sq. m
	(Note: In a detailed estimate of a building the measurements of this item are included in the item of boarding in floors.)						
6.	Providing teak wood 3 cm × 25 cm moulded wall string including plugging to wall and housing for treads and risers Length: First flight						
	$\sqrt{(2.34)^2 + (1.5)^2} = 2.66\,\text{m}$	1	2.66			2.66	
	Second flight						
	$\sqrt{(0.90)^2 + (0.60)^2} = 1.08\,\text{m}$	1	1.08			1.08	
	Third flight						
	$\sqrt{(2.08)^2 + (1.35)^2} = 2.04\,\text{m}$	1	2.04			2.04	
						5.78	5.78 m
7.	Providing teak wood 3 cm × 25 cm cut and mitred string including housing to newels (Length as calculated above)						
	First and third flights	2	2.04			4.08	
	Second flight	1	1.08			1.08	
						5.16	5.16 m
8.	Providing 8 cm × 10 cm teak wood carriage including notching at top to receive the treads and risers Lengths as calculated in item no. 5						
	First flight	1	2.66			2.66	
	Second flight	1	1.08			1.08	
	Third flight	1	2.04			2.04	
						5.78	5.78 m

(*continued*)

Example 12.1   (Continued)

Item No.	Description	No.	Length in m	Breadth in m	Depth or height in m	Quantity	Total
9.	Providing 8 cm × 10 cm teak wood handrail rounded and hollowed on both sides and housed to newels						
	First floor landing horizontal length	1	1.90			1.90 m	1.90 m
10.	Providing 8 cm × 10 cm teak wood handrail rounded and hollowed on both sides fixed along the flights						
	Length assumed equal to cut outer string	2	2.04			4.08 m	
		1	1.08			1.08 m	
						5.16 m	5.16 m
11.	Providing 2.5 cm × 2.5 cm teak wood balusters 0.90 m long including housing both ends						
	Number = 2 × no. of treads						
	= 2 × (9 + 3 + 8) = 40						
	On first floor, balusters are provided @10 cm cc						
	Number × 1.90 ÷ 0.10 = 19						
	Total no. = 40 + 19 = 59						59 no.	59 no.
12.	Providing 12.5 cm × 12.5 cm teak wood framed newels	1	1.50			1.50 m	
		1	2.70			2.70 m	
		1	3.30			3.30 m	
		2	1.15			2.30 m	
						9.80 m	9.80 m
13.	Extra for moulded solid caps to newels	4				4.00 no.	4.00 no.

***Example* 12.2**   An RCC stair

Brief Specifications

Width of stair	110 cm
Width of landing	100 cm
Rise	15 cm
Tread	30 cm
Concrete block below first step 30 cm × 30 cm	
Thickness of waist slab	15 cm
Thickness of landing slab	15 cm
Thickness of parapet	10 cm
Height of parapet	85 cm

Reinforcement

Stair

Main bars	10 no. 16mm dia. (5 straight and 5 bent up)
Distribution	10 mm dia. 30 cm centre to centre

Parapet

Main bars	10 mm dia. 30 cm centre to centre
Distribution	6 mm dia. 30 cm centre to centre

Concrete block below first step

Main bars	3 nos. 16 mm dia.
Stirrups	6 mm dia. 30 cm centre to centre

***Example* 12.2**   An RCC stair (Refer to Fig. 12.4)

MEASUREMENT SHEET

Item No.	Description	No.	Length in m	Breadth in m	Depth or height in m	Quantity	Total
1.	Providing formwork to risers, treads and waist slab Waist slab, bottom Length $= \sqrt{(3.90)^2 + (1.95)^2}$ $= 4.36$ m Width $= 1.0 + 0.10 = 1.10$ m						
	Width $= 1.0 + 0.10 = 1.10$ m	2	4.36	1.10		9.59	
	Sides	2	4.36		0.27	2.35	
	Risers	25	1.10		0.15	4.13	

(*continued*)

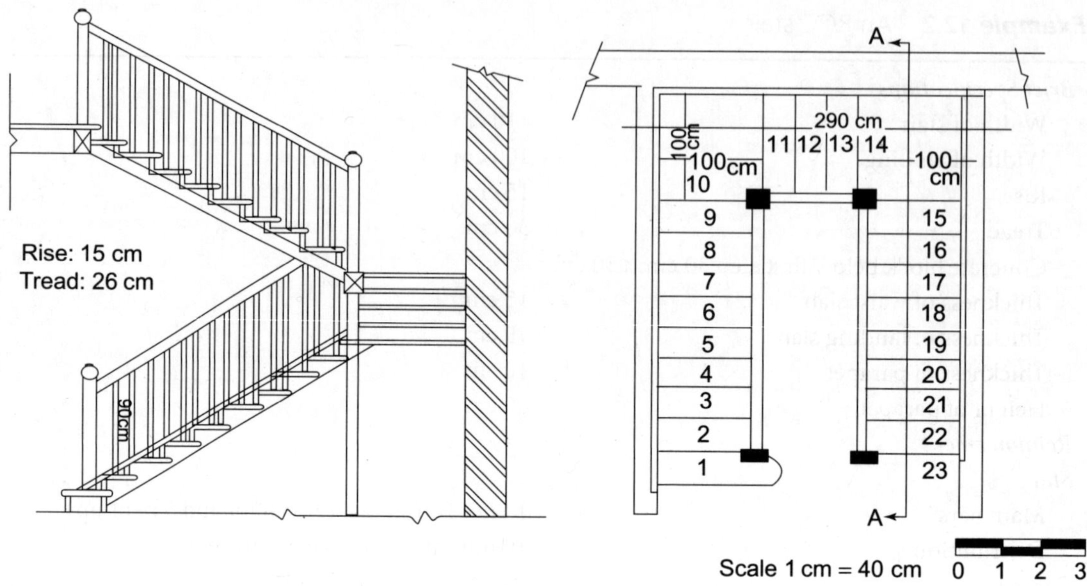

Fig. 12.3

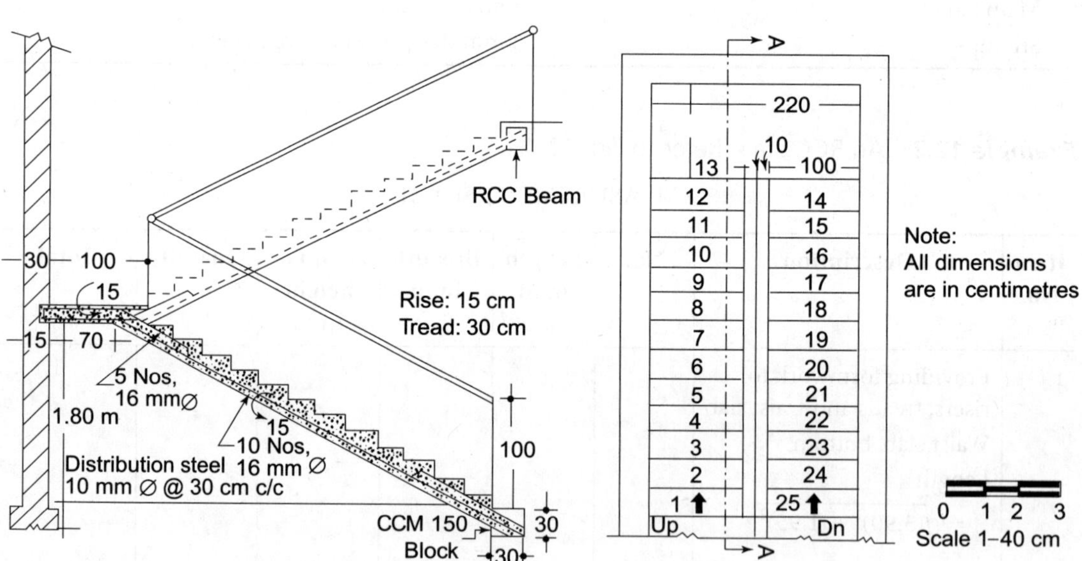

Fig. 12.4   Example 12.2: An RCC stair

***Example* 12.2**   (Continued)

Item No.	Description	No.	Length in m	Breadth in m	Depth or height in m	Quantity	Total
	CC block below first step: Sides	2	1.10		0.30	0.66	
	End	1	0.30		0.30	0.09	
						16.82 m^2	
	Deduct for triangular portion at sides	$\frac{1}{2} \times 26$	0.30		0.15	0.58 m^2	
						16.24 m^2	16.24 sq. m
2.	Providing formwork to RCC slab (landing) Bottom sides	1	2.20	0.70		1.54	
	Length = 2.20 + 0.30 (for bearing) = 2.50 m	1	2.50		0.15	0.38	
	Length = 1.0 + 0.15 for bearing = 1.15 m	2	1.15		0.15	0.35	
						2.27 m^2	2.27 sq. m

(Note: In an estimate of a building these measurements are included in the item of formwork to floor slabs.)

Item No.	Description	No.	Length in m	Breadth in m	Depth or height in m	Quantity	Total
3.	Providing formwork to RCC parapet Sides	2×2	3.60		0.85	12.24	
	Ends	2×2	0.10		0.85	0.34	
	On first floor	2×1	1.10		0.85	1.87	
	Add for Δlar portions	$2 \times 24\left(\dfrac{0.30 \times 0.15}{2}\right)$				1.08	
						15.53 m^2	15.53 sq. m
4.	Providing, cutting, bending, hooking, conveying, placing in position and binding m.s reinforcement as per design	—As per bar bending schedule given in the next page—				2.85	2.85 q

(*continued*)

***Example* 12.2**   (Continued)

Item No.	Description	No.	Length in m	Breadth in m	Depth or height in m	Quantity	Total
5.	Providing and laying in position c.c M 150 in steps including ramming and curing						
	Waist slab	2	4.36	1.10	0.15	1.44	
	Steps	26	1.10	$\left(\dfrac{0.30 \times 0.15}{2}\right)$		0.64	
	Block below first step	1	1.10	0.30	0.30	0.10	
						2.18	2.18 cu. m
6.	Providing and laying in position c.c in M 150 in landing including ramming and curing						
	Length = 2.20 + 0.30 for bearings = 2.50 m						
	Width = 1.00 + 0.15 for bearing—0.30 tread at landing level = 0.85 m	1	2.50	0.85	0.15	0.32 m³	0.32 cu. m

(Note: In an estimate of a building these measurements are included in the item of R C C floor slabs.)

Item No.	Description	No.	Length in m	Breadth in m	Depth or height in m	Quantity	Total
7.	Providing and laying in position c.c M 150 for parapet including ramming, curing, and finishing	2	3.60	0.10	0.85	0.61	
	On first floor	1	1.10	0.10	0.85	0.09	
	Add for Δlar portions	24	$\left(\dfrac{0.30 \times 0.15}{2}\right)$ 0.10			0.05	
						0.75 m³	0.75 cu. m

***Example* 12.2** An RCC stair

BAR BENDING SCHEDULE

Type and shape of bar	No.	Length of each bar in m	Total length in m	Weight per m in kg	Weight in kg
Bottom straight, 16 mm dia.					
Length:					
Bearing on wall 0.15 m					
Straight portion in					
Landing 0.70 m					
Inclined length 4.36 m					
Two hooks					
$= 2 \times 9 \times 0.16$ 0.29 m					
Total length 5.50					
Two flights and alternate bars hence no. of bars $= 2 \times 5$	10	5.50	55.00	1.58	86.90
Bottom bent up, 16 mm dia.					
Length: 5.50 m (as above) $\frac{1}{2} \times$ (depth of slab; for increase in length due to one bend) + 0.15 m anchoring in bottom beam = $5.50 + 0.08 + 0.15$ = 5.73 m					
No. of bars 2×5 as above	10	5.73	57.30	1.58	90.53
Secondary steel, straight bars, 10 mm dia.					
Length = 1.05 m					
In flights, alternate bars no. = $2(4.36/0.30 + 1) = 32$					
In landing at top and bottom = $2 \times (0.85/0:15 + 1)$ = 14					
For bent up portions					
No. of bars = $2 \times \left(\dfrac{1.0}{0.15} + 1 \right) = 16$					
Total no. of bars = $32 + 14 + 16 = 62$	62	1.05	65.10	0.62	40.36
Bars taken up in parapet					
Length = 0.85 + 0.15 (rise) + 0.12 (vertical portion in waist slab) + 1.05 (horizontal portion in waist slab) + 9×0.01 (for one hook) = 2.27 m less 2 cm cover = say 2.25 m					
Number = $2 \times (4.36/0.30 + 1) = 32$	32	2.25	72.00	0.62	44.64

(continued)

***Example* 12.2**   (Continued)

Type and shape of bar	No.	Length of each bar in m	Total length in m	Weight per m in kg	Weight in kg

At first floor level
  Length:
  Vertical   0.80 m
  Anchorage in beam   0.40 m
  Hooks   0.18 m
  Total length =   1.38 m
No. of bars $= \dfrac{1.10}{0.30} + 1 = 5$
	5	1.38	6.90	0.62	4.28

6 mm dia. secondary bars in parapet. In flights, length $= 4.36 - 0.05$ for clear cover $= 4.31$ m No. of bars $= \dfrac{1.00}{0.30} + 1 = 5$ per flight					
Total no. of bars $= 2 \times 5$ On first floor Length $= 1.05$ m	10	4.31	43.10	0.22	9.48
No. of bars 5 as above Concrete block below first step Main bars, 16 mm dia.	5	1.05	5.25	0.22	1.16
Length $= 1.05 + 2 \times 9 \times 0.16$ (for two hooks) $= 1.34$ m Stirrups 6 mm dia. 30 cm c/c	3	1.34	4.02	6.35	6.35
Length: $4 \times 0.25 + 2 \times 9 \times 0.06$ for hooks $= 1.11$ m Number of stirrups $= 1.05/2.25 + 1 = 5$	5	1.11	5.55	0.22	1.22
	Total	weight		That is,	284.92 2.85 quintals

Plumbing, Drains and Sanitary Fittings

13.1 GENERAL

Taking out quantities of plumbing, drainage and sanitary fittings requires a considerable amount of skill and experience. The information available about this section of work, at the time the estimate is being prepared, is scanty. The estimator may have to prepare a layout of these services. The layout should be so prepared that all the necessary services are provided with the minimum amount of material and labour.

The measurements may be taken out in the following order:

1. connection to water-main,
2. pipework,
3. pipe fittings,
4. water-storage tanks, and
5. sanitary appliances.

13.2 CONNECTION TO WATER-MAIN

The connection with the water-main and provision of the communication pipe from the main to the boundary of the site together with the charges of the water and local authorities in connection therewith, are enumerated. The size and length of the connecting pipe are mentioned in the description.

If the cost of carrying out this bulk of the work cannot be estimated with a fair degree of accuracy, a provisional sum is included for the work.

13.3 PIPEWORK

Pipework is classified according to the size of a pipe, thickness of metal, method of jointing and fixing. Excavation and refilling in trenches are either included with the item or measured separately.

All pipework is measured in running metres. Length is measured along the centre line of the pipe line. Pipes not exceeding one metre in length are kept separate. Similarly pipes laid in trenches are measured separately from pipes fixed to walls, ceilings and so on.

13.4   PIPE FITTINGS

All pipe fittings such as bends, offsets, T-junctions, reducers, elbows, tees and so on are fully described and enumerated.

13.5   WATER-STORAGE TANKS

Water-storage tanks are enumerated. The description should include the material of tanks, their type, actual capacity and also the number, types and sizes of connections for pipes and mountings.

13.6   SANITARY APPLIANCES

The appliances commonly provided and their units of measurement are given below.

	Sanitary appliance	Unit of measurement
1.	WC pan	number
2.	Flushing tank	number
3.	Wash-basin	number
4.	*Nahni* tap	number
5.	Gullies	number
6.	Ventilation pipe	metre run
7.	Inspection chamber and manhole:	
	(i) Up to 2.1 m deep	number
	(ii) Exceeding 2.1 m but not exceeding 4.2 m deep	number
	(iii) Exceeding 4.2 m but not exceeding 6 m deep	number
	(iv) More than 6 m deep	according to various trades

13.7   DRAINS

Drains are classified according to material of which they are made such as:

1.   cast iron pipes,
2.   glazed stoneware pipes, and
3.   concrete pipes.

Drainage pipes are measured in running metres. All fittings, specials and so on are enumerated as 'extra-over'.

Concrete beds, haunching and coverings, formwork and so on, are fully described and measured in running metres.

***Example* 13.1**   Plumbing, drains and sanitary fittings (Reference Fig. 13.1)

MEASUREMENT SHEET

Item No.	Description	No.	Length in m	Breadth in m	Depth or ht in m	Quantity	Total
1.	Providing connection to water-main and laying 2 cm dia. pipe line from main to the boundary of the site including making good main road pavement and all works disturbed	1.0				1.00	1.00 no.
2.	Providing stopcock pit 20 cm × 20 cm internally and 0.75 m deep, consisting of 10 cm thick brickwork on 10 cm c.c. (1:4:8) prop. 0.75 m × 0.75 m bed including all excavation, part filling, ramming and so on	1.0				1.00	1.00 no.
3.	Providing and connecting in position 10 cm c.i square stopcock box with hinged cover bedded and pointed in cement mortar	1.0				1.00	1.00 no.
4.	Providing and connecting in position 2 cm brass stopcock with square head and loose key	1.0				1.00	1.00 no.
5.	Providing and connecting in position water meter including the construction of brick chamber as per specification	1.0				1.00	1.00 no.
6.	Providing and connecting 2 cm dia. G.I pipe including necessary excavation and returning, filling and ramming selected material and removing surplus material away Approximate length	1.0	15.00			15.00 m	15.00 m

(*continued*)

Back lane
7.1 m
1.75 m
Drainage
line
Bedroom
3 m × 4 m
Kitchen
3.5 m × 3.75 m
Livingroom
3.5 m × 3.5 m
Proposed
water
line
Bedroom
3 m × 4 m
Verandah
2 m
Water
meter
3.0 m Stop cockpit
Gate
Compound
2.0 m
Water-main
Road

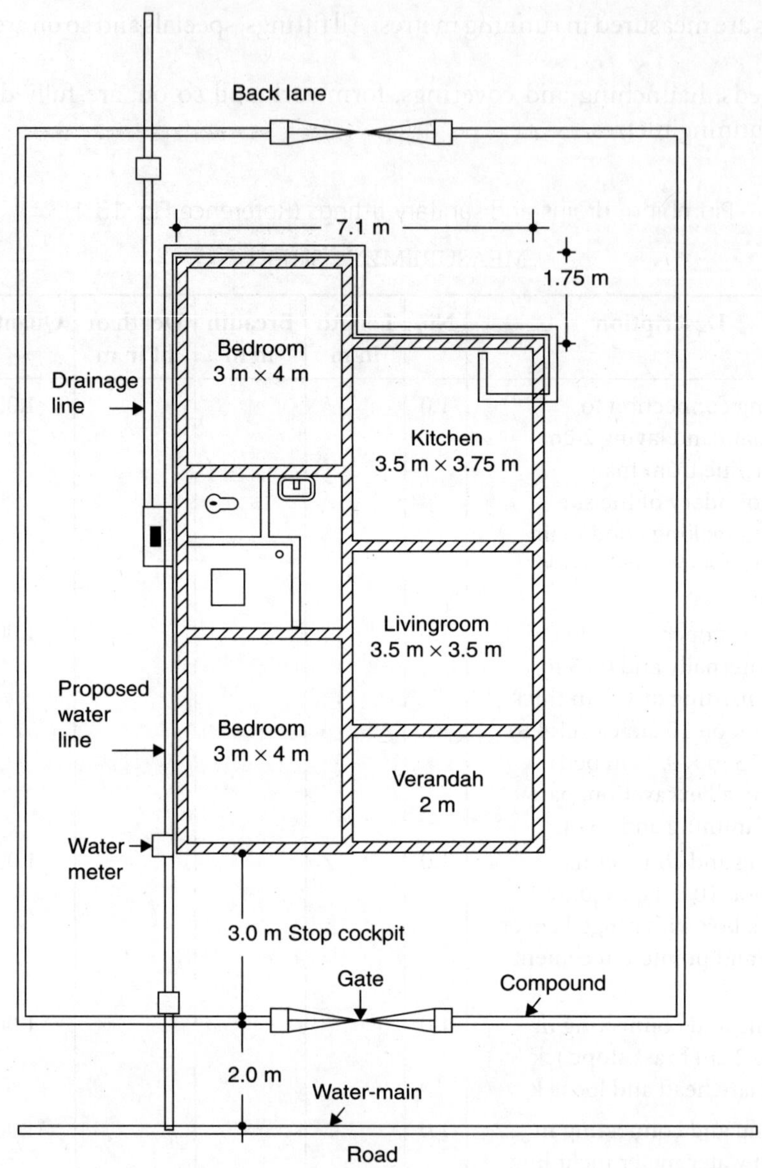

Fig. 13.1

***Example* 13.1**   (Continued)

Item No.	Description	No.	Length in m	Breadth in m	Depth or ht in m	Quantity	Total
7.	Providing and connecting 2 cm dia. G.I pipe line including making holes in masonry and making good masonry and plaster on completion of the work Approximate length: Bathroom 4.0 m Washbasin 2.0 m WC 3.0 m Sink in kitchen 15.0 m Total length 24.0 m	1.0	24.00			24.00 m	24.00 m
8.	Providing and connecting 2 cm dia. bends	7				7	7 nos
9.	Providing and connecting 2 cm dia. tees.	3				3	3 nos
10.	Providing and connecting bibcock of approved quality	1				1	1 nos
11.	Providing and connecting stop cocks of approved type, make and quality	4				4	4 nos
12.	Providing and connecting wash basin including all necessary connections	1				1	1 no.
	Sanitary Fittings						
13.	Providing and laying in position 67.5 cm size WC pan including all necessary connections	1				1	1 no.
14.	Providing and laying in position foot rests of approved type, make and quality	2				2	2 nos
15.	Providing and erecting on bracket 14 litres capacity CI flushing tank including all necessary connections	1				1	1 nos
16.	Providing and connecting as directed *Nahani* trap of approved make, type and quality	2				2	2 nos

(*continued*)

***Example* 13.1**   (Continued)

Item No.	Description	No.	Length in m	Breadth in m	Depth or ht in m	Quantity	Total
17.	As above gully traps	3				3	3 nos
18.	Providing and constructing 0.75 m × 0.75 m inspection chamber consisting, of brickwork in cm and cement plaster to inside face, including excavation, concrete bed, part refilling and removal of surplus excavated material	1				1	1 no.
19.	Providing and concreting intercepting sewer trap of approved make, type and quality as directed.	1				1	1 no.
20.	Providing and laying 10 cm dia. stoneware pipe including necessary excavation and refilling	1	30.00			30.0 m	30.00 m
21.	Providing mica flap valve complete with connection up to manhole	1				1	1 no.

CHAPTER 14 | Electrical Work

14.1 GENERAL

It is undoubtedly scientific and accurate to have detailed wiring plans ready before estimating the cost of electrical work. However, at an early stage in the construction of a work, the detailed plans and specifications for the electrical work are rarely available. In such cases, it is advisable for the estimator to study the general plans and specifications very carefully and note the locations of all outlets, such as switches, meters, plugs, fuse boxes and so on.

The scope of this chapter is limited to the measurement of electrical installations in a building.

14.2 SERVICE CONNECTIONS

Service connections may be of (i) insulating cables, or (ii) bare cables. In the former case, the connections are fully described and measured in running metres. In the latter case, the size of the cable is specified and the measurements are taken out in kilograms.

Wall and service brackets are described and enumerated. Bearer wire and cradles are measured by weight in kilograms and their size, is stated in the description.

14.3 POLES FOR OVERHEAD LINES

Poles and struts for overhead lines are enumerated. The description of the item should state the type, sectional sizes, girth or diameter at top and bottom (if round or tapered) and the total length.

Anticlimbing devices of barbed wire are described and measured by weight in kilograms.

Steel brackets, cross arms, clamps and so on, fixed to poles, and struts, are fully described and enumerated.

14.4 UNDERGROUND CABLES

The measurements of underground cables have to be classified according to their size and type.

The unit of measurement is running metre. The actual length fixed is measured. The description should include the method of laying and jointing and the necessary filling.

Terminal and joint boxes of various kinds are enumerated and measured separately according to size and type.

Manholes are fully described and enumerated.

14.5 CONTROL GEAR

Switch board and each item of control gear such as insulators, fuse boards and so on are fully described and separately enumerated stating the method of fixing.

14.6 POINT WIRING

As specified in the SMM of BWIS: 1200-1965, point wiring includes all work necessary in complete wiring of a tumbler switch circuit of any length from the tapping point on the distribution circuit to the following via the switches:

1. ceiling rose,
2. back plate,
3. socket outlet,
4. lamp holder, and
5. call bell buzzer via the bell push or ceiling rose.

Point wiring may be in any one of the following systems:

1. tough rubber sheathed (TRS),
2. vulcanised insulated rubber in casing and capping (VIR), and
3. vulcanised insulated rubber in conduit. The conduit may be (i) concealed, or (ii) surface.

Wirings in different systems are fully described and measured separately by enumerating the respective points. Provision for scaffolding is deemed to be included in the item and need not be measured separately. It is obvious that the cost of wiring per point will depend upon the distance or length between the tumbler switch and ceiling rose, connector, back plate, socket outlet of lamp holder, and also upon the fitting. For this reason, wiring points are classified as under and measured separately.

1. Short points: Length not exceeding 3 m.
2. Medium points: Length exceeding 3 m but not exceeding 6 m.
3. Long points: Length exceeding 6 m but not exceeding 10 m.
4. Special points: Length exceeding 10 m.

14.7 METERS, FANS, LAMPS AND SO ON

The point wiring is not to include the lamp holders, lamps, meters, fans, brackets, shades, regulators, fuses, call bell, buzzers and so on. These are fully described and enumerated separately.

***Example* 14.1**   Electrical work (Reference: Fig. 14.1)

MEASUREMENT SHEET

Item No.	Description	No.	Length in m	Breadth in m	Depth ht in m	Quantity	Total
1.	Providing service connections with insulating cables	1	10 m			10 m	10 m
2.	Providing and erecting on TW boards of required size DPP switches with fuses for 10 amp. of approved type, make and quality.	1				1 no.	1.00 no.
3.	Providing and erecting TW cupboard of required size to accommodate 1 meter and a fuse	1				1 no.	1 no.
4.	Providing and connecting meter of approved make and of 230V, 50 cyc. capacity	1				1 no.	1 no.
5.	Providing and connecting fuses for 10 amp and of approved type, make and quality	1				1 no.	1 no.
6.	Providing 2 pin plug and socket for 5 amp erected on TW boards as per directions with all necessary accessories of approved type, make and quality including wiring with VIR in casing and capping; one each in kitchen, living and bedroom	4				4 nos	4 nos
7.	Providing and erecting pendant or batten holders or OC bracket light points and fan points as directed including wiring with VIR in casing and capping						

(continued)

Example **14.1**   (Continued)

Item No.	Description	No.	Length in m	Breadth in m	Depth ht in m	Quantity	Total
	Short points up to 3 m Living room and front bed,	4				4 nos	
	verandah and outside of front wall	2				2 nos	
						6 nos	6 nos
8.	—as above—	3				3 nos	
	Medium points up to 6 m, one each in WC, bath and passage						
	kitchen	2				2 nos	
	One each on outside of side wall	2				2 nos	
						7 nos	7 nos
9.	—as per item no. 7— Long points						
	Rear bed	1				2 nos	
	Outside of rear wall	1				1 no.	
						3 nos	3 nos
10.	Providing and connecting in position lamp holders and lamps 25 to 60 watts of approved make, type and quality	12				12 nos	12 nos
11.	Providing and connecting in position 25 cm dia. EI shades of approved make, type and quality	12				12 nos	12 nos
12.	Providing and erecting on TW boards of required size fan regulators of approved make and quality	4				4 nos	4 nos
13.	Providing ceiling fans of approved size, make, type and quality	4				4 nos	4 nos
14.	Providing call bell including necessary wiring and connections	1				1 no.	1 no.

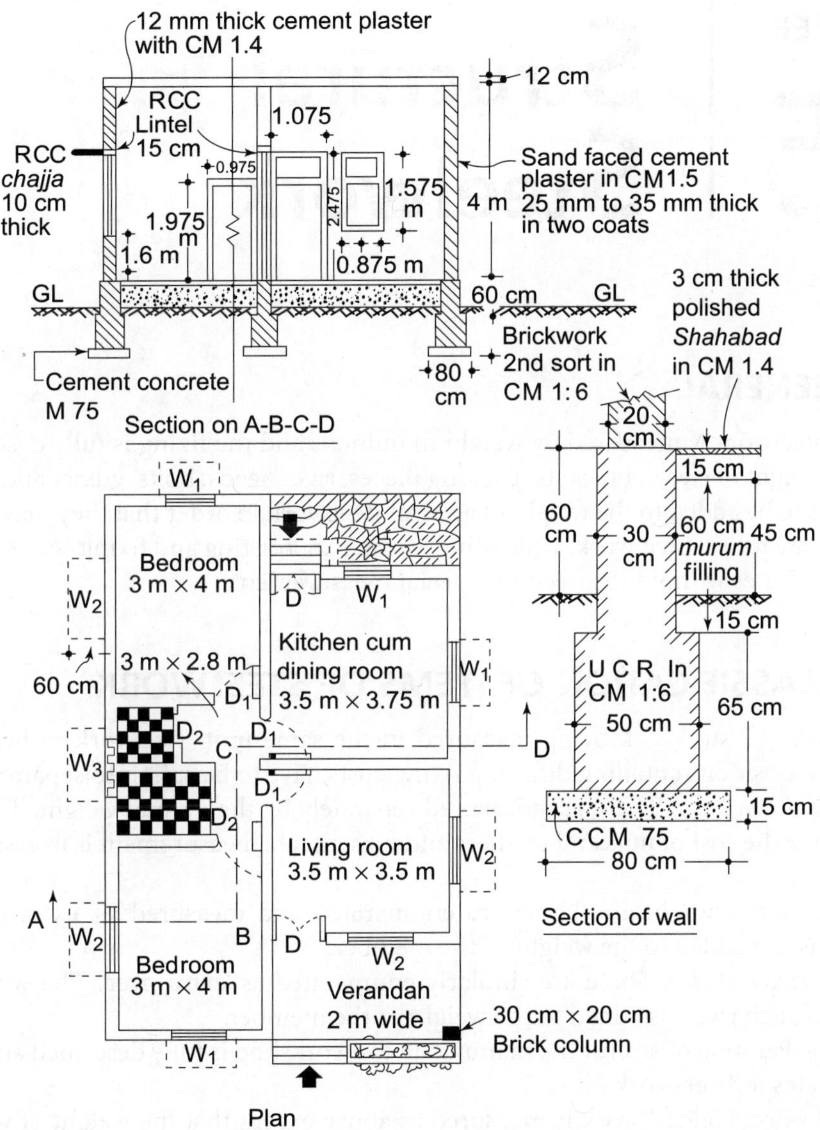

Fig. 14.1 Example 14.1: Electrical work

CHAPTER 15

Structural Steelwork

15.1 GENERAL

Structural steelwork is measured by weight in quintals and the fixing is fully described. The additional weight of cleats, brackets, packing pieces, rivet heads, bolts, gusset and fish plates, and so on, is to be added to the weight of general steelwork, in order that they may be covered by the unit rate for the steelwork. Unloading, getting in, hoisting and fixing is also covered by the steelwork rates without the need for special measurement.

15.2 CLASSIFICATION OF ITEMS OF STEELWORK

1. *Steelwork*: All steelwork being measured in the same unit, steelwork to be erected in different positions entailing differing fixing costs have to be split into separate items.
2. *Holding-down bolts*: These are measured separately by the quintal weight. The rates are to include the cost of nuts and washers and no separate measurement is necessary for the same.
3. *Counter-sunk rivet heads*: These are enumerated and measured as 'extra-over'. Their weight is not added to the weight of the member.
4. *Rivets driven at site*: These are similarly enumerated as 'extra-over'. The weight of the heads of such rivets is added to the weight of the member.
5. *Painting*: Painting of steel at the manufacturer's workshop is fully described and included in the rates for steelwork.
6. *Welded work*: Welded work is measured as above except that the weight of welds is not added to the weight of general steel work.

***Example* 15.1**   A steel roof truss (Fig. 15.1)

MEASUREMENT SHEET

Item No.	Description	No.	Length in m	Breadth in m	Depth or height in m	Quantity	Total
1.	Steelwork in roof truss including erecting and fixing the roof truss in position						
	Double angle 50 × 50 × 6 mm						
	Member ABCDEFG Single angle 50 × 50 × 6 m	2 × 2	8.97				
	Member BLKH	1	15.76				
	CL	2	0.71				
	LD	2	1.74				
	DK	2	2.22				
	KE	2	2.15				
	KG	2	4.56				
	EJ	2	1.79				
	JF	2	0.78				
	Fixing angles	2 × 2	0.40				
	Total length…		81.04 m				
	Weight		81.04 m @ 4.5 kg per metre		364.48		
	M.S flat 50 × 6 mm						
	Member HG	1		3.66 m @ 2.40 kg per metre			
	Gusset plate 6 mm thick (Note: Weight of gusset plate per sq. m is worked out by assuming weight of steel per cu m as 78 q)				8.78		
					Area		
	Gusset plate at joint:						
	B	2	0.40	0.30	0.24		
	C	2	0.20	0.15	0.06		
	D	2	0.35	0.20	0.14		
	E	2	0.35	0.20	0.14		
	F	2	0.20	0.15	0.06		

(continued)

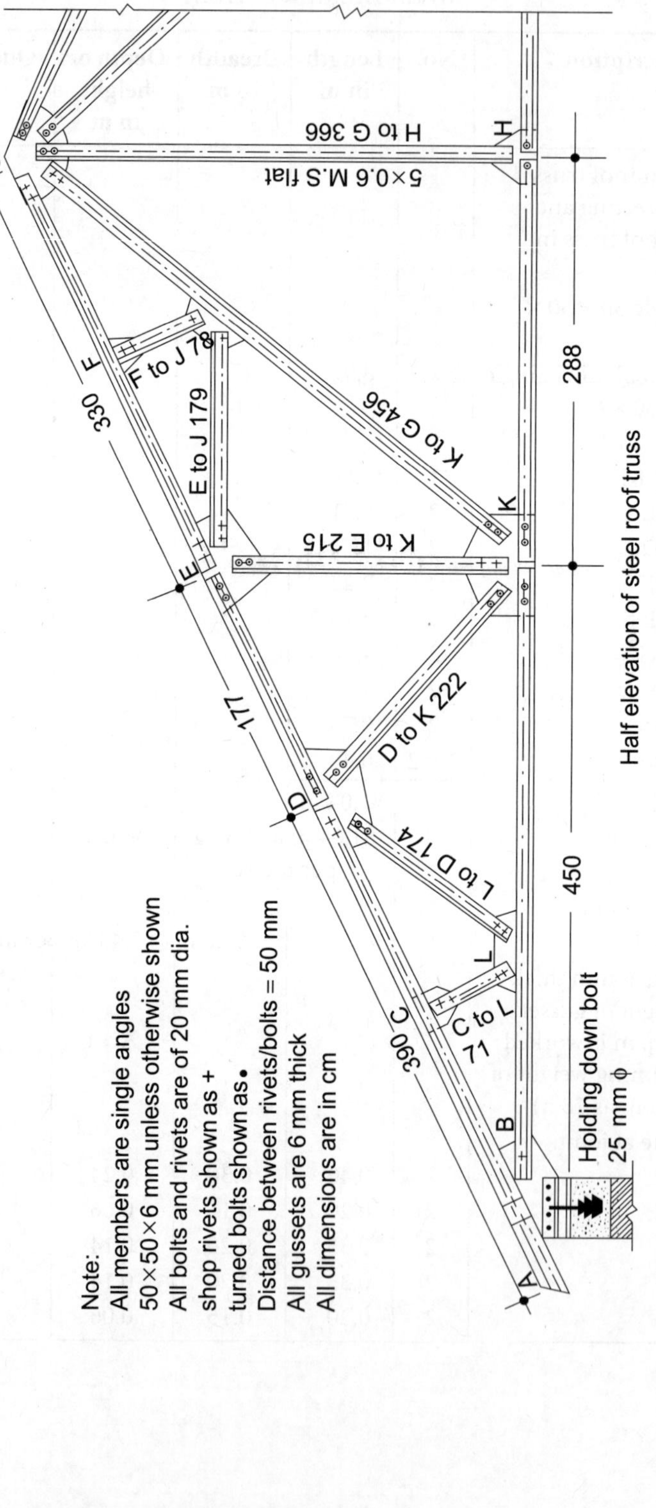

Fig. 15.1 Example 15.1 steel roof truss

***Example* 15.1** (Continued)

Item No.	Description	No.	Length in m	Breadth in m	Depth or height in m	Quantity	Total
	G	1	0.45	0.25	0.11		
	H	1	0.20	0.15	0.03		
	J	2	0.30	0.20	0.12		
					Area		
	K	2	0.30	0.20	0.12		
	L	2	0.20	0.25	0.10		
	Total area				1.12 m^2		
	Weight		@ 46.8 kg per sq. m			52.42 kg	
	Fixing plate 12 mm thick	2	0.40	0.40	0.32	29.95 kg	
			0.32 sq. m @ 93.60 kg per sq. m				
	Site rivet heads 20 mm dia. (Volume of hemisphere $V = \frac{2}{3}\pi r^3$						
	Dia. of rivet head $= 1.6 \times$ dia of rivet						
	Wt. of steel per cu m 78 quintals)	38	@ 0.067 kg per rivet head				
						2.55 kg	
	Total weight					458.18 kg	458 kg
							4.58 q
2.	Providing and fixing in position holding down bolts 25 mm dia. 25 cm long	2 × 2	@ 5.0 kg per no.			20 kg	20 kg
							0.20 q
3.	Providing and fixing in position counter-sunk rivets	2 × 7				14 nos	14 nos
4.	Fixing in position site rivets	38				38 nos	38 nos
5.	Painting in 3 coats with paint of approved colour and shade						
	Member ABCDEFG	2 × 2	8.97				
	BLKH	1	15.76				

(*continued*)

***Example* 15.1** (Continued)

Item No.	Description	No.	Length in m	Breadth in m	Depth or height in m	Quantity	Total
	CL	2	0.71				
	LD	2	1.74				
	DK	2	2.22				
	KE	2	2.15				
	KG	2	4.56				
	EJ	2	1.79				
	JF	2	0.78				
	Fixing angles	2×2	0.40				
			81.04 m	0.20	(girth)	16.21 m^2	
	M.S flat GH	1	3.66 m	0.10	(girth)	0.37 m^2	
	Gusset plates at joints						
	B	2×2	0.40	0.30		0.48	
	C	2×2	0.20	0.15		0.12	
	D	2×2	0.35	0.20		0.28	
	E	2×2	0.35	0.20		0.28	
	F	2×2	0.20	0.15		0.12	
	G	2×1	0.45	0.25		0.22	
	H	2×1	0.20	0.15		0.06	
	J	2×2	0.30	0.20		0.24	
	K	2×2	0.30	0.20		0.24	
	L	2×2	0.20	0.25		0.20	
	Fixing plates	2×2	0.40	0.40		0.64	
						19.46 sq. m	19.46 sq. m

CHAPTER 16 | Measurement of Earthwork

16.1 GENERAL

Almost all civil engineering projects involve earthwork. The knowledge of methods employed to calculate the volume of embankment or cutting is therefore essential to an estimator. The methods of finding out volume of excavation in trenches or over areas with the help of spot levels are explained in Chapter 5. This chapter deals with the methods employed for finding out volumes of cutting and embankment required in works like roads, railways, canals and so on.

It is important to note that the degree of accuracy of the final result will depend upon the accuracy of the field work and the method of calculation adopted, if the arithmetical errors are eliminated. Also, the methods of calculation based on field measurements involve assumptions regarding the geometrical form of the solid which may further impair the accuracy of the result.

16.2 METHODS OF MEASUREMENT

Measurement of earthwork from cross-sections is a universally applicable method. In this method, the total volume is divided into a series of solids by the planes of cross-sections at representative spacings. The cubic content of each of these solids is then determined separately, the addition of those in the cutting giving total volume of excavation, and similarly for banking.

The following two methods based on cross-sections are commonly used to calculate the volume:

1. prismoidal formula, and
2. method of end areas.

These methods are explained below.

16.3 PRISMOIDAL FORMULA

The solids between sections take very nearly the form of a prismoid and hence they are considered as such in this method. The formula for the volume of a prismoid, namely the 'prismoidal formula' given below, is used to find the volume of earthwork.

$$V = \frac{L}{6}(A_1 + A_2 + 4A_m)$$

where L is the perpendicular distance between end sections, A_1 and A_2 are the areas of those end sections, A_m is the area of the cross-section in a plane parallel to the end planes and midway between them and V is the volume of the prismoid.

It must be understood that A_m is not the average of the end areas A_1 and A_2, except in the special case where it may actually work out to be so, that is, where the solid is composed of prisms and wedges only.

If there are more than three cross-sections and all of them are at equal intervals, every alternate sectional area may be treated as a mid-area.

Thus, if L is the distance between the sections, A_1, A_2, A_3 ... A_n are the cross-sectional areas, the volume of the first prismoid of length $2L$ will be

$$V_1 = \frac{2L}{6}(A_1 + A_3 + 4A_2)$$

and that of second prismoid will be:

$$V_2 = \frac{2L}{6}(A_3 + A_5 + 4A_4)$$

Similarly, the volume of the last prismoid of length $2L$ will be:

$$V = \frac{2L}{6}(A_{n-1} + A_n + 4A_{n-2})$$

Summing up, we have the total volume:

$$V = \frac{L}{3}(A_1 + 4A_2 + 2A_3 + A_4 \ldots\ldots 2A_{n-2} + 4A_{n-1} + A_n)$$

which is known as Simpson's rule for volume. This rule can be stated as below:

$$V = \frac{L}{3}\{(A\,\text{first} + A\,\text{last}) + 4(\Sigma A\,\text{even}) + 2(\Sigma A\,\text{odd})\}$$

where   A first   =   area of first cross-section,

                A last   =   area of last cross-section,

                ΣA even   =   sum of areas of even cross-sections,

and ΣA odd   =   sum of areas of odd cross-sections.

It is obvious that Simpson's rule will not yield good results due to the fact that solids of length $2L$ may differ very considerably from a true prismoid.

16.4   THE METHOD OF END AREAS

It is seen from the above discussion that the prismoidal formula cannot be used where mid area is not known. Thus, when only end areas are known, the following rule is used to determine the volume of the prismoid. Using the same notations the rule states:

$$V = L\left(\frac{A_1 + A_2}{2}\right)$$

This formula, though very convenient, has the limitation that it yields correct results only if the mid-area is the mean of the end areas. However, for most practical purposes the results, though slightly higher, may be accepted as sufficiently accurate.

16.5   FORMULAE FOR CROSS-SECTIONS

In the above methods of finding out volume of cutting or embankment, the areas of cross-sections are used. One must, therefore, know the formulae for obtaining cross-sectional areas from the known data. Below are listed formulae for the dimensions of cuttings and embankments for some of the simplest cases.

Let b = formation width

$S : 1$ = side slopes

$n : 1$ = transverse slope of original ground

d = depth of earthwork on the centre line

w and w_1 = the horizontal distances from the centre line to the intersection of the side slopes with the original surface

W = $w + w_1$

d_1 and d_2 = the vertical distances from formation level to the intersections of the side slopes with the original surface, namely,

$$d_1 = \frac{w - b/2}{s} \text{ and } d_2 = \frac{w_1 - b/2}{s}$$

b_1 = the formation width of excavation in side-hill section

b_2 = the formation width of filling in side-hill section

A = the sectional area

Case I: Level section

$$w = w_1 = \frac{b}{2} + sd$$

$$A = d(b + sd)$$

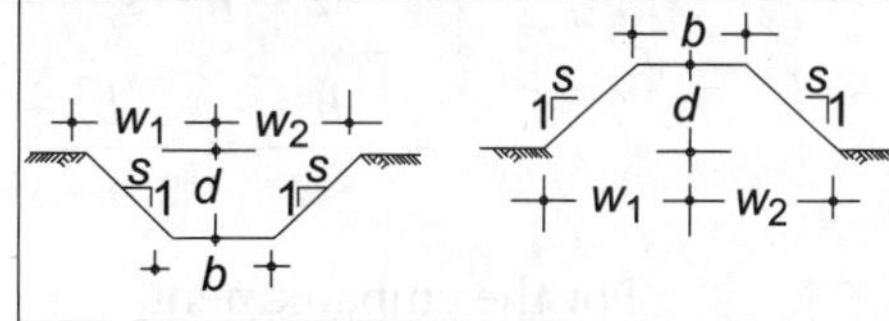

Fig. 16.1

Case II: Two-level section

$$w = \left(d + \frac{b}{2s}\right)\left(\frac{ns}{n+s}\right)$$

$$w_1 = \left(d + \frac{b}{2s}\right)\left(\frac{ns}{n-s}\right)$$

$$A = \frac{ww_1}{s} - \frac{b^2}{4s}$$

or $A = sd_1d_2 + \dfrac{b}{2}(d_1 + d_2)$

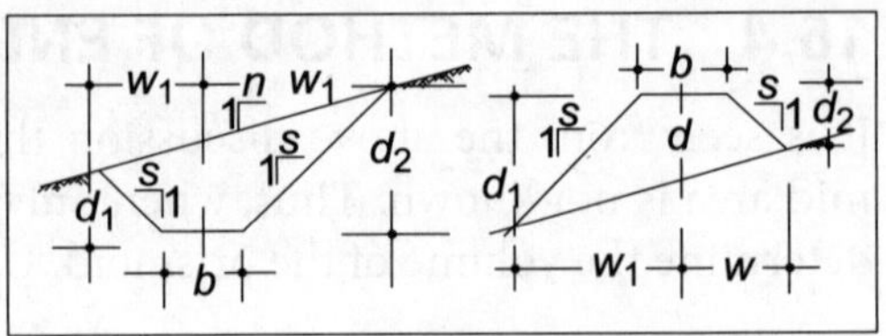

Fig. 16.2

Case III: Three-level section

$$w = \left(d + \frac{b}{2s}\right)\left(\frac{ns}{n+s}\right)$$

$$w_1 = \left(d + \frac{b}{2s}\right)\left(\frac{n_1 s}{n_1 - s}\right)$$

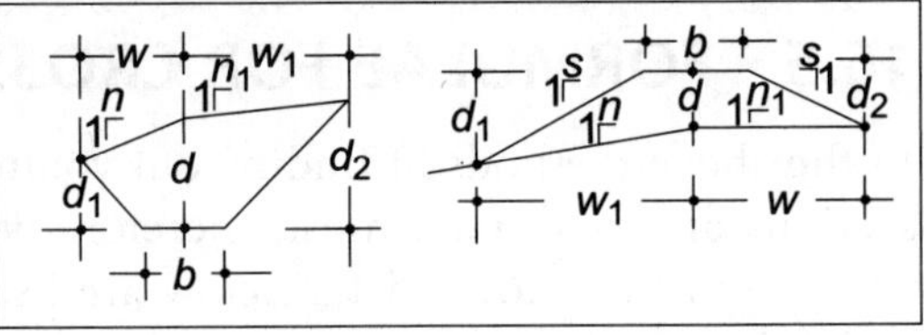

Fig. 16.3

The expression for w or w_1 may apply to both side widths according as the ground falls or rises from the centre to both sides.

$$A = \frac{w + w_1}{2}\left(d + \frac{b}{2s}\right) - \frac{b^2}{4s} = \frac{d(w + w_1)}{2} + \frac{b}{4}(d_1 + d_2)$$

Case IV: Side-hill two-level section
For the cutting,

$$b_1 = \frac{b}{2} + nd$$

$$w = \left(d + \frac{b}{2s}\right)\left(\frac{ns}{n-s}\right)$$

or $= \dfrac{b}{2} + \dfrac{b_1 s}{n - s}$

$$A = \frac{b_1 d_1}{2} = \frac{b_1^2}{2(n-s)}$$

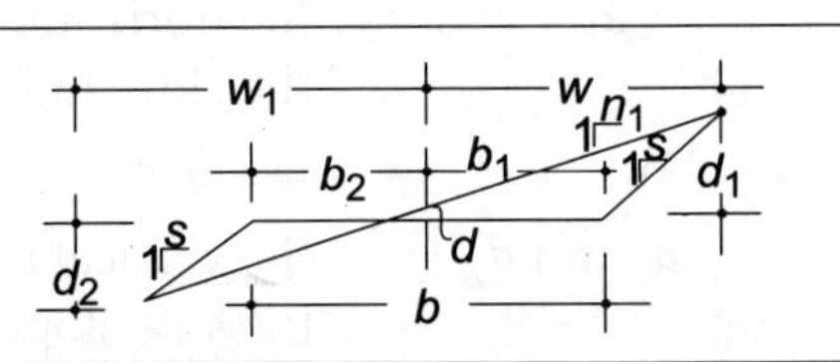

Fig. 16.4

For the embankment,

$$b_2 = (b - b_1) \text{ and}$$

$$w_1 = \left(\frac{b}{2s} - d\right)\left(\frac{ns}{n-s}\right)$$

$$A = \frac{b_2 d_2}{2} = \frac{b_2^2}{2(n-s)}$$

Where the centre line is in embankment, $b_1 = \left(\dfrac{b}{2} - nd\right)$, and the formulae again apply.

Example 16.1

Calculate the volume of earthwork in a railway embankment with the following data.

Chainages in m	Height of embankment at the centre in m	Transverse slope of ground
1000	1.50	8:1
1060	1.00	10:1
1120	1.25	12:1

Formation width = 10 m
Sides slopes of embankment = 2 : 1

Solution:

Cross-sectional areas would be found first by using the formulae given in the preceding section.
Cross-section at chainage 1000.00

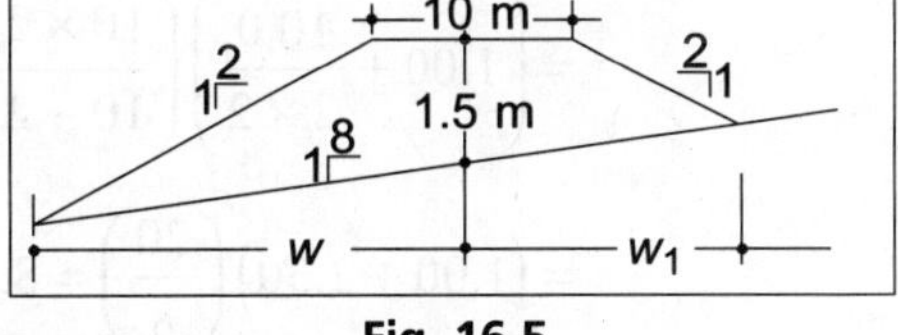

$$w = \left(d + \frac{b}{2s}\right)\left(\frac{ns}{n-s}\right)$$

Substituting the values from data:

$$w = \left(1.5 + \frac{10}{2 \times 2}\right)\left(\frac{8 \times 2}{8 - 2}\right)$$

$$= (1.5 + 2.50)\left(\frac{16}{6}\right) = (4.00)(2.67)$$

$$= 10.98 \text{ m}$$

$$w_1 = \left(d + \frac{b}{2s}\right)\left(\frac{ns}{n+s}\right)$$

Substituting the values from the data

$$w_1 = \left(1.5 + \frac{10}{2 \times 2}\right)\left(\frac{8 \times 2}{8 + 2}\right)$$

$$= (1.50 + 2.50)\left(\frac{16}{10}\right)$$

$$= (4).(1.6) = 6.4 \text{ m}$$

$$\text{Cross-sectional area} = A_1 = \frac{ww_1}{s} = \frac{b^2}{4s}$$

$$= \frac{6.4 \times 10.68}{2} - \frac{10 \times 10}{4 \times 2}$$

$$= 34.18 - 12.50$$

$$= 21.68 \text{ sq. m}$$

Cross-section at chainage 1060

$$w = \left(d + \frac{b}{2s}\right) = \left(\frac{ns}{n - s}\right)$$

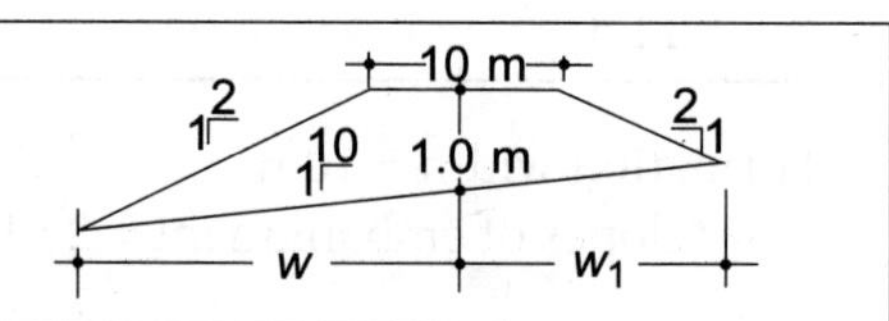

Fig. 16.6

$$= \left(1.00 + \frac{10.0}{2 \times 2}\right)\left(\frac{10 \times 2}{10 - 2}\right)$$

$$= \left(1.00 + 2.50\right)\left(\frac{20}{8}\right) = 8.75 \text{ m}$$

$$w_1 = \left(d + \frac{b}{2s}\right) = \left(\frac{ns}{n + s}\right)$$

$$= \left(1.00 + \frac{10.00}{2 \times 2}\right)\left(\frac{10 \times 2}{10 + 2}\right)$$

$$= \left(1.00 + 2.5\right)\left(\frac{20}{12}\right)$$

$$= (3.50)(1.67) = 5.85 \text{ m}$$

$$\text{Area} = A_m = \frac{ww_1}{s} - \frac{b^2}{4s}$$

$$= \frac{8.75 \times 5.85}{2} - \frac{10 \times 10}{4 \times 2} = 13.09 \text{ sq. m}$$

Cross-section at chainage 1120

$$w = \left(d + \frac{b}{2s}\right) = \left(\frac{ns}{n-s}\right)$$

$$= \left(2.25 + \frac{10}{2 \times 2}\right)\left(\frac{12 \times 2}{12 - 2}\right)$$

$$= \left(2.25 + 2.50\right)\left(\frac{24}{10}\right)$$

$$= (4.75)\,(2.40) = 11.40 \text{ m}$$

$$w_1 = \left(d + \frac{b}{2s}\right) = \left(\frac{ns}{n+2}\right)$$

$$= \left(2.25 + \frac{10}{2 \times 2}\right)\left(\frac{2 \times 2}{12 + 2}\right)$$

$$= \left(2.25 + 2.50\right)\left(\frac{24}{14}\right)$$

$$= (4.75)(1.71) = 8.12 \text{ m}$$

$$A_2 = \frac{ww_1}{s} - \frac{b^2}{4s}$$

$$= \frac{11.40 \times 8.12}{2} - \frac{10 \times 10}{4 \times 2}$$

$$= 46.29 - 12.50 = 33.79 \text{ sq. m}$$

Finally, by using the prismoidal formula volume of earthwork would be determined

$$\text{Length} = L = \text{Ch:}1120 - \text{Ch:}1000 = 120 \text{ m}$$

$$V = \frac{L}{6}\,(A_1 + 4A_m + A_2)$$

$$= \frac{120}{6}\,(21.68 + 4 \times 13.09 + 33.79) = 2156.60 \text{ cu. m}$$

Example 16.2

Calculate the quantity of earthwork of a hill road with the following data.

    Length of the road   =   50 m

    Formation width   =   10 m

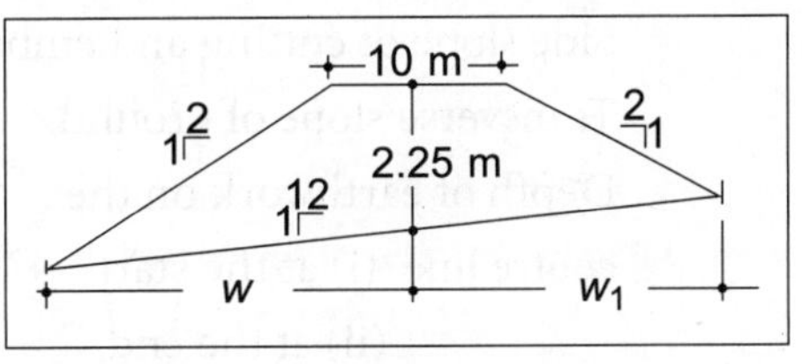

Fig. 16.7

Side slope of cutting and embankment $\quad = \quad$ 2:1

Transverse slope of ground $\quad\qquad\qquad = \quad$ 6:1

Depth of earthwork on the

centre line (i) at the start $\qquad\qquad = \quad$ 0.50 m

$\qquad\qquad\qquad$ (ii) at the end $\qquad\qquad = \quad$ 0.30 m

Solution:

(i) Cross-sectional areas would be determined first.

Cross-section at the start

Using the formula from Section 16.4

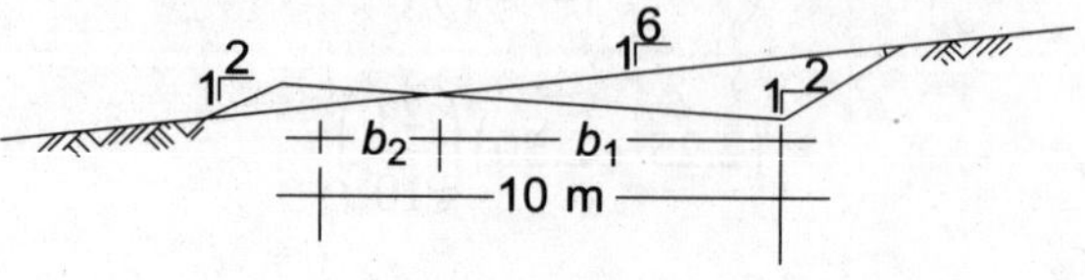

Fig. 16.8

$$b_1 = \frac{b}{2} + nd = \frac{10}{2} + 6 \times 0.5$$

$$= 5.0 + 3.00 = 8.00 \text{ m}$$

$$\text{Area of cutting} = A_1 = \frac{b_1^2}{2(n-s)}$$

$$= \frac{(8.00)^2}{2(6-2)} = 8.00 \text{ sq. m}$$

Area of embankment

$$b_2 = (b - b_1)$$

$$= 10 - 8.00 = 2:00 \text{ m}$$

$$\text{Area} = A_1 = \frac{b_2^2}{2(n-s)} = \frac{(2.00)^2}{2(6-2)} = 0.50 \text{ sq. m}$$

Cross-section at the end

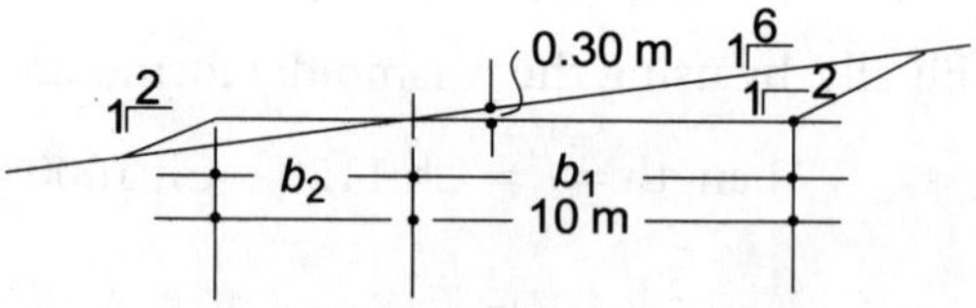

$$b_1 = \frac{b}{2} + nd = \frac{10}{2} + 6 \times 0.30$$

$$= 5.00 + 1.80 = 6.80 \text{ m}$$

$$\text{Area of cutting} = A_2 = \frac{b_1^2}{2(n-s)}$$

$$= \frac{(6.80)^2}{2(6-2)} = 5.78 \text{ sq. m}$$

Fig. 16.9

Area of embankment

$$b_2 = b - b_1 = 10.00 - 6.80 = 3.20 \text{ m}$$

$$\text{Area } A_2 = \frac{b_2^2}{2(n-s)} = \frac{(3.20)^2}{2(6-2)} = 1.28 \text{ sq. m}$$

(ii) Volume of earthwork in cutting and embankment would be determined by using the end area method:

$$V = L\left(\frac{A_1 + A_2}{2}\right)$$

Volume of earthwork in cutting:

$$V = 50\left(\frac{8.00 + 5.78}{2}\right)$$

$$= 50 \times (6.89) = 344.50 \text{ cu. m}$$

Volume of earthwork in embankment:

$$V = 50\left(\frac{0.50 + 1.28}{2}\right)$$

$$= 50 \times (0.89) = 44.50 \text{ cu. m}$$

(iii) Excess volume of earthwork would be:
Volume of cutting − volume of embankment

$$= 344.50 - 44.50 = 300 \text{ cu. m}$$

Squaring the Dimensions

17.1 GENERAL

After the quantities are taken out, the next step in the preparation of an estimate is to square the dimensions. The term 'squaring dimensions' refers to the calculation of number, length, areas, and volumes and their entry in the last two columns on the measurement sheet (in PWD method) or third column on the dimension paper (in English method).

There are three methods of squaring dimensions.

1. decimal system,
2. duodecimal system, and
3. fractional part and practice.

These methods are explained below with the help of examples.

17.2 DECIMAL SYSTEM

The decimal system, in which the calculations are done in 'tenths' and 'hundredths' is used in the PWD method of taking out quantities.

Example 17.1

Multiply 12.20 m × 7.70 m to find area of surface dressing.

$$
\begin{array}{ll}
12.20 & \text{(length)} \\
\times\, 7.70 & \text{(width)} \\
\hline
85400 & \\
8540\text{xx} & \\
\hline
93.9400 & \\
\times\, 1 & \text{(number)} \\
\hline
93.94\ \text{m}^2 & \text{(area)}
\end{array}
$$

If the dimensions are in feet units, inches are expressed in tenths and hundredths part of feet as below.

Example 17.2

$$3 \times 12' - 9'' \times 3' - 6'' \times 2' - 3''$$

First, $\quad 12' - 9'' - -12.75$ (length)

$\quad\quad\quad 3' - 6'' -- 3.50$ (width)

$$63750$$
$$3825xx$$
$$44.625 \text{ m}^2 \quad \text{(area)}$$

Then

$$44.625$$
$$\times 2' - 3'' - - 2.25 \quad \text{(depth)}$$
$$2231250$$
$$892500x$$
$$892500xx$$
$$100.40625 \text{ m}^3 \quad \text{(volume)}$$
$$\times 3 \quad \text{(number)}$$
$$301.23 \text{ m}^3 \quad \text{(total volume)}$$

17.3 DUODECIMAL SYSTEM[*]

In this method the calculations are made by working in twelfth and twelfth parts of such twelfths instead of tenths and hundredths as mentioned above in the decimal system. The system of duodecimal is used in English method of taking out quantities where the dimensions are in feet and inches.

The following simple example will illustrate this method.

$$8.9$$
$$4.3$$

---

[*]Duo means twelfths.

$$
\begin{array}{lll}
& & 8.9 \\
& & 4.3 \\
\hline
8 \times 4 = -\; - & -\quad - & 32.0 \\
9 \times 4 = 36 \div 12 & -\quad - & 3.0 \\
8 \times 3 = 24 \div 12 & -\quad - & 2.0 \\
9 \times 3 = 27 \div 12 & -\quad - & 0.2.3 \\
\hline
& & 37.2.3 \\
\hline
\end{array}
$$

That is, 37 feet square, 2 twelfths and 3 parts.

It will be noticed from the above example that:

$$\text{Feet} \times \text{Feet} = \text{Feet}^2$$

$$\text{Feet} \times \text{Inches} = \text{Twelfth and Twelfth} \div 12 = \text{feet}^2$$

$$\text{Inches} \times \text{Inches} = \text{parts and, parts} \div 12 = \text{twelfths}$$

Where the dimensions are 'timesed' the result obtained by squaring is multiplied by the 'timesing' or combination of timesing and dotting on.

For example,

$$
\begin{array}{rl}
^{2\!/\!_{3}.2/}8.9 & \\
4.3 & \\
\hline
37.2.3 & \quad \text{(as detailed above)} \\
\times 2(3+2) = 10 \qquad 10 & \\
\hline
371.10.6 & \quad \text{that is, } 371\dfrac{10\frac{1}{2}}{12}\ \text{ft}^2 \\
\hline
\end{array}
$$

It is generally simpler and quicker to multiply one of the dimensions by the timesing first. Thus:

$$
\begin{array}{ll}
& 8.9 \\
& \times 10 \\
\hline
& 87.6 \\
& \times 4.3 \\
\hline
87 \times 4 = \ \dots\dots \quad \dots \ \dots \ \dots & 348.0 \\
6 \times 4 = \ 24 \div 12 \quad \dots \ \dots \ \dots & 2.0 \\
87 \times 3 = 261 \div 12 \quad \dots \ \dots \ \dots & 21.9 \\
6 \times 3 = \ 18 \div 12 \quad \dots \ \dots \ \dots & 0.1.6 \\
\hline
& 371.10.6 \qquad \text{that is, } 371\dfrac{10\frac{1}{2}}{12}\ \text{ft}^2 \\
\hline
\end{array}
$$

The following worked example will illustrate the application of the above principles to larger dimensions.

4/2·1/	69.9			
	3.6		Exc. fdn. tr. n.e. 5′ dp.	
	4.3	12451.2		

First,   3.6

$4(2 + 1)$ ...   12.00

$3 \times 12 =$ 36.0

$6 \times 12 = 72 \div 12$ 6.0

                                                                                 42.0

Then   69.9

                                                                                 42.0

$69 \times 42 =$ ...   2898.0

$9 \times 42 = 378 \div 12$ ...   31.6

                                                                               2929.6

and further,   2929.6

                                                                                   4.3

$2929.0 \times 4 =$ 11716.0

$6 \times 4 =$ $24 \div 12$ 2.0

$2929 \times 3 = 8787 \div 12 =$ 732.3

$6 \times 3 =$ $18 \div 12 =$ 0.1.6

                                                                             12450.4.6

that is, $12451 \, \tfrac{2}{12} \, \text{ft}^3$

17.4   FRACTIONAL PART AND PRACTICE

In this method the dimensions to be squared are expressed as fractions. For example,

(i)   4.9

                                                                                 3.3

$4' - 9'' \times 3' =$ 14.3

$4' - 9'' \times 3/12 =$ 1.2.1

                                                                               15.5.1

(ii)

$$17' \times 11' =$$

	17.3
	11.6
$17' \times 11' =$	187.0
$\times 6'' = \frac{1}{2}$ of $17' - 3''$	8.7.6
$\times 3'' = \frac{1}{4}$ of $11' - 0''*$	2.9.0
	198.4.6

*$11' - 0''$ and not $11' - 6''$ as the inches have been already multiplied in the step immediately preceding.

(iii)

	47.1
	19.1½
	2.6
First	47.1
	19.1½
$47 \times 19 =$	893.0
$\times 1½ = 1/8$ of $47' - 1''$	5.10.1*
$\times 1'' = 1.12$ of $19' - 0''$	1.07.0
	900.5.7
Then	900.5.7
	2.6
$\times 2' =$	1800.11.2
$\times 6'' = 1/2$ of $900.5.7 =$	450.2.9
	2251.1.11

set as 2251.2 that is, $2251\frac{2}{12}$ ft^3

*It is not necessary to work more accurately than to the nearest part.

Method of working out by practice for the second example above is given below:

	17.3	
	11.6	
$17' - 3'' \times 10' =$	172.6	
$\times \ 1' =$	17.3	
$\times \ 6'' = 1/2 \times 17' - 3'' =$	8.7.6	
	198.4.6	that is, $198\frac{5}{12}$ ft^2

Similarly,	27.9
	21.3
$27' - 9'' \times 10' =$	277.6
$\times 10' =$	277.6
$\times\ \ 1' =$	27.9
$\times\ \ 3' - (= 1/4 \times 27' - 9'') -$	9.2.3
	591.11.3

set as 59.11 or 591 $\tfrac{11}{12}$ ft^2

It will be noticed in the above examples that it is at times difficult to separate the fractional part and practice method from the duodecimal system. Often, the most convenient process is the combination of both systems.

17.5   CHECKING THE SQUARING

All squared dimensions and waste calculations must be checked by a different person, each result being ticked preferably in red or other coloured ink. Where necessary, corrections are to be made in coloured ink and all such corrections must be cross-checked.

Abstracting and Bill of Quantities

18.1 GENERAL

The last stage in the preparation of an estimate in the PWD system of taking out quantities is preparation of an abstract. In the English system, the corresponding operation is writing a bill of quantities. Abstracting in the English system is an intermediate stage between taking out dimensions and writing the bill of quantities.

18.2 ABSTRACT—PWD METHOD

Abstracting in the PWD method is the operation of transferring all the items from the measurement sheets to the abstract sheets without changing the serial order. The ruling of the abstract sheet is shown below.

The dimensions from the measurement sheets of Example 5.1, are transferred here for illustration (see table on the next page).

In the first column marked 'quantity'—total quantity of each item as shown in the last column of the measurement sheet is entered and its unit of measurement is mentioned in the second column. In the third column, item number and full particulars of the work have to be written. In the measurement sheets, description of an item of work may be short and abbreviated but in the abstract sheets, all major factors that are required to make a realistic assessment of the unit rate must be described. The rate is entered in the fourth column and the unit quantity of work for which it stands is written under column five. It is preferable to use the same unit in the second and fifth columns. The cost of an item of work is then found out by multiplying the quantity of work in the first column and the unit rate in the fourth column (and dividing the result by column five if the units in column two and five are not identical). The amount so found out is written in the last column.

The total of the amount column is taken at the end of each page and carried over to the next page. First entry on each new page would be the total 'brought forward' from the last page. Thus, at the end of the last page the total cost of all the items would be obtained. To the total cost so obtained a percentage is usually added (5% in the Maharashtra State PWD) to allow for possible unforeseen expenditure (called 'contingencies'). To the 'estimated cost' so worked out, a certain percentage is added (1½% to 2%) for provision of tools and plant and work-charged establishment.

Quantity	Unit	Description	Rate ₹	Per	Amount
95.89	m²	1. Surface dressing 15 cm thick and removing material 50 m away	21.00	m²	2,013.69
22.42	m³	2. Excavation in trenches in ordinary soil including disposal of excavated material, lift 1.5 m, lead 50 m	207.00	m³	4,640.94
13.45	m³	3. Excavation in fdn. trenches in hard soil including disposal of excavated material, lift 1.5 m, lead 50 m and returning, filling and ramming the selected material around foundation	365.00	m³	4,909.25
56.97	m²	4. Providing planking and strutting up to 1.5 m deep	40.00	m²	2,278.80
6.72	m³	5. Providing and laying CC–M 10 in foundation trenches, including ramming and curing	5830	m³	39,177.60

The ruling of the abstract sheet shown above is used in the public works department. It does not contain the column of item number.

The form may be improved as below:

Name of work .... Page no.

Item	Particulars of work	Quantity	Unit	Rate ₹	Amount

The main advantage of the improved form is that the quantity unit rate and amount columns are all adjacent to one another which greatly assist in pricing and checking priced bills of quantities.

18.3   ABSTRACT IN ENGLISH METHOD

In the English method of taking out quantities, the items of work are entered on the dimension papers in order of construction, whereas, in the last stage, while billing, they are grouped according to trade. In simple cases bills may be prepared directly from the dimension paper; but in large and complicated cases, an intermediate step of grouping all items together as they would appear in the bill is necessary. This operation of collecting similar items together and classifying them into sections with a view to their being in a suitable order for writing of the bill is called 'abstracting'.

The abstract paper is usually a double sheet ruled in columns about an inch wide. Sometimes columns are subdivided with an additional line to separate the main figures (of quantity) from the references (page numbers of dimension papers).

The items are grouped in sections. For example, in the case of a building work the major sections would be:

1.   *Carcase*: foundation, concrete work, masonry, brickwork, roof and so on.
2.   *Finishing*: carpentry, joinery, plumbing, plastering, glazing, drainage, fencing and so on.

As to the order of items in each trade section, the usual practice is to adopt the order of cubic, square, linear and enumerated items. The provisional sums come last in the section to which they apply.

Writing down the items: The items are entered in the abstract in the same order in which they will appear in the bill. Headings of the section or subsection are written at the top, covering two or three columns. Similarly, descriptions are also spread over two or three columns. In the first column, quantities under all times of that section or subsection (collected from wherever they may be on the dimension papers) and measured in cubes are entered. In the same column, against each quantity is written the page number of the dimension paper from which that quantity is transferred. These items will be followed by items under the same trade but measured in area, length, and so on in the order mentioned earlier.

Concrete work

(Note: The abstract is normally prepared across double-foolscap width sheet)
(1:2:4 Concrete)

400	2	Ddt.		
200	6	150	3	
50	13	20	4	
650		170	10	
− 340		340		
310				
9) 310	(34.44			
3) 34.44	(11.48			
	cu. yds			

Deductions are entered in the second column under the main heading of the item under consideration as shown above. It may be noted in the above example that cubic contents have been reduced to cubic yards being the correct unit of measurement for this class of work.

As each item is entered on the abstract, the description of the appropriate item on the dimension sheet is crossed through with a vertical line having short horizontal lines at its extremities, as shown below:

2/	52.6		Cone.	(1:2:4)
	3.6		in fdns. and n. e.	ex 6" 12" th
	0.9			
2/	52.6		Hor. d. p. c. in	
	3.6		c. c. (1: 2: 4) and tar felt	
2/	51.6		1 B wall bks	in common
	2.6		in (1:6)	ct mortar

This enables the abstractor to pick out items from the dimensions in any order he likes, and at the same time leave no doubt as to which item he has dealt with and which he has not.

When all the items are transferred to the abstract sheets all entries are checked, cast, deductions made and the totals 'reduced' to the recognised units of measurements used in the bill. All the latter work is checked and the totals finally transferred to the bill of quantities. In order to indicate that the deductions have been made and the adjusted quantity transferred to the bill, the dimensions are crossed as shown below.

Concrete work

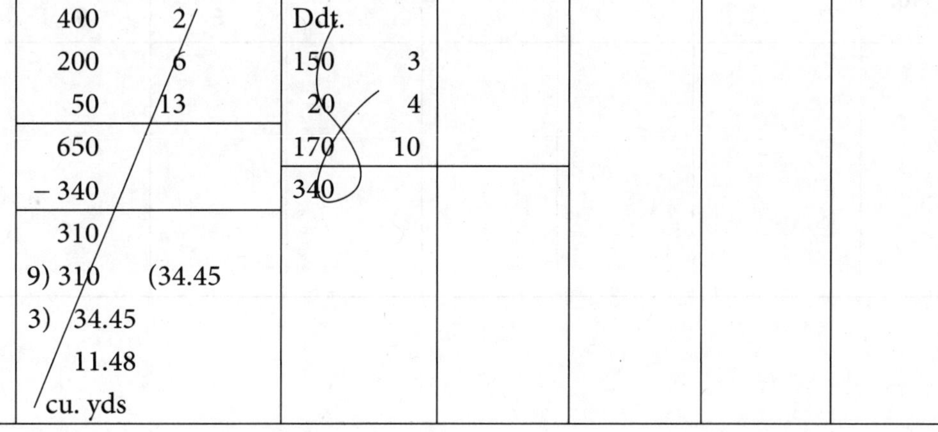

18.4   BILL OF QUANTITIES

'Billing' in the English system of taking out quantities is theoretically copying out the quantities and descriptions from the abstract in the form of a schedule or list on a specially ruled paper with money columns for pricing. The only difference between 'abstract' in the PWD system and the 'bill of quantities' in the English system is that in the latter, the items of work are grouped according to trades whereas in the former they are in the order of construction.

Two forms of ruling of bill of quantities are shown below. The first is used in India and the second in Great Britain and USA.

FORM I

Item No.	Estimated quantity	Item of work	Tendered rates			Unit	Total amount
			in figures		in words		
			₹	P.			

FORM II   Bill of quantities

Item No.	Description	Quantity	Unit	Rate	Amount		
					£	$	d/c

CHAPTER 19

Factors Affecting Cost of any Work

19.1 GENERAL

A comparative statement of tenders received for a work often shows considerable differences among the tendered sums, and between tendered sums and the estimated cost of the work. This, even after allowing for the higher tenderers not wanting the job, still indicates that the prices have not been built up on a sound and scientific basis. Lack of thorough knowledge of the various factors affecting the price of an item of work must be considered as the chief cause for such differences although many other factors such as cut-throat competition to keep staff and machinery going, rushed tenders, firms of different standards and so on are also responsible to some extent.

It is essential that the estimators, architects, engineers and contractors must have a sound practical understanding of all the factors involved in building up a price.

The various factors affecting the unit price of an item of work may be grouped under the five main component parts of which the unit price is built up. These are:

1. material,
2. labour,
3. plant,
4. overheads, and
5. profit.

The most important considerations in each of the above sections are dealt with in the subsequent articles.

19.2 MATERIALS

As a generalisation it may be stated that the cost of materials in a unit price of an item of work is nearly fifty per cent of the total price. Any savings effected or excess amount spent on the materials would, thus, decrease or increase the total cost of the work by a considerable margin. Various factors which would affect the cost of materials are listed below:

1. *Quantity of materials*: The estimate of the amount of materials required must be accurate. Due allowance must be made for waste and for extra material for laps and so on, where

measurements are net. Any error in such estimates is bound to reflect in the unit price worked out.

If materials on the site of work are found to be in excess of the requirement, such materials have to be disposed of often at a loss. Even when such materials are used on other works, the transportation charges may be heavy and double handling may increase the waste. On the other hand, if materials purchased and transported on site are found to be short of the requirement it may result in:

(i) stoppage of the work, and

(ii) purchase and transportation, of material often at a higher cost than the cost involved in bulk purchase and transportation.

It is obvious from the above considerations that any mistake in estimating the amount of materials required for executing a work is bound to affect the actual cost of the work.

2. *Quality of materials*: Materials of the specified quality should be purchased. Any cheap rate would be more than offset if the materials of inferior quality brought on the site are rejected by the architect or engineer.

3. *Large quantities*: Materials, if purchased in large quantities, can be had at cheaper rates than when purchased in small quantities. Bulky materials such as bricks, aggregate and so on, purchased directly from the producers may cost less than if purchased from builder's merchants or suppliers.

It is generally preferable to purchase materials from a few specialist builders' merchants. The advantages in so doing are:

(i) efficient service,

(ii) cheap rates due to bulk purchase, and

(iii) reduction in the number of accounts.

4. *Conditions of supply*: Quotations should be obtained for materials and supplies of all kinds. For bulky materials, quotations should be for materials delivered to site. While considering the different quotations, rates alone should not be the criterion. Conditions of the supply stated by different suppliers must be carefully studied. For example, a low price for a large quantity may be far from economic if it is tagged with one or more of the conditions mentioned below:

(i) acceptance of the whole quantities at once,

(ii) no guarantee of deliveries as and when required, and

(iii) dumping the whole material at the entrance.

19.3 LABOUR

Labour is the most uncertain factor in ascertaining the estimated cost of any unit of work. An obvious approach to determine the cost of labour is to assume hourly or daily output of a workman and find out the total number of man-hours or days required for the given work. These are multiplied by the wages per man-hour or per labour per day. Uncertainty in labour cost is due to the following reasons:

1.   Output of different artisans varies considerably according to their skill.
2.   Output of an individual craftsman also varies from day to day depending upon such varied things as the weather and his domestic life.
3.   Efficient site organisation and management increases the labour output whereas faulty or slack organisation will decrease it.
4.   Higher wages, incentive payments, and provision of other facilities such as clean and healthy labour camps, less working hours, improve the output.

Peculiarities of the job and site may have some effect on the output of the workmen. Extra rates have to be paid to workers engaged on work entailing extra discomfort, inconvenience or risk.

Past experience coupled with a thorough study of the peculiarities of the proposed work should enable an estimator to determine a realistic cost of the labour involved. The rates of wages should include the provision for the following:

1.   insurance of various kinds,
2.   holidays with pay,
3.   workmen's compensation,
4.   welfare obligations, and
5.   extra rates for overtime and for workmen on night gangs.

19.4   PLANT

Owing to the shortage of labour, the use of plants is increasing and is likely to continue to do so. However, it must not be assumed that the use of a mechanical plant is warranted in all circumstances. In fact, the cost of work will depend upon the decision of the estimator with respect to two basic questions namely, whether a plant can be used and, if a plant can be used, whether it is cheaper to purchase or to hire the plant.

On a large extensive work, there may be ample opportunity to use a mechanical plant and if human labour is used instead, the cost of the work would increase. On the other hand, if there is no space sufficient for the movement of a plant or if the amount of work involved is small, the extensive use of a plant may prove impracticable or wasteful.

Purchase of a plant would be economic only if the plant is to operate at its maximum capacity for a large portion of each year of its life. Only general purpose machines which can be used to carry out a number of building operations are economic to purchase. Purchase of specialised plants may prove to be economic only if there is possibility of sufficient work ahead where it can be used. Invariably it may prove economic to hire a specialised plant, if the contractor has undertaken a work outside his normal activities.

Whether a plant is owned or hired, to keep its cost to the minimum, it must operate as near its peak performance as possible. To achieve this, the labour force and other plants have to be matched with it. This is a difficult task that the site management has to perform and upon its efficiency will depend the cost of the plant.

To calculate the cost of operating a plant owned by the contractor or to determine its hire charges, the following items should be taken into consideration:

1. depreciation,
2. interest on capital outlay,
3. maintenance and repairs,
4. fuel and lubricating oil,
5. operators' wages, their travelling expenses and lodging allowances, and
6. transport of plant to and from site.

The costs under first three of the above items are known as 'standing charges' and those of items four and five are called 'running costs'.

The cost of a plant worked out as above should be allocated to those items of work for which it is used. In the case of minor tools and the plant, it would be difficult to allocate their use to individual items of work; so, their cost can more conveniently be included under 'overheads'.

19.5 OVERHEAD CHARGES

Under this heading are included all costs which cannot be specifically attributed to any particular contract. Examples of such costs are office expenses, including erecting office sheds, rents, rates, lighting and heating, stationery, postage, telephone accounts, salaries of staff and so on. In a similar manner, the costs of small plants, such as iron baskets, ladders, ropes and so on, are also included.

Estimating overheads: Every firm maintains the account of expenses under these headings. At the end of each financial year, it is therefore possible to find out the percentage these costs bear to the total turnover of the firm. For example, if the overhead charges of a firm amounted to ₹900,000 in a year, and if the turnover of the firm in that year was ₹1,00,00,000, then the percentage would work out to be 9.

The firm would allow 9% to the net cost of all items to cover overheads.

This percentage figure will vary from firm to firm depending upon the size of the firm and efficiency of its site organisation and management. An average value of 10% is taken while analysing the rates of a few items of work in Chapter 20.

19.6 PROFIT

The margin of profit a contractor may expect depends upon many factors. If he has contracts almost to the limit of his capacity, he will be prepared to take on new work only at a comparatively high figure. Another contractor, in need of work to keep his workmen and equipment fully occupied, may expect only a nominal profit margin.

The cost of a work is another factor on which the profit margin will depend. More the cost of a work, smaller is the percentage of profit. The percentage varies from 15% on small jobs to 5% on big ones. Overheads and profit may be accounted for in any of the following ways:

1. To the cost of materials, labour and plants, add say 10% for overheads and then to the total add say another 10% for profit.
2. To the cost of materials, labour and plants, add for overheads, and profit say $(10 + 10 =)$ 20%.

It may be noted that in the first case 10% profit is expected on the overheads also.

The Rates and Cost Committee's report, published in 1957 by the Government of India, has recommended that an allowance of 10% would be adequate for the contractor's actual expense on overheads and an allowance of 10% of the prime cost as the contractor's profit would be reasonable.

CHAPTER 20 | Analysis of Rates

20.1 GENERAL

The various factors affecting the price of any item of work are discussed in Chapter 19. In this chapter, it is proposed to frame a few examples of analysis for the reader's guidance. It is obvious that while framing such examples, certain assumptions regarding the factors affecting the price of an item of work have to be made. Also the costs of materials and labour rates constantly undergo changes from time to time and from place to place. It must therefore be clearly understood that the prices of various items of work analysed in the following sections represent the general rates for the respective items of work described under the assumed conditions. They are based on the average costs of materials and labour rates during the year 2014–15.

20.2 EXCAVATION

A typical excavating job will usually require loosening and digging, transportation and disposal. The methods commonly used for these operations are:

1. hand labour,
2. excavation by hand labour and transportation and disposal by power equipment, and
3. excavation, transportation and disposal by power equipment.

The method that is best suited to a given job will depend upon a number of factors. Generally it may be correct to say that the first method would be suitable and economic for works involving small quantities of excavation. If the haulage distance is more than 100 m, the second method would be advantageous. For works involving large quantities of excavation, the third method may be the most economic and suitable.

The information that is useful in analysing the price per unit of excavating and removal of stuff under different conditions is given in the tables. The use of tables is illustrated by analysing prices of typical items of work.

20.2.1   Excavation by Hand Labour

Memoranda

Table 20.1 Labour in excavating and so on: Average quantity excavated and thrown out by an excavator working 8 hours per day in a small gang

Particulars	Units per day	Hours per unit
1. Surface dressing, average 15 cm deep in common top soil	30 sq. m	0.25
2. Excavation over areas in		
(i) ordinary soil	5.50 cu. m	1.50
(ii) hard soil	4.00 cu. m	2.00
(iii) ordinary rock	0.60 cu. m	13.00
(iv) hard rock (requiring blasting)	0.60 cu. m	13.00
(v) hard rock (chiselling)	0.20 cu. m	40.00
3. Excavation in trenches, in		
(i) ordinary soil	4.00 cu. m	2.00
(ii) hard soil	2.00 cu. m	4.00
(iii) ordinary rock	0.50 cu. m	16.00
(iv) hard rock (blasting)	0.75 cu. m	10.00
(v) hard rock (chiselling)	0.15 cu. m	54.00
4. Removing excavated material		
(i) vertical removal: lift 15 m	8.00 cu. m	1.00
(ii) removing through a distance not exceeding 50 m	10.00 cu. m	0.80
5. Return and fill in layers	15.00 cu. m	0.55
6. Ramming around foundations in 30 cm layers	15.00 cu. m	0.55
7. Filling in carts or trucks	10.00 cu. m	0.80

Table 20.2 Bulk when excavated

	Type of soil excavated	Increase in bulk percentage
1.	Earth and clay	25
2.	Sand and gravel	10 to 13
3.	Chalk	33
4.	Rock	50

Table 20.3   Capacity of vehicles, dumpers

	Particulars of vehicles	Capacity
1.	10- to 12-tonne lorry	6.00 cu. m
2.	$\frac{1}{2}$ tonne (25 hp) dumper	0.375 cu. m
3.	1 tonne dumper	0.75 cu. m
4.	Two-bullock cart	0.33 to 0.75 cu. m
5.	Ordinary wheelbarrow	0.56 cu. m
6.	Basket	0.035 cu. m

Analysis

Item 1.   Surface dressing average 15 cm deep and removing excavated material 50 m away.

Labour for: 10 sq. m	Hours	Rate ₹	Per	Amount ₹
(Refer to Table 20.1) Mazdoors for excavating $0.25 \times 10 = 2.50$ Mazdoors for removing 50 m Increase in bulk 25% $10 \times 0.15 \times 1.25 = 1.875$ cu. m	2.50	82.00	Hour	205.00
$1.875 \times 0.80 = 1.50$ hour	1.50	82.00	Hour	123.00
Add 20% establishment charges and profit Price per 10 sq. metre	Total			328.00 65.60
				393.60

Price per sq. metre say ₹12 to 15.

Item 2.   Excavation over area for a basement in hard soil, depth 1.5 m and removing the material through a distance of 50 m.

Labour per 1 cu. m	Hours	Rate ₹	Per	Amount ₹
1. Mazdoors for excavating (refer to Table 20.1)	2.00	82.00	Hour	164.00
2. Mazdoor for lifting $= 1.25 \times 1 = 1.25$	1.25	82.00	Hour	102.50
3. Mazdoors for removing $= 1.25 \times 0.80$	1.00	82.00	Hour	82.00
	Total			348.50
Add 20% establishment charges and profit				69.70
				418.20

The rate varies from ₹400 to ₹500 per cu. m.

Item 3. Excavation in trenches in ordinary rock including returning and filling in layers. Lift 1.5 m.

Labour for 1 cu. m	Hours	Rate ₹	Per	Amount ₹
1. Mazdoors for excavating (Table 20.1)	16.00	82.00	Hour	1,312.00
2. Mazdoor for lifting $1.50 \times 1 = 1.55$	1.50	82.00	Hour	123.00
3. Mazdoors for removing $1.50 \times 0.80 = 1.20$	1.20	82.00	Hour	98.40
4. Mazdoors for return and fill in layers $1.50 \times 0.55 = 0.83$	0.83	82.00	Hour	68.06
		Total		1,601.46
Add 20% for establishment charges and profit				320.29
				1,921.75

Price per cubic metre say ₹1900 to 2000.

20.2.2 Earth-moving Equipment

Where large quantities of excavation are involved, the use of one or more types of earth-moving machines is essential on grounds of economy. The more generally used types of earth-moving equipment, their approximate capacities, economic hauls, loading and unloading times are given in Table 20.4.

Further details of each of the equipment described in the table are given in Table 20.5 to 20.9.

Table 20.4 Earth-moving equipment data (approximate)

	Equipment	Capacity in cu. m	Economical haul in m	Time in minutes		Speed m/min	
				Load	Unload	Loaded	Empty
1.	Bulldozer Angledozer	0.70 to 7.00	up to 100	—	—	30 to 50	50 to 100
2.	Scraper Tractor	3.00 to 10.50	60 to 500	0.25 to 1.50	0.20 to 0.50	70 to 115	100 to 130
	Motorised	up to 18	300 to 900	0.25 to 2.00	0.20 to 1.00	70 to 130	100 to 130
3.	Power shovel	0.38 to 1.53	—	—	—	—	—
		Upto 30.00	—	—	—	—	—
		exceptionally	—	—	—	—	—
4.	Trucks and dumpers	0.5 to 19.00	over 200	1.00 to 20.00	0.26 to 2.00	3 to 12	5 to 15

20.2.3   Method of Working out Hourly Ownership and Operating Cost of Construction Equipment

There are several methods of determining the probable cost of owning and operating construction equipment. However, there is no method that gives exact costs under all operating conditions.

The cost of owning and operating a plant may include the following items:

1.   *Depreciation*: Straight line method may be used to determine depreciation per unit time. For example, a given unit of equipment, whose original cost is ₹5,00,000 may have a useful life of 3000 hours per year for 5 years and a salvage value of ₹50,000. The amount of depreciation is determined as follows:

$$\text{Total depreciation} = ₹5,00,000 - 50,000 = ₹4,50,000$$

$$\text{Annual depreciation} = 4,50,000/5 = ₹90,000$$

$$\text{Hourly rate of depreciation} = 90,000/3000 = ₹30.00$$

2.   *Cost of maintenance and repairs*: As an approximation these costs may be taken 100 per cent of the amount of depreciation.
3.   *Investment costs*: These include interest, insurance, taxes and storages. The cost under each of these factors is expressed as a per cent of the average value of equipment. The average value of the equipment as a per cent of the original cost may be determined by the following formula:

$$\text{Average value, \% of original cost} = \frac{(1+n)\times 100}{2n}$$

For example, a plant with five years of useful life will have:

$$\text{Average value} = \frac{(1+5)\times 100}{10} = 60\% \text{ of the original cost.}$$

The following percentage figures may be used as a guide in determining the total investment cost per annum.

Interest	14%
Insurance	1%
Taxes	2.5%
Storage	2.5%
Total	20.0%

Running costs: These include the cost of the fuel consumed, lubricating oil and so on, and will vary with the type of equipment, the conditions under which it is used, and location. It is possible to estimate cost with a reasonable degree of accuracy for a given project.

The costs per working hour of a few types of equipment are worked out by the above method and tabulated in Table 20.5.

Table 20.5   Hourly ownership and operating cost of construction equipment

Equipment	Hourly hire charges
Bulldozer	₹3,966.00
Scraper	₹15,000.00
Tractor operated shovel	₹2,000.00
Dumper	₹1,000.00 to 1,100.00
Concrete mixer 280 NT	₹534.00

Table 20.6   Blade capacities and bulldozer output, in cubic metres (bank measure)

Blade size		Tractor drawbar hp	Speed		Blade capacity in cu. m	Output, cu. m per h			
Length in m	Height m		Forward	Reverse		Haul distance in m			
						30	60	90	120
4.09	1.33	160	30	60	7.20	160	175	185	200
3.43	1.16	130	45	100	3.65	140	80	56	43
3.12	1.16	80	36	100	3.35	115	65	45	35
2.90	0.96	65	36	100	2.23	74	41	28	22
2.49	0.96	65	36	100	1.82	64	35	25	19
2.18	0.83	50	45	50	1.14	35	20	13	10
1.73	0.70	35	45	55	0.68	22	20	8	7

Note: Bulldozer output per trip may be worked out as below:

1. Volume of earthwork per trip

$$V = \frac{L \times nH^2}{2}$$

where L = length of blade; H = height of blade; n = side slope of earth moving in front (normally assumed 2:1).

2. Output in cu. yd per hour

$$V = \frac{\text{Drawbar hp of the tractor}}{\text{Dozing distance in hundreds of ft}}$$

Table 20.7 Representative hourly output for crawler and wheel type tractor scrapers in cubic metres (bank measure)

Haul distance, one way, in metres	Crawler tractor	Wheel tractor
60	90	80
120	73	70
180	60	60
240	52	55
300	45	50
450	34	40
600	27	34
900	20	26
1200	16	20
1500	13	17

Table 20.8 Average hourly output of power shovels, in cu. m per h for 90° swing, optimum depth and no delays

Class of material	Size of shovel (cu. m)			
	0.4	0.6	0.8	1.5
1. Ordinary soil	80	125	152	215
2. Hard soil	57	80	110	160
3. Well blasted rock	45	72	95	135
4. Poorly blasted rock	19	39	57	85

20.2.4 Transportation of Excavated Material

Dumpers, trucks and tractor-pulled wagons may be used to haul the materials excavated by power shovels. The capacity of a hauling unit is normally expressed in cubic metre as struck or heaped. However, it is advisable to ascertain the actual capacity of a unit by measuring the volume of earth in several representative loads.

If the volume of excavated material is expressed as bank measure, loose volume should be calculated thus:

$$\text{Loose volume} = \text{Bank volume} \times (1 + \text{increase in bulk expressed in decimal})$$

The capacity of vehicles is given in Table 20.3 and increase in bulk when excavated in different soils is also given in Table 20.3.

Table 20.9 gives the number of trips per hour needed to be taken by trucks for various haul distances and speeds.

20.2.5 Analysis

(*Note*: In the following analysis, the costs of delivery of the machines to, and removal from, the site are not included.)

Item 4.   Excavate to reduce level by a bulldozer with a haul distance of 60 m.

Output of bulldozer per hour (from Table 20.6): 20 cu. m
Analysis

One hour of bulldozer with operator (From Table 20.5)	₹3,966.00
Add 20% establishment charges and profit	₹793.20
Total cost for 20 cu. m	₹4,759.20

Rate per cubic metre = say ₹238.00

Item 5.   Excavate over site with 12 cu. m capacity wheel tractor scraper, hauling about 300 m and deposit on site.

In medium soil the average output for 300 m haul distance from
Table 20.7   ...   ... = 50 cu. m
Analysis

One hour of scraper with operator (Table 20.5)	₹15,000.00
Add 25% establishment charges and profit	₹3,750.00
Total cost for 50 cu. m	₹18,750.00

Rate per cubic metre = 18750/50 = ₹375.00

Table 20.9   Number of trips per hour for trucks

One way distance in km	Time in minutes			Operating factor[2]	Trips per hour
	Fixed time	Travel round trip	Round trip time		
		(1) Speed 15 km per hour			
1	7	8	15	0.83	3.32
2	7	16	23	0.83	2.16
3	7	24	31	0.83	1.62
4	7	32	39	0.83	1.28
5	7	40	47	0.83	1.06

(*continued*)

Table 20.9   (Continued)

One way distance in km	Time in minutes			Operating factor[2]	Trips per hour
	Fixed time	Travel round trip	Round trip time		
(2) Speed 20 km per hour					
1	7	6	13	0.83	3.85
2	7	12	19	0.83	2.63
3	7	18	25	0.83	2.00
4	7	24	31	0.83	1.62
5	7	30	37	0.83	1.32
(3) Speed 30 km per hour					
1	7	4	11	0.83	4.55
2	7	8	15	0.83	3.33
3	7	12	19	0.83	2.63
4	7	16	23	0.83	2.18
5	7	20	27	0.83	1.84
(4) Speed 40 km per hour					
1	7	3	10	0.83	4.98
2	7	6	13	0.83	3.85
3	7	9	16	0.83	3.11
4	7	12	19	0.83	2.63
5	7	15	22	0.83	2.26

Note: [1]*Fixed time: 4 min. at the shovel and 3 min. at the dump.*
   [2]*Actual operating time is assumed to be 50 min. per hour.*

Item 6.   Excavate in ordinary soil with a 1.135 cu. m capacity tractor-operated shovel and haul 2 kilometres in 6 cu. m dumpers.

Output of shovel per hour (from Table 20.8) 215 cu. m solid

Dumper capacity, 6 cu. m loose = 6/1.25

(Refer to Table 20.3) = 5 (say) cu. m solid

Therefore, dumper loads and journeys = 215/5 = 43

Total round trip time = 15 min. ($\because$ from Table 20.9, assuming dumper speed 30 km per hour)

Loads per hour = 60/15 = 4

Dumpers required = 43/4 = say, 11

Analysis

One hour of shovel and operator (from Table 20.5)	₹2,200.00
11 hours dumpers and operators at ₹650.00 per dumper, per hour	₹11,000.00
Total	₹13,000.00
Add 20% establishment charges and profit	₹2,600.00
Total cost for 215 cu. m =	₹15,600.00

Rate per cubic metre = ₹15,600/215 = (say) ₹72.55 say ₹70 to 75.

20.2.6   Schedule of Rates

Item	Rate ₹	Per
1. surface dressing average 15 cm thick, in ordinary soil	21.00	sq. m
2. excavation over areas lift 1.5 m lead 50 m in		
(i) ordinary soil	207.00	cu. m
(ii) hard soil	365.00	cu. m
(iii) ordinary rock	429.00	cu. m
(iv) hard rock (blasting)	1,033.00	cu. m
3. Excavation in trenches for foundation, lift 1.5 m, lead 50 m		
(i) ordinary soil	207.00	cu. m
(ii) hard soil	365.00	cu. m
(iii) ordinary rock	429.00	cu. m
(iv) hard rock (blasting)	1,033.00	cu. m
(v) hard rock (chiselling)	1,322.00	cu. m
4. Filling in plinth with approved excavated material	120.00	cu. m
-do- Material brought from outside	599.00	cu. m
5. Hand packed rubble stone soling	1,382.00	cu. m

20.3   RATES OF MATERIALS AND WAGES OF LABOUR

Rates of materials and hourly and daily wages of labour used in the analysis of rates in the following sections are given in Tables 20.10 and 20.11, respectively.

Table 20.10   Rates of materials

Material	Rate ₹	Per
Cement (open market)	6,000.00	t
Sand	1,670.00	cu. m
Broken stone 25 mm to 37.5mm	1,050.00	cu. m
Broken stone 12 mm to 25 mm	1,050.00	cu. m
Mild steel	61,000.00	tonne
Headers	70.00	no.
Khandli	88.00	no.
Bricks	9,000.00	1,000 nos
Teak wood (good quality)	1,18,981.00	cu. m
Battens	51.00	m
Manglore tiles	24,000.00	1,000 no.
Manglore ridge tiles	42.00	no.
Asbestos cement sheets		
(i) Corrugated	149.00	sq. m
(ii) Plain	257.00	sq. m
Galvanised corrugated and plain 24 BWG	49,300.00	q
Shahabad stone (rough, 4 cm thick)	283.00	sq. m
Shahabad stone (polished, 2.5 cm thick)	292.00	sq. m
Ceramic tiles	509.00	sq. m
Marble chips	56.00	kg
Granite	2,656.00	sq. m
Rubble	653.00	cu. m

Table 20.11   Labour rates

Kind of labour	Rate in ₹	
	Per day	Per hour
Mazdoor		
(i) Male	615	82
(ii) Female	615	82
Bhisti	648	85
Mason	677	90
Fitter	677	70
Helper to fitter	615	45
Carpenter	677	70
Helper to carpenter	615	45
Tile layer	615	65
Painter	677	60

20.4   TRANSPORT OF MATERIALS

The bulk and weight of building materials and plant make it necessary to consider carefully the cost of transport. Materials may be transported by road or rail. Table 20.12 gives useful information for calculating the lead charges of materials while analysing the unit price of an item of work. The procedure of calculating lead charges with the help of Table 20.12 is illustrated in an example. The analysis of rates in the following sections include lead charges for an initial lead up to 5 km.

Table 20.12   Data for calculating transportation charges of materials

Material	Capacity		Labour hours	
	Bullock cart **0.75 to 1 t**	**Truck** **3 to 5 t**	**Bullock** **cart**	**Truck**
Crushed stone Gravel	0.85 cu. m	3.00 cu. m	0.75 to 1.00	3 to 6
Sand	0.85 cu. m	3.00 cu. m	0.75 to 0.50	1.50 to 1.00
Steel bars (bundled)	0.75 t to 1.00 t	3.00 t to 5.00 t	0.40 to 1.25	1.50 to 3.75
Rubble	0.33 cu. m	0.95 cu. m	0.30 to 0.50	2.00 to 3.00
Brick $19 \times 9 \times 9$ cm	2.50 to 3.00	1000 to 1200	1.00 to 1.20	2.00 to 2.50
Mangalore tiles nos	250 to 300	1000 to 1200	1.20 to 1.50	2.5 to 3.0

Note: Average speed of bullock cart-3 km per hour.
Average speed of truck-45 km per hour.

Example 20.1

Estimate the total cost and cost per tonne of loading, unloading, and hauling 1000 tonnes of cement from railway wagons to a work site 4 km distant. Trucks will haul 12 tonnes per load and can maintain an average speed of 20 km per hour allowing for lost time. Four mazdoors will be stationed at the wagon and two at the job to unload the trucks. A mazdoor will handle about three tonnes per hour.

Analysis

Rate of unloading the wagons, $4 \times 3 = 12$ tonnes per hour

Rate of loading the truck, 4×3 $= 12$ tonnes per hour

Time to load a truck, 12/12   =   1 hour

Time to unload a truck, same   =   1 hour

Travel time round trip, 4 km

@ 20 km per hour, (Table 20.9)   =   0.50 hour

Round trip time   =   2.50 hour

Trips per hour per truck $= 1/2.50 = 0.40$

Cement hauled per hour by one truck $12 \times 0.40 = 4.80$ t

No. of trucks needed to haul 12 t per hour $= 12/4.8 = 2.5$

Using two trucks, the time required will be $= 1000/9.6 =$ say 104 hours

The cost will be:

Trucks with drivers $2 \times 104 = 208$ hrs @ 600.00 =	₹	1,24,800.00
Labourers $4 \times 104 = 416$ h @ ₹82.00	= ₹	34,112.00
Total cost	= ₹	1,58,912.00
Handling and transporting cost per tonne	= ₹	158.91

If three trucks are used, the time required will be $= 1000/14.4 =$ say 70 hours.

The cost will be:

Trucks $3 \times 70 = 210$ hours @ ₹600 per hour	= ₹	1,26,000.00
Labourers $4 \times 70 = 280$ hours @ ₹82 per hour	= ₹	12,600.00
Total cost	= ₹	1,71,920.00
Handling and transporting cost per tone	= ₹	171.92

It appears economical to use three trucks.

20.5   CONCRETE

Memoranda.

20.5.1   Quantities per Metre Cube

When the proportions of mix are given by volume, the quantities required per cubic metre may be approximately computed as below:

Let a, b, c be the proportions by volume of cement, fine aggregate, and coarse aggregate, respectively. The proportion of cement a is usually taken as unity. If V is the void percentage (expressed in decimal) in aggregate, then

$$\text{Quantity of cement per cu. m of concrete} = (1 + V)a/a + b + c$$

$$\text{Quantity of fine aggregate per cu. m of concrete} = (1 + V)b/a + b + c$$

$$\text{Quantities of coarse aggregate per cu. m of concrete} = (1 + V)c/a + b + c$$

If, for example, broken stone aggregate is found to have about 45% voids, the volume of dry materials required for one cubic metre of concrete would be: $1 + 0.45$ cu. m.

1:3:6 mix has 10 parts $(1 + 3 + 6)0.145$

Table 20.13 Materials required per cubic metre of cement concrete (based on Portland cement weighing 1442 kg per cubic metre and assuming 45 per cent voids in aggregate)

Mix	Nominal mix	Proportions for one bag batch			Gross volume of materials per batch	Yield per bag of cement	Minimum size of mixer for 50 kg batch	Materials per cubic metre of concrete				
		Cement	Fine aggregate	Coarse aggregate				Cement	Fine aggregate		Coarse aggregate	
Unit	–	kg	cu. m	cu. m	cu. m	cu. m	litres	t	cu. m*	t	cu. m*	t
M 200	$1:1\frac{1}{2}:3$	50	0.052	0.104	0.191	0.128	200 T or 200 NT	0.395	0.40	0.585	0.81	1.17
M 150	$1:2:4$	50	0.07	0.14	0.245	0.16	280 NT	0.315	0.44	0.63	0.88	1.26
	$1:2\frac{1}{2}:5$	50	0.088	0.175	0.298	0.193	400 NT	0.259	0.46	0.65	0.91	1.30
M 100	$1:3:6$	50	0.105	0.21	0.35	0.222	400 NT	0.225	0.47	0.675	0.94	1.35
	$1:4:8$	50	0.138	0.277	0.45	0.29	800 NT	0.172	0.48	0.69	0.96	1.38
	$1:5:10$	50	0.174	0.347	0.556	0.35	800 NT	0.143	0.50	0.715	1.00	1.43

(*Worked to the nearest of 0.01 cu. m)

$$1 \text{ part cement} = \frac{(1+0.45)\times 1}{1+3+6} = \frac{1.45}{10} = 0.145 \text{ cu. m}$$

$$3 \text{ parts fine aggregate} = \frac{(1+0.45)\times 3}{1+3+6} = 0.435 \text{ cu. m}$$

$$6 \text{ parts coarse aggregate} = \frac{(1+0.45)\times 6}{1+3+6} = 0.870 \text{ cu. m}$$

Quantity of water required for mixing concrete can easily be found from the 'water–cement ratio'.

Quantities measured by weight: The computation required when the proportions are given by weight are quite simple. For example, in a 1:2:4 mix by weight, there would be 1 kg of cement for every 2 kg of fine aggregate and 4 kg of coarse aggregate.

Table 20.13 and 20.14 give approximate quantities of materials required for one cu. m of concrete for some of the more common mixes.

Table 20.14   Water–cement ratios

DATA	
1 cubic metre of cement weighs	1442 kg
1 bag of cement weighs	50 kg
1 litre of water weighs	1 kg
1 cubic metre of water weighs	1000 kg

Water–cement ratio by weight	Quantity of water in litres per 50 kg cement
0.31	15.50
0.36	18.00
0.40	20.00
0.45	22.50
0.49	24.50
0.54	27.00
0.58	29.00
0.63	31.50
0.67	33.50
0.71	35.50
0.76	38.00
0.80	40.00

20.5.2   Labour

The number of labour hours required for mixing, placing, compacting or consolidating and curing 1 cu. m of concrete varies greatly according to the nature of the job, conditions of the work, skill of the workers and efficiency of site organisation and management.

Table 20.15 gives approximate values of labour hours for structural concrete.

Table 20.15   Approximate labour required for mixing, placing, compacting and curing concrete

Kind of work	Labour hours per cu. m			
	Mazdoor		Mason	Bhisti
	Male	Female		
Hand mixing	5	–	–	0.50
Machine mixing (conveying aggregate, cement and water)	3	–	–	0.50
Placing in foundations	1.5	3.5	0.5	0.50
Placing in columns and thin walls	2 to 2.5	4.5	0.50	0.50
Placing in thick walls	1.5 to 2	3.5	0.50	0.50
Placing in stairs	2 to 3	3.5	0.50	0.50
Compacting concrete	0.50	–	–	–
Curing concrete	–	–	–	10

20.5.3   Plant

The plant required for a structural concretework may vary from a small mixer and minor tools and plant such as *phawaras*, iron baskets, wheel barrows and so on, to a big mixer, weighing and heating devices, cranes or derricks and others, depending upon the size of the job. The plant commonly used on small to moderate sized jobs include mixers and vibrators.

Table 20.16 gives approximate values of output in metre cubes for mixers of different sizes.

Table 20.16   Average output of mixers per 8 hour day (based on approximately 3 minutes period for filling, mixing and discharging)

Size of mixer (litres)	Output in cu. m (hand loading)
200 T or 200 NT	20/22
280 NT	28/30
400 NT	32/35
300 NT	45/50

T: tilting; NT: non-tilting

Analysis

Item 7.　Providing and laying plain cement concrete (M:10) nominal mix in foundation trenches including compacting and curing.

Analysis

Materials for 1.00 cu. m	Quantity or number	Rate ₹	Per	Amount ₹
(Reference to Table 20.13)				
Coarse aggregate	0.96 cu. m	1,050.00	cu. m	1,008.00
Fine aggregate	0.48 cu. m	1,670.00	cu. m	801.60
Cement	0.1725 t	6,000.00	T	1,035.00
Water (Table 20.14 water–cement ratio 0.49)	Lump	–	–	50.00
Sundries	Lump			50.00
		Material cost		2,944.60
Labour				
For 1.00 cu. m				
Machine mixing (Reference: Table No. 20.15)				
Mazdoors for conveying aggregate and cement	3 hours	82.00	hour	246.00
Male mazdoors for placing	1.5 hours	-do-	hour	123.00
Female mazdoors for placing	3 hours	82.00	hour	246.00
Male mazdoors for compacting	0.5 hours	82.00	hour	41.00
Bhisti (including curing)	1.5 hours	85.00	hour	127.50
		Labour cost		783.50
Plant				
Mixer				
280 NT	0.3 hours	534.00	hour	160.20
		Plant cost		
Total cost of material, labour and plant			₹	3,887.70
Add 20% establishment charges and profit			₹	777.54
Total cost			₹	8,553.54

Rate per cubic metre say, ₹5000

Item 8.   Providing and laying M 20 (nominal mix 1:2:4) cement concrete for RCC work in beams and slabs including compacting and curing.

Analysis

Materials for : 1.00 cu. m	Quantity or number	Rate ₹	Per	Amount ₹
(Reference to Table 20.13)				
Coarse aggregate	0.88	1,050.00	cu.m	924.00
Fine aggregate	0.44	1,670.00	cu.m	734.80
Cement	0.315	6,000.00	T	1890.00
Water	lump			50.00
Sundries	lump			50.00
		Material cost		3648.80
Labour for: 1.00 cu. m				
(Reference to Table 20.15)				
Male mazdoor for conveying aggregate and cement	3 h	82.00	hour	246.00
Male mazdoor for placing (lift 3 m to 4 m)	3 h	82.00	hour	246.00
Female mazdoor for placing	3.5 h	82.00	hour	287.00
Male mazdoor for compacting	0.5 h	82.00	hour	41.00
Bhisti	2 h	85.00	hour	170.00
		Labour cost		990.00
Plant for: 1.00 cu. m				
Mixer 280 NT	0.30 h	534.00	hour	160.00
Needle vibrator	0.30 h	179.00	hour	179.00
		Plant cost		339.20
Total cost of material, labour and plant				4,978.00
Add 20% establishment charges and profit				995.60
Total cost say,				5,973.60

Rate per cubic metre ₹5,900 to ₹6,000

Item 9. Providing, cutting, bending, hooking, conveying, placing in position and binding mild steel reinforcement.

Analysis

Material for: 1 quintal	Quantity or number	Rate ₹	Per	Amount ₹
Mild steel bars including 5% waste	1.05 q	6,100.00	q	6,405.00
16 SWG wire	1.00 kg	74.00	kg	74.00
		Material cost		6,479.00
Labour cost for: 1 quintal				
Fitter	8 h	677.00	Day	677.00
Helper	8 h	615.00	Day	615.00
		Labour cost		1,292.00
Total cost of material and labour				7,771.00
Add 20% establishment charges and profit				1,554.00
Total cost				9,325.00

Cost per quintal say ₹9,350.00

20.5.4 Schedule of rates

Item	Rate (₹)	Per
1. Providing formwork to:		
(i) columns	750.00	sq. m
(ii) walls	750.00	sq. m
(iii) stairs	800.00	sq. m
(iv) beams	750.00	sq. m
(v) floors or roofs	750.00	sq. m
2. Providing, cutting, bending, conveying, placing in position and binding ms reinforcement	9,350.00	quintal
3. Providing and laying in foundation cement concrete:		
(i) 1 : 3 : 6 proportion (M:15)	6,359.00	cu. m
(ii) 1 : 4 : 8 proportion (M:10)	5,830.00	cu. m
4. Providing and laying in floors cement concrete 1 : 5 : 10 proportion	3,600.00	cu. m
5. Providing and laying cement concrete M 150 (nominal mix 1 : 2 : 4)		
(i) in column footings, sills	6,359.00	cu. m
(ii) lintels and beams	6,721.00	cu. m

(continued)

(Continued)

Item	Rate (₹)	Per
(iii) in columns	6,359.00	cu. m
(iv) in walls	6,359.00	cu. m
(v) in chajja and stairs	6,721.00	cu. m
(vi) in floor slabs and roofs	6,359.00	cu. m

20.6 STONE MASONRY

20.6.1 Materials

Materials required per cubic metre of stone masonry are given in Tables 20.17 and 20.18.

Table 20.17 Materials required for one cubic metre of stone masonry

Kind of masonry cu. m	Mortar cu. m	Stonem		
		Rubble	Khandki no.	Headers no.
Uncoursed rubble masonry	0.30 to 0.40	1.25	—	7 to 10
Random rubble masonry	0.30 to 0.35	1.25	—	7 to 10
Coursed rubble masonry (*khandi* facing and UCR backing)	0.25 to 0.30	0.50	50	7 to 10
Ashlar masonry	0.20 to 0.23	0.50	50	7 to 10

Water required may vary from 125 litres to 600 litres per m^3 of mortar. An allowance of ₹20.00 per cu. m of mortar is usually satisfactory.

Table 20.18 Materials required for 1 cu. m of mortar (weight of sand per cu. m = 1.42 t; Weight of lime per cu. m = 0.650 t)

Mix	Proportions by volume			Proportions by weight		
Cement : sand and lime : sand	Cement bags	Lime m^3	Sand m^3	Cement *t*	Lime *t*	Sand *t*
1 : 1	26.0	0.9	0.90	1.298	0.585	1.30
1 : 2	13.34	0.465	0.93	0.670	0.60	1.32
1 : 3	9.57	0.33	1.00	0.476	0.65	1.42
1 : 4	7.25	0.25	1.00	0.360	0.16	1.42
1 : 5	5.80	0.20	1.00	0.288	0.13	1.42
1 : 6	4.81	0.167	1.00	0.248	0.02	1.42

20.6.2   Labour

Labour hours required per cubic metre of stone masonry are given in Table 20.19.

Table 20.19   Labour hours required for 1 cu. m of stone masonry

Kind of masonry	Mason		Mazdoors	
	Dressing	Setting	Unskilled	Bhisti
1. UCR masonry in:				
(i) foundation and plinth	–	6 to 7	12 to 14	2 to 3
(ii) superstructure (lift 3 to 4 m)	–	7 to 8	14 to 16	2 to 3
2. Random rubble masonry in:				
(i) foundation	2	7 to 8	14 to 16	2 to 3
(ii) superstructure (lift 3m to 4m)	2	8 to 9	16 to 18	2 to 3
3. Coursed rubble masonry in: superstructure (lift 3 to 4 m)				
(i) first sort	10	11 to 12	16 to 18	2 to 2.5
(ii) second sort	8	9 to 10	16 to 18	2 to 2.5
4. Ashlar masonry in superstructure (lift 3 to 4 m)	12 to 15	13 to 15	16 to 18	2

20.6.3   Analysis

Item 11.   Uncoursed rubble masonry in cement mortar (1:6) in foundation and plinth including curing.

Material for 1.0 cu. m	Quantity or number	Rate ₹	Per	Amount ₹
(Refer to Tables 20.17 and 20.18)				
Rubble	1.25 cu. m	653.00	cu. m	816.25
Headers	7 no.	70.00	no.	490.00
Sand	0.35 cu. m	1,670.00	cu. m	584.50
Cement	0.087 t	6,000.00	t	522.00
Water	Lump	–	–	50.00
Sundries	Lump	–	–	50.00
		Material cost ₹		2,512.75
Labour				
For 10 cu. m (refer to Table 20.19)				
Masons	6 h	90.00	hour	540.00
Male mazdoors	6 h	82.00	hour	492.00

(*continued*)

(Continued)

Material for 1.0 cu. m	Quantity or number	Rate ₹	Per	Amount ₹
Female mazdoors	6 h	82.00	hour	492.00
Bhisti	2 h	85.00	hour	510.00
Curing	1 h	85.00	hour	85.00
		Labour cost ₹		2,119.00
Cost of material and labour				4,631.75
Add 20% establishment charges and profit				926.35
Cost for 1 m^3				5,558.10

Rate per cubic metre ₹5,500 to ₹5,600

Item 12. Coursed rubble masonary first sort (*khandki* facing and UCR backing in cm (1:5) in superstructure. Height above ground 3 m to 4 m.

Material for 1.0 cu. m	Quantity or number	Rate ₹	Per	Amount ₹
(Refer to Tables 20.17 and 20.18)				
Rubble	0.5 cu. m	652.00	cu. m	326.00
Khandki	50 no.	88.00	no.	4,400.00
Through stones	7 no.	102.00	no.	714.00
Sand	0.25 cu. m	1,670.00	cu. m	417.50
Cement	0.072 t	6,000.00	tonne	432.00
Water				50.00
Sundries				50.00
		Material cost		6,389.50
Labour for				
1.00 cu. m				
(Refer to Table 20.19)				
Masons for dressing	10 h	85.00	hour	850.00
Masons for setting	12 h	90.00	hour	1,080.00
Bhisti	2 h	85.00	hour	170.00
Male mazdoors	16 h	82.00	hour	1,312.00
Female mazdoors	16 h	82.00	hour	1,312.00
Blacksmith for sharpening chisels	1 h	85.00	hour	85.00
Bellows boy	1 h	82.00	hour	82.00

(continued)

(Continued)

Material for 1.0 cu. m	Quantity or number	Rate ₹	Per	Amount ₹
Scaffolding	lump			100.00
Curing (Bhisti)	1 h	85.00	hour	85.00
		Labour cost		5,076.00
Cost of materials and labour say,			₹	11,465.50
Add 20% establishment charges and profit			₹	2,293.10
Total ₹				13,758.60

Rate per cubic metre ₹13,800.00

20.6.4   Schedule of rates

	Item	Rate (₹)	Per
1.	Uncoursed rubble masonry in cement mortar (1:6):		
	(i) in foundation and plinth	4,734.00	cu. m
	(ii) in superstructure up to 4 m height	5,025.00	cu. m
2.	Random rubble masonry in superstructure in cement mortar (1 : 6). Height up to 4 m	5,423.00	cu. m
3.	Coursed rubble masonry in cement mortar (1 : 4) in superstructure:		
	(i) first sort, (a) in walls	12,427.00	cu. m
	(b) in pillars	12,485.00	cu. m
	(ii) second sort, (a) in walls	11,703.00	cu. m
	(b) in pillars	12,303.00	cu. m
4.	Ashlar dressed stone masonry in cement mortar (1 : 5)	24,201.00	cu. m
5.	Cut stone masonry in cement mortar (1 : 3) in sills, copings, steps, string course and so on.		
	(i) fine dressed	12,427.00	cu. m
	(ii) medium dressed	11,703.00	cu. m

20.7   BRICKWORK

20.7.1 Materials

Number of bricks required per cu. m of brickwork:

Standard size of brick: 19 cm × 9 cm × 9 cm

Volume of one brick = 0.001539 cu. m

Volume of bricks per metre cube of brickwork: assuming 30% mortar (1.00 − 0.30)
= 0.70 cu. m

$$\text{Number of bricks per cu. m} = 0.70/0.001539 = \quad 455 \text{ no.}$$
$$\text{Wastage say 5\%} = \quad \underline{25 \text{ no.}}$$
$$\text{Total} \quad \underline{480 \text{ no.}}$$

Table 20.20 gives the number of bricks and volume of mortar required per unit quatity of
brickwork.

Table 20.20

Kind of brickwork	Quantity of mortar in cu. m	Number of brick, (including 5% wastage)
1. Brickwork 1st sort in cm (1:4)	0.20 to 0.24	525 to 550 per cu. m
2. Brickwork 2nd sort in cm or lime mortar	0.30 to 0.35	480 to 500 per cu. m
3. 10 cm thick brickwork in cm (1:3)	0.02	50 to 55 per sq. m

20.7.2 Labour

Table 20.21 Labour hours required for unit quantity of brickwork using lime/cement mortar
and common bond

Kind of brickwork	Hours		per
	Brick layer	Mazdoor	
1. Brickwork 1st sort in cement mortar	4	4	cu. m
2. Brickwork 2nd sort in cement	3	3	cu. m
3. 10 cm thick brickwork	2	2	sq. m

20.7.3 Analysis

Item 13. Brickwork II sort in cm (1 : 6) in superstructure, 4 m height.

Materials for 1 cu. m	Quantity or number	Rate ₹	Per	Amount ₹
Bricks (size 19 × 9 × 9 cm) (Table 20.20)	480 no.	9,000.00	1000 no.	4,320.00
Sand (Table 20.18 and 20.20)	0.2 cu. m	1,670.00	cu. m	334.00
Cement (Table 20.18)	0.057 t cu. m	6,000.00	t	342.00
Water	Lump			30.00
Sundries	Lump			20.00
		Material cost ₹		5,046.00

(continued)

(Continued)

Labour 1.00 cu. m	No.	Rate ₹	Per	Amount ₹
Bricklayers	3 hours	90.00	hour	270.00
Male mazdoors for mixing mortar	3 hours	82.00	hour	246.00
Female mazdoors	3 hours	82.00	hour	246.00
Bhisti (including curing)	1.6 hours	85.00	hour	136.00
Scaffolding	Lump			50.00
		Labour cost		948.00
Total cost of material and labour				5,994.00
Add 20% establishment charges and profit				1,198.80
Total, say,				7,192.80

Rate per cubic metre ₹7,200.00

Item 14.   10 cm thick brickwork in cm (1 : 3) (Excluding reinforcement)

Materials for 10 sq. m	Quantity or number	Rate ₹	Per	Amount ₹
Bricks (size 19 × 9 × 9) cm	500 no.	9,000.00	1000	4,500.00
Cement (Table 20.20)	0.10	6,000.00	tonne	600.00
Sand (Table 20.20)	0.20 cu. m	1,670.00	cu. m	334.00
Water	Lump	25.00	hour	50.00
Sundries	Lump	12.50	hour	50.00
		Material cost ₹		5,534.00
Labour for 10 sq. m				
Bricklayers	16 h	90.00	hour	1,440.00
Male mazdoor	16 h	82.00	hour	1,312.00
Female mazdoor	16 h	82.00	hour	1,312.00
Bhisti	4 h	85.00	hour	340.00
Scaffolding	Lump	–	–	100.00
		Labour cost ₹		4,504.00
Total cost of material and labour				10,038.00
Add 20% establishment charges and profit ₹				2,007.60
Cost per 10 sq. m				12,045.60

Rate per sq. m say ₹1,205

20.7.4   Schedule of Rates

Item	Rate ₹ Ps.	Per
1. Brickwork II sort in lime mortar ($1.1\frac{1}{2}$) or cm (1 : 6) in foundation and plinth	7,755.00	cu. m
2. Brickwork II sort in lime mortar ($1:1\frac{1}{2}$) or cm in superstructure (height 3 to 4 m)	7,994.00	cu. m
3. Brickwork 1st sort in cm (1:6) or lime mortar ($1:1\frac{1}{2}$) in superstructure (height 3 to 4 m)	9,525.00	cu. m
4. 10 cm thick brickwork in cm (1:4) (height 3 m to 4 m)	802.00	sq. m
5. Cement concrete hollow block $100 \times 200 \times 400$ mm in cm 1:6	1,231.00	sq. m
6. Aerated autoclave cellular block in cm 1:6	1,026.00	sq. m

20.8   DOORS AND WINDOWS

20.8.1   Frames

Item 15.   Providing TW frames for (a) doors 87.5 cm × 1.975 m and (b) windows 87.5 cm × 107.5 cm. Size of frame members: 9.5 cm × 7 cm.

(a)   Door frame

Materials

	No.	Length in m	Breadth in m	Depth in m	Quantity
1. Timber: Verticals (197.5 cm + 2 cm)	2	2.00	0.07	0.095	0.0266
Head (87.5 cm + 30 cm horns)	1	1.17	0.07	0.095	0.0078
				Total	0.0344

Analysis

Materials For: 0.0344 cu. m	Quantity or number	Rate ₹	Per	Amount ₹
Timber	0.0344 cu. m	1,18,981.00	cu. m	4,092.00
		Materials cost		4,092.00
Carpenter for joinery and erection	4 h	90.00	hour	360.00
Helper	2 h	82.00	hour	164.00
		Labour cost		524.00
Cost of labour and materials				4,616.94
Add 20% establishment charges and profit				923.39
Total				5,540.33

Rate per cubic metre say ₹1,60,000.00 to 1,65,000.00

(b) Window frame

Material

	No.	Length in m	Breadth in m	Depth in m	Quantity
Vertical posts Head and sill	2	1.08	0.07	0.095	0.0144
(87.5 + 3 cm for horns)	2	1.17	0.07	0.095	0.0156
				Total	0.0300

Analysis

Material for 0.03 cu. m	Quantity or number	Rate ₹	Per	Amount ₹
Timber	0.03 cu. m	1,18,981.00	cu. m	3,569.43
		Material cost		3,569.43
Labour for 0.03 cu. m				
Carpenter for joinery and erection	4.00 h	90.00	hour	360.00
Helper	2.00 h	82.00	hour	164.00
		Labour cost		524.00
Total cost of material and labour				4,093.43
Add 20% establishment charges and profit				818.69
Total				4,912.12

Rate per cu. m say ₹1,64,000

The rate per cu. m of frames for doors, windows, cupboards and so on, may be taken uniform as ₹1,10,000.

20.8.2   Shutters

Door shutter: Width = overall width – 2 × width of vertical posts + 2 × 12 mm for rebate.

$$\text{Size of door} = 87.5 \text{ cm} \times 197.5 \text{ cm}$$

$$\text{Size of shutter} = 87.5 \text{ cm} - 2 \times 7 \text{ cm} + 2.4 \text{ cm}$$

$$= 75.9 \text{ say, } 76 \text{ cm}$$

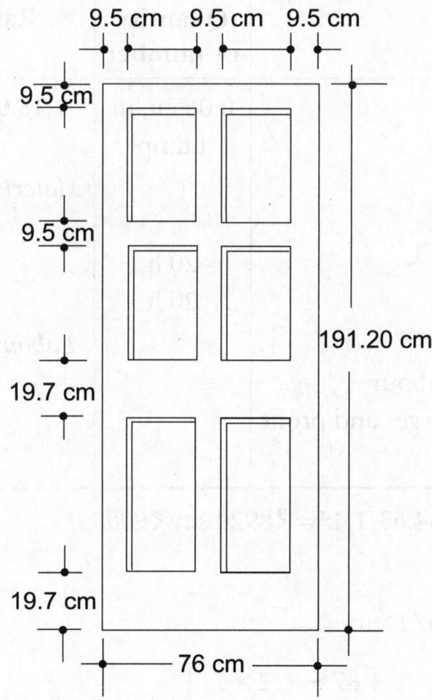

Fig. 20.1   Door shutter

$$\text{Height} = \text{overall height} - \text{width of head(s)} - 0.05 \text{ cm clearance}$$

$$+ 1.2 \text{ cm for rebate (twice for doors with sills)}$$

$$= 197.5 \text{ cm} - 7.00 - 0.05 \text{ cm} + 1.20 \text{ cm} = 191.7 \text{ cm}$$

Sizes of timbers required

Particulars	No.	I	B	D	Quantity
Styles	3	1.910	0.095	0.035	0.0190
Top and frieze rails	2	0.76	0.095	0.035	0.0050
Lock and bottom rails	2	0.76	0.197	0.035	0.0105
Width of panel (00.76 − 3 × 0.095 + 0.07 for bearing in style)	2	0.505	0.36	0.035	0.0250
Top panels	2	0.48	0.36	0.035	0.0250

Total .. 0.0595 cu. m, say 0.06 cu. m

Analysis

Materials for 1.45 sq. m	Quantity or number	Rate ₹	Per	Amount ₹
Timber	0.06 cu. m	1,18,981.00	cu. m	7,138.86
Sundries	Lump			200.00
		Material cost ₹		7,338.86
Labour for 1.45 sq. m				
Carpenters	20 h	90.00	hour	1,800.00
Helpers	20 h	82.00	hour	1,640.00
		Labour cost ₹		3,440.00
Total cost of material and labour				10,778.86
Add 20% establishment charges and profit				2,155.77
Total ₹				12,934.63

Rate per metre square = 12934.63/1.45 = ₹8920 *say* ₹9000

Window Shutter

Size of window: 87.5 cm × 107.5 cm

$$\text{Width of each shutter} = \left[\frac{87.5 - (2 \times 7)}{2}\right] + 1.25 \text{ cm for rebate} = 38 \text{ cm}$$

Height of each shutter = 107.5 − 2 × 7 + 2(1.25) for rebate = 93.5 + 2.5 = 96.0 cm

Size of glass panel

Width = 38 − 2 × 7 + 0.02 cm for bearing = 26 cm, that is, 0.26 m

$$\text{Height} = 96 - \left[\frac{(7 + 9.5) - (3 \times 3.5)}{4}\right] + 2 \text{ cm for bearing} = 96.50 \text{ cm}$$

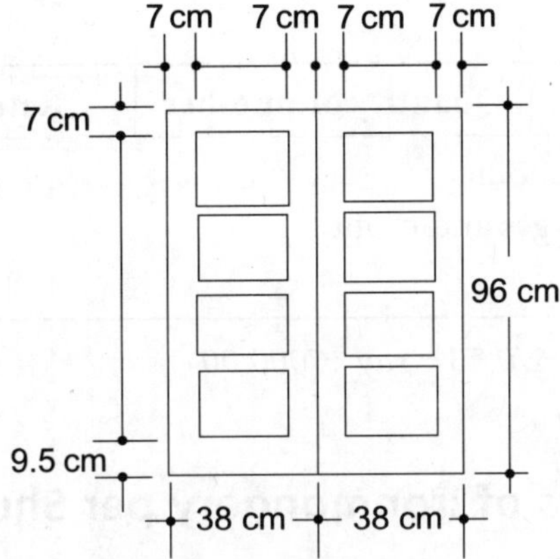

Fig. 20.2   Window shutter

Materials required

Particulars	No.	L	B	D	Quantity
Timber					
Styles	2 × 2	0.96			
			0.07	0.035	0.0113
Top rail	1	0.76			
Bottom rail	1	0.76	0.095	0.35	0.0025
Sash bars	6	0.38	0.035	0.035	0.0028
				Total	0.0116 cu. m
					say, 0.012 cu. m
Glass panels	8	0.26		0.22	0.46 sq. m

Analysis

For 0.76 × 0.96 = 0.73 sq. m	Quantity or number	Rate ₹	Per	Amount ₹
Timber	0.012 cu. m	1,18,981.00	cu. m	1,427.77
Glass panels	0.46 cu. m	1,200.00	sq. m	552.00
	Lump			50.00
		Material cost		2,029.77
Labour				
for: 0.73 sq. m				
Carpenter	12 h	90.00	hour	1,080.00
Helper	12 h	82.00	hour	984.00
Glazier	2 h	82.00	hour	164.00
		Labour cost		2,228.00

(*continued*)

(Continued)

For 0.76 × 0.96 = 0.73 sq. m	Quantity or number	Rate ₹	Per	Amount ₹
Total cost of materials and labour				4,257.77
Add 20% establishment charges and profit				851.55
			Total cost	5,109.32

Rate per square metre ₹5109.32/0.73 = say ₹7000.00

20.8.3   Requirements of Ironmongery per Shutter

Table 20.22   Minimum number of fixtures required for doors and windows

	Particulars	Door	Window
1.	Holdfasts	6 no.	4 no.
2.	Hinges (single shutter)	3 no.	2 no.
	(double shutters)	6 no.	4 no.
3.	Handle per shutter	1 no.	1 no.
4.	Tower bolts per shutter	2 no.	2 no.
5.	Stoppers per shutter	1 no.	—
6.	Eye and hook per shutter	—	1 no.

20.8.4   Labour Required for Fixing or Hanging Shutter to Frames

Table 20.23   Labour hours required to set the frame, install hardware and fit and hang door or window shutters

Type of door	Size	Labour hours	
		Carpenter	Helper
1. Single shutter	67.5 cm × 187.5 cm	2	2
	97.5 cm × 207.5 cm	2.5	2.5
2. Double shutter	97.5 cm × 187.5 cm	3	3
	117.5 cm × 207.5 cm	4	4
3. Window shutters	47.5 cm × 97.5 cm	$1\frac{1}{2}$	$1\frac{1}{2}$
	107.5 cm × 117.5 cm	2	2

20.8.5 Schedule of Rates

Item	Rate	Per
1. Supplying and fixing in position TW frames for doors, cupboards and so on	₹1,60,000.00	cu. m
2. Supplying ledged and battened shutters for doors	₹ 7000.00	sq. m
3. Supplying ledged, braced and battened shutters for doors	₹ 7,200.00	sq. m
4. Supplying framed, ledged, braced and battened shutters for doors	₹ 7,600.00	sq. m
5. Providing framed and panelled door shutter (6 raised panels 3.5 cm thick)	₹ 8,750.00	sq. m
6. Providing partly panelled and partly glazed shutters for doors	₹ 6,282.00	sq. m
7. Providing flush type shutters for doors	₹ 6,548.00	sq. m
8. Providing rolling shutter fabricated with 18 gauge steel laths with side guides, bottom rail, brackets, doors suspension shafts, housing, shafts at top	₹ 5,308.00	sq. m
9. Providing collapsible gate with channel gates, pivoted flat bars, top and bottom slides, rollers, stoppers, handles and mechanical gear arrangement	₹ 4,011.00	sq. m
10. Providing fully panelled window shutter (6 raised panels, 3.5 cm thick)	₹ 5,800.00	sq. m
11. Providing fully glazed window shutter	₹ 4,804.00	sq. m
12. Providing partly panelled and partly glazed window shutter	₹ 4,500.00	sq. m
13. Providing steel windows	₹ 3,600.00	sq. m
14. Providing louvred windows shutter with mild steel bars between louvers	₹ 4,000.00	sq. m
15. Ventilators fully glazed	₹ 7,740.00	sq. m
Cupboards	₹	
16. Providing fully panelled shutter for cupboards	₹ 3,000.00	sq. m
17. Providing partly panelled shutter for cupboards	₹ 4,800.00	sq. m
18. Providing grillwork for windows, ventilators, 35 kg/sq. m	₹ 3,258.00	sq. m
19. *Irommongery*	₹	
(i) Providing fixtures and fastenings to double leaf shutters with iron (chromium plated)	₹ 741.00	sq. m
(ii) Single leaf	₹ 558.00	sq. m
Labour only		
20. Fixing shutters to frames	₹ 750.00	sq. m
21. Hanging shutters to frames with hinges and so on	₹ 800.00	sq. m
Painting	₹	
22. Three coats of oil painting of approved colour	₹ 110.00	sq. m

20.9 ROOFS

20.9.1 Timber Roof Trusses

The quantity of timber required for a roof truss can be determined from the drawing. For example, quantity of timber in a king post roof truss of 6 m span and 2 m rise may be estimated as below:

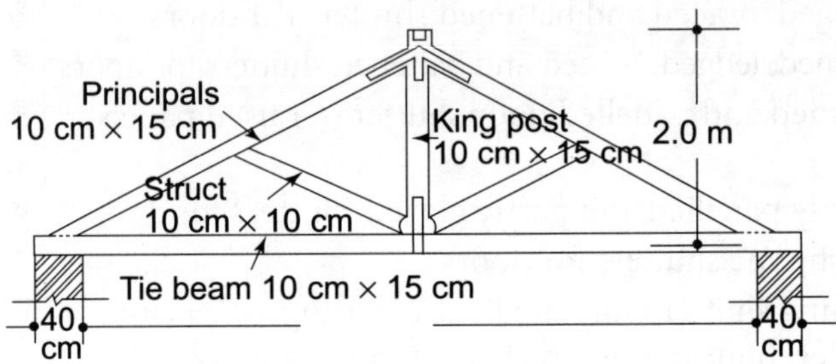

Fig. 20.3   Roof truss

Member	No.	Length (m)	Breadth (m)	Depth (m)	Quantity
Tie beam	1	6.80	0.10	0.15	0.102 cu. m
Principal rafters length $= \sqrt{(3.40)^2 + (2)^2} = 3.94\ \text{m}$	2	3.94	0.15	0.10	0.118 cu. m
King post	1	2.00	0.10	0.15	0.003 cu. m
Struts	2	1.60	0.10	0.10	0.032 cu. m
			Total		0.255 cu. m

Analysis

Material for one truss	Quantity or number	Rate ₹	Per	Amount ₹
Timber 0.255 cu. m	0.255 cu. m	1,18,981.00	cu. m	30,340.15
Hoop iron and strap	10 kg	100.00	kg	1,000.00
12 mm diameter bolts	25 nos	80.00	no.	2,000.00
		Material cost ₹		33,340.00
Labour for one truss				
Carpenter	40 h	90.00	hour	3,600.00
Helper	40 h	82.00	hour	3,280.00
Fitter	6 h	82.00	hour	492.00
		Labour cost		7,372.00

Total cost of material and labour	40,712.00
Add 20% establishment charges and profit	8,142.40
(say)	48,854.40

Cost per cu. m = 48,854/0.255 = ₹1,91,584.00

Rate per cu. m ₹1,90,000 to ₹2,00,000.00

20.9.2 Woodwork in Purlins, Common Rafters, Wall Plates, Ridge Plates and so on

Analysis

Material for 1 cu. m	Quantity or number	Rate ₹	Per	Amount ₹
Timber 1 cu. m	1 cu. m	1,18,981.00	cu. m	1,18,981.00
Nails	lump			250.00
		Material cost ₹		1,19,231.00
Labour for 1 cu. m				
Carpenter	100 h	90.00	hour	9,000.00
Helper	100 h	82.00	hour	8,200.00
		Labour cost		17,200.00
Cost of material and labour			₹	1,36,431.00
Add 20% establishment charges and profit			₹	27,286.20
			Total cost say,	1,63,717.20

Rate per cubic metre, ₹1,60,000.00 to 1,65,000.00

20.9.3 Roof Coverings

Memoranda	1.	Mangalore tiled roof	
		Number of tiles/sq. m	150 no.
		Ridge tiles per 10 sq. m	10 no.
		Quantity of battens per sq. m of roof (spacing 35 cm)	3 m run
		Nails per 10 sq. m approximately	2.5 kg

    2.   *GI sheets*

GI Sheets are available in lengths of 1.8288 m, 2.1336 m, 2.4384 m, 2.7432 m, 3.048 m, having a standard width of 0.8128 m

    3.   *AG sheets*

AC sheets are also available in lengths as per

GI sheets. The standard width is 1.0541 m

End overlap	15 cm
Side overlap	16.93 cm say 17 cm

Analysis

Materials for 10 sq. m	Quantity or number	Rate ₹	Per	Amount ₹
Mangalore tiles	150 no.	24,000.00	1000 no.	3,600.00
Mangalore ridge tiles	10 no.	42.00	no.	420.00
Wooden batten (4.5 cm × 2.50 cm)	30 m	124.00	m run	3,720.00
Nails	2.5 kg	75.00	kg	187.50
Sundries				50.00
		Material cost ₹		7,977.00
Labour for 10 sq. m				
Carpenter	8 h	90.00	hour	720.00
Male mazdoor	8 h	82.00	hour	656.00
Tile layer	8 h	82.00	hour	656.00
Female mazdoor	8 h	82.00	hour	656.00
		Labour cost		2,688.00

Cost of materials and labour	10,665.00
Add 20% establishment charges and profit	2,133.00
Total ₹	12,798.00

Rate per square metre, say ₹1,280.00

20.9.4   Schedule of Rates

Item	Rate (2014) ₹	Per
1. Providing country teak woodwork in trusses including fixtures	1,60,000.00	cu. m
2. Providing country teak woodwork in purling, rafters and so on	1,30,000.00	cu. m
3. Providing and applying three coats of oil paint to woodwork	100.00	sq. m
4. Providing and fixing Mangalore tiled roofing on country teak wood battens	800.00	sq. m
5. Providing and fixing Mangalore ridge and hip tiles	350.00	m

(*continued*)

(Continued)

Item	Rate (2014) ₹	Per
6. Providing and fixing 6 mm thick asbestos cement corrugated or semi-corrugated sheet roofing with ridges including galvanised iron 'J' bolts, washers, bolts, nuts	450.00	sq. m
7. Providing and fixing corrugated galvanised iron sheets of 24 BWG for roofing including 'J' bolts and lead washers	850.00	sq. m
8. Providing and fixing 12 mm thick laminated board plank ceiling with necessary fixtures without form work	2,435.00	sq. m

20.10 FLOORS

Memoranda

20.10.1 Materials

Table 20.24 Quantities of materials for one sq. m flooring

Type of flooring	Sand or concrete bedding, cu. m	Cement concrete (1:2:4)	Cement mortar cu. m	Shahabad stone or tile or bricks or mobile chips
1. Indian patent stone flooring (4 cm thick c.c. 1:2:4 prop. over 10 cm thick c.c. 1:4:8	0.1	0.04	—	—
2. *Shahabad* stone flooring set in cm over 10 cm thick sand/or c.c (1:5:10) or (1:4:8) prop	0.1	—	0.2	1.00 sq. m
3. Tile flooring set in cm (1:3) over 10 cm thick c.c (1:4:8) bed	0.01	0.1	—	1.00 sq. m
4. Terrazo flooring 12 mm to 20 mm thick over 25 mm thick c.c M150 (1:2:4) over 10 cm thick c.c (1:5:10) bed	0.1	0.025	—	15–30 kg marble chips
5. Brick on edge flooring set in cm (1:6) over 10 cm thick c.c (1:5:10)	0.1	—	0.025	58 no. bricks

20.10.2   Labour

Table 20.25   Labour hours required for laying 1 sq. m floor finish

(Note: Laying bed concrete sand is measured separately and hence not included in this section)

Type of floor	Pavior	Male mazdoors	Bhisti
1. Indian patent stone flooring	1	1	$\frac{1}{2}$
2. *Shahabad* stone flooring	1	1	$\frac{1}{2}$
3. Tile flooring	2	2	$\frac{1}{2}$
4. Mosaic flooring	1	4	$\frac{1}{2}$
5. Brick on edge flooring	1.2	1.2	$\frac{1}{2}$

20.10.3   Analysis

Item: Providing 2.5 cm thick rough Shahabad stone flooring in cu. m (1:3)

Materials for 10 sq. m	Quantity or number	Rate ₹	Per	Amount ₹
Cement mortar (i) Sand	0.2 cu. m	1,670.00	cu. m	334.00
(ii) Cement	0.095 t	6,000.00	tonne	570.00
(iii) Water	Lump			50.00
Shahabad stones	10 sq. m	283.00	sq. m	2,830.00
		Material cost		3,784.00
Labour for 10 sq. m				
Pavior (Table 20.25)	10 h	90.00	hour	900.00
Male mazdoor	10 h	82.00	hour	820.00
Bhisti	5 h	85.00	hour	425.00
		Labour cost		2,475.00

Cost of material and labour, say,	5,929.00
Add 20% establishment charges and profit	1,185.80
Total cost per 10 sq. m ₹	7,114.80

Rate per square metre ₹710 to 750

The rate analysis for other items can be carried out on the same lines. Prevailing unit rates for a few common items of flooring are listed below:

20.10.4 Schedule of rates

Item	Rate per sq. m (₹)
1. Providing and laying Indian patent stone flooring (33 mm thick) with 1:2:4 cement concrete laid to proper level and slope in alternate bays including compaction, marking lines to give appearance of tiles, finishing smooth in any colour as directed	1,000.00
2. Providing 5 cm thick rough Shahabad stone flooring in cu. m (1:3)	900.00
3. Providing 2.5 cm thick rough Shahabad stone flooring in cu. m (1:3)	731.00
4. Providing 2.5 cm thick polished Shahabad stone flooring in cu. m (1:2)	826.00
5. Providing ceramic tile flooring in cm (1:2)	1,392.00

20.11 PLASTERING AND POINTING

Memoranda

20.11.1 Materials

Table 20.26 Quantities of mortar required for 1 sq. m of plaster

(no waste included)

Thickness of plaster in cm	1	2	2.5	3	4	5
Quantity of mortar in cu. m	0.01	0.02	0.025	0.03	0.04	0.05

Table 20.27 Quantities of materials required for 1 cu. m of finished mortar

Proportions by volume, cement: sand	Quantity	
	Cementing material in tones	Sand in cu. m
1:2	0.58 to 0.65	0.08 to 0.90
1:3	0.43 to 0.48	0.9 to 1.0
1:4	0.36	1.00
1:5	0.29	1.00
1:6	0.23	1.00

Finishing

1. *Sand faced plaster:* About 1.5 to 23 kg of cement per sq. m of surface over and above the quantity of cement mentioned above.
2. *Cement faced plaster:* About 2.1 to 3 kg of cement per sq. m of surface over and above the quantity of cement mentioned above.
3. *Neeru finish:* About 4 to 23 kg of *neeru* per sq. m of surface.

20.11.2 Labour

Table 20.28 Approximate labour hours required per 10 sq. m of plastering

Class of work	Labour hours			
		Mazdoor		
	Plasterer	Male	Female	Bhisti
1. Preparation of surface	—	$\frac{1}{2}$	—	$\frac{1}{2}$
2. Rough cost three-coat work	4	4	4	2
3. Floating coat, three-coat work	3	3	3	2
4. Rough coat two-coat work	4	4	4	2
5. Finishing coat	1	1	1	1
6. Curling	—	—	—	10

20.11.3 Analysis

Item: Cement plaster 2 cm thick in cm (1:6) to brickwork with sand finish (external plastering)

Materials for 10 sq. m	Quantity or number	Rate ₹	Per	Amount ₹
Sand (screened)	0.20	1,670.00	cu. m	334.00
Cement 0.2 × 0.23	0.046 t	6,000.00	tonne	276.00
Water	Lump			50.00
Sundries				50.00
		Material cost ₹		710.00
Labour for 10 sq. m				
Plasterer	5 h	90.00	hour	450.00
Male mazdoor	5 h	82.00	hour	410.00
Female mazdoor	5 h	82.00	hour	410.00

(continued)

(Continued)

Materials for 10 sq. m	Quantity or number	Rate ₹	Per	Amount ₹
Bhisti (including curing)	13 h	85.00	hour	1,105.00
Scaffolding	Lump			100.00
		Labour cost ₹		2,475.00
Cost of material and labour			₹	3,158.00
Add 20% establishment charges and profit				637.00
			Total cost	3,822.00

Rate per sq. m ₹380.00 to 400.00

Pointing

Memoranda

20.11.4 Materials

Approximate quantity of mortar required per 10 sq. m of pointing is 0.025 cu. m.

Table 20.29 Approximate quantities of materials required per 10 sq. m of pointing

Proportion of mortar by volume	Quantity	
	Cement tones	Sand cu. m
1:1	0.036	0.025
1:2	0.018	0.020
1:3	0.012	0.025

20.11.5 Labour

Table 20.30 Approximate labour hours required per 10 sq. m of pointing

Class of work	Labour hours			
	Mason	Mazdoor		
		Male	Female	Bhisti
1. Preparation of surface	—	1/2	—	1/2
2. Pointing	2 to 3	1/2 to 1	2 to 3	1/2
3. Curing	—	—	—	5

20.11.6   Analysis

Item: Cement pointing (1:2) to brickwork

Materials for 10 sq. m	Quantity or number	Rate ₹	Per	Amount ₹
Cement	0.018 t	6,000.00	tonne	108.00
Sand (screened)	0.025 cu. m	1,670.00	cu. m	41.75
Water	Lump	—	—	40.00
Sundries	Lump	—	—	35.00
		Material cost ₹		224.00
Labour for 10 sq. m				
Mason	2 h	90.00	hour	180.00
Male mazdoor	1 h	82.00	hour	82.00
Female mazdoor	2 h	82.00	hour	164.00
Bhisti	6 h	85.00	hour	510.00
Scaffolding up to 4 m				150.00
		Labour cost ₹		1,086.00
Cost of material, labour and so on			₹	1,310.75
Add 20% establishment charges and profit			₹	262.15
		Total cost per 10 sq. m		1,572.90

Rate per sq. m say ₹157.00

20.11.7   Schedule of rates

	Item	Rate per sq. m ₹
1.	Providing cement plaster 12 mm thick in cu. m (1:4) proportion without *neeru* finish to concrete or brickwork	380.00
2.	—do— 20 mm in two coats	574.00
3.	—do— 25 mm in two coats	574.00
4.	Providing pointing with cement mortar (1:1) for stone masonry	182.00
5.	—do— Cement mortar (1:2)	179.00
6.	—do— Cement mortar (1:3)	173.00
7.	Providing pointing with cement mortar (1:1) proportion for brickwork	182.00
8.	—do— cement mortar (1:2)	179.00
9.	—do— cement mortar (1:3)	173.00

20.12   WHITE WASH, COLOUR WASH, DISTEMPERING MATERIALS

Memoranda

20.12.1   Materials

Table 20.31   Materials required per sq. m of white washing

Coat	Lime (kg)	Gum (g)
First coat	1.35	56.70
Second coat	0.68	28.35
Third coat	0.56	28.35

Materials required per sq. m of distemper 0.8 to 1.0 litre.

20.12.2   Labour

Table 20.32   Labour hours required for white washing and so on for 10 sq. m of surface

	Item	Painter	Mazdoor
1.	White washing—one coat	$\frac{1}{2}$	$\frac{1}{2}$
2.	White washing—two coats	1	1
3.	White washing—three coats	$1\frac{1}{2}$	$1\frac{1}{2}$
4.	Colour washing	$\frac{1}{2}$	$\frac{1}{2}$
5.	Distempering one coat	$1\frac{1}{3}$	$1\frac{1}{3}$

20.12.3   Analysis

Item: Providing white wash—two coats.

Materials for 10 sq. m	Quantity or number	Rate (₹)	Per	Amount (₹)
Lime: 1.65				
0.68				41.94
2.33 kg	2.33 kg	18.00	kg	
Gum	0.085 kg	100.00	kg	8.50
Water, sundries	Lump			5.00
		Material cost		55.44

(*continued*)

(Continued)

Materials for 10 sq. m	Quantity or number	Rate (₹)	Per	Amount (₹)
Labour				
For: 10 sq. m				
Painter	1.5 h	85.00	hour	127.50
Mazdoor	1.5 h	82.00	hour	123.00
Scaffolding	Lump			10.00
		Labour cost		260.50

Total cost of material and labour: 315.94

Add 20% establishment charges and profit: 63.19

Total cost per 10 sq. m   379.13

Rate per square metre ₹37 to 40/-

20.12.4   Schedule of rates

	Item	Rate per sq. m (₹)
1.	Providing and applying white wash in two coats to old surfaces	38.00
2.	Providing and applying white wash in three coats to new work	53.00
3.	Providing and applying colour wash of approved colour	60.00
4.	—do— three coats	85.00
5.	Providing and applying washable (dry) distemper of approved colour in two coats	85.00
6.	—do— three coats	85.00
7.	Providing and applying oil-bound distemper of approved quality (two coats)	85.00
8.	—do— three coats	110.00

A Two-Storeyed Residential Building

Name of Work: A Residential Building

MEASUREMENT SHEET

Item No.	Description	No.	Length in m	Breadth in m	Depth or ht in m	Quantity	Total
1.	*Surface dressing* Average depth 10 cm C to C						
	length: 3.40						
	+5.50						
	+3.40						
	12.30 m						
	Add half the width of trench on either side						
	$(0.50 + 0.50)$ = 1.00						
	13.30 m						
	C to C width 4.40						
	+4.40						
	+1.75						
	10.55 m						
	Add half the width of trench on either side						
	$(0.50 + 0.50) = 1.00$ m						
	11.55 m	1	13.30	11.55		153.63	
	Add below steps:						
	Front	1	1.80	0.50		0.90	
	Side	1	1.35	0.50		0.68	

(continued)

Schedule of openings

Type	Particulars	No	Size in cm
D	Fully panelled door, double shutter with ventilator fully glazed	2	117.5×247.5
D_1	Flush door, single shutter	13	97.5×197.5
D_2	Ledged, braced and framed door single shutter	4	77.5×187.5
W_1	Fully panelled window double shutter with ventilator fully glazed	3	207.5×157.5
W_2	,,		107.5×157.5
W_3	Louvred	4	67.5×97.5
W_4	Precast cement concrete jali	1	70×300
G	CI grill	4	150×150
C	Cupboard with fully panelled shutters	8	97.5×197.5

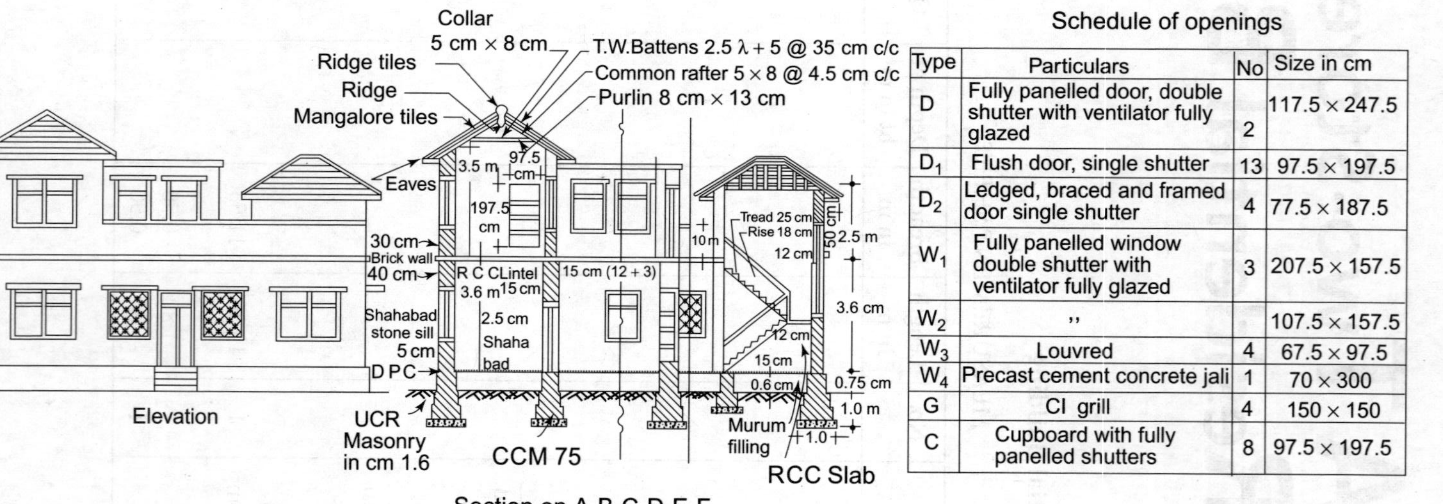

Estimates 1   A residential building

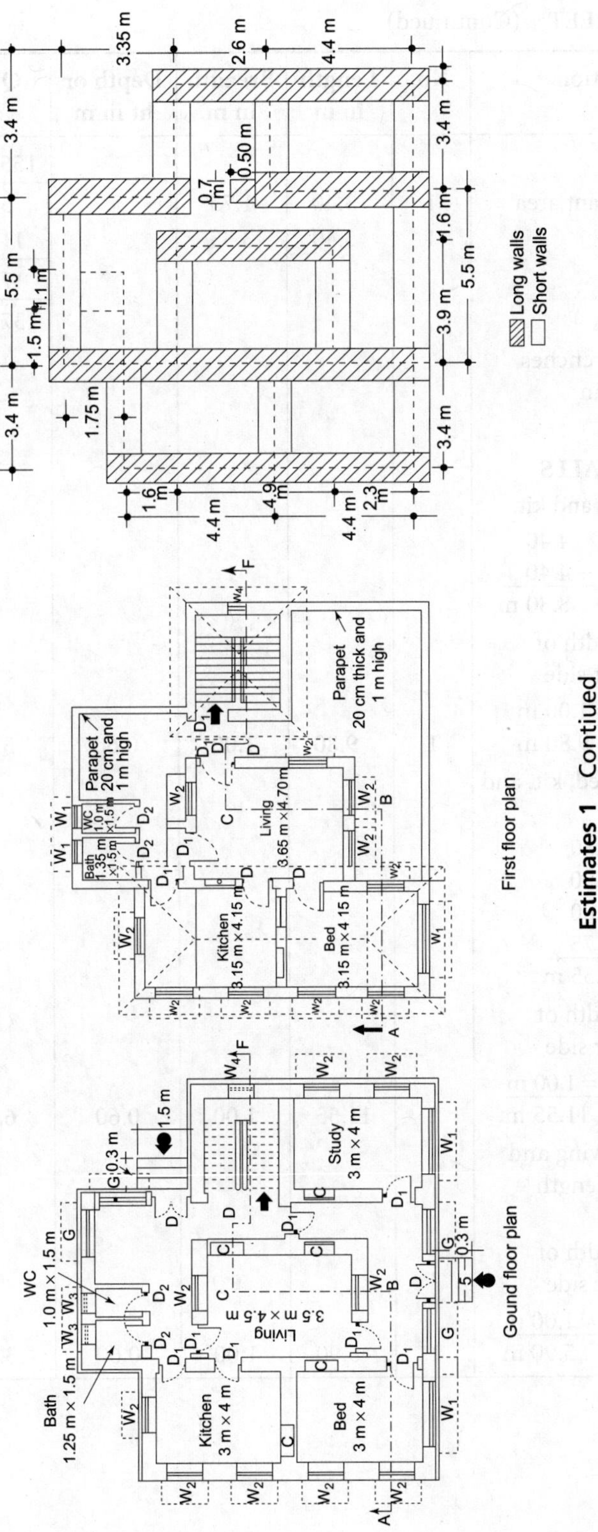

3.35 m
2.6 m
4.4 m
3.4 m
0.50 m
0.7 m
3.4 m
1.6 m
5.5 m
5.5 m
3.9 m
1.5 m
1.1 m
3.4 m
1.75 m
3.4 m
1.6 m
4.4 m
4.9 m
4.4 m
2.3 m
Long walls
Short walls
Parapet 20 cm and 1 m high
Parapet 20 cm thick and 1 m high
WC 1.0 m ×1.5 m
Bath 1.35 m ×1.5 m
Living 3.65 m ×4.70 m
Kitchen 3.15 m ×4.15 m
Bed 3.15 m ×4 15 m
First floor plan
1.5 m
0.3 m
WC 1.0 m ×1.5 m
Bath 1.25 m ×1.5 m
Living 3.5 m ×4.5 m
Study 3 m × 4 m
Kitchen 3 m × 4 m
Bed 3 m × 4 m
0.3 m
Gound floor plan
Estimates 1 Contiued

MEASUREMENT SHEET   (Continued)

Item No.	Description	No.	Length in m	Breadth in m	Depth or ht in m	Quantity	Total
						155.21	
	Deduct redundant area	1	3.40	1.75		5.95	
		1	3.40	3.35		11.39	
						17.34	
						137.87 sq. m	137.87 sq. m
2.	Excavation in trenches for foundation in ordinary soil.						

LONG WALLS

Side wall of bed and kit.

C to C length   4.40
　　　　　　　+ 4.40
　　　　　　　　8.80 m

Add half the width of trench on either side

(0.50 + 50)   = 1.00 m
　　　　　　　　9.80 m

		1	9.80	1.00	0.60	5.88	

Wall between bed, kit. and living. C to C

Length: 2.30
　　　　　+4.90
　　　　　+1.60
　　　　　+1.75
　　　　　10.55 m

Add half the width of trench on either side

(0.50 + 0.50)   = 1.00 m
　　　　　　　　11.55 m

		1	11.55	1.00	0.60	6.93	

Wall between living and passage C to C length = 4.90 m

Add half the width of trench on either side

(0.50 + 0.50)   = 1.00 m
　　　　　　　　5.90 m

		1	5.90	1.00	0.60	3.54	

(continued)

MEASUREMENT SHEET   (Continued)

Item No.	Description	No.	Length in m	Breadth in m	Depth or ht in m	Quantity	Total
	Wall between study and passage C to C length = 4.40 m Add half the width of trench on either side (0.50 + 0.50) = 1.00 m 5.40 m	1	5.40	1. 00	0.60	3.24	
	Sidewall of rear verandah C to C length = 3.35 m Add half the width of trench on either side (0.50 + 0.50) = 1.00 m 4.35 m	1	4.35	1.00	0.06	2.61	
	Wall below first step of stair At plinth level: Clear length = 1.10 m Half width of wall = 0.25 m add offset = 0.15 m 1.50 m Deduct half width of trench already considered 0.50 m 1.00 m	1	1.00	0.70	0.60	0.40	
	Side wall of study and stair C to C length = 4.40 +2.80 m 7.20 m Add half the width of trench on either side (0.50 + 0.50) = 1.00 m 8.20 m	1	8.20	1.00	0.60	4.92	
	SHORT WALLS Of bed, kit. study of stair C to C length = 3.40 m						

(continued)

MEASUREMENT SHEET   (Continued)

Item No.	Description	No.	Length in m	Breadth in m	Depth or ht in m	Quantity	Total
	Deduct half the width of trench on either side $(0.50 + 0.50)\ \underline{= 1.00\ m}$ $\ 2.20\ m$	6	2.40	1.00	0.60	8.64	
	Of front and rear verandah C to C length $= 5.50\ m$ Deduct half the width of trench on either side $(0.50 + 0.50) = 1.00\ m$	2	4.50	1.00	0.60	5.40	
	of living room: C to C length $= 3.90\ m$ Deduct half the width of trench on either side: $(0.50 + 0.50)\ \underline{= 1.00\ m}$ $\ 2.90\ m$	2	2.90	1.00	0.60	3.48	
	Of passage: C to C length $= 1.60\ m$ Deduct half the width of trench on either side: $(0.50 + 0.50)\ \underline{= 1.00\ m}$ $\ 0.60\ m$	1	0.60	1.00	0.60	0.36	
	Below steps:						
	Front	1	1.80	0.50	0.15	0.14	
	Side	1	1.35	0.50	0.15	0.10	
						40.10	
						45.66 cu. m	45.66 cu. m
3.	Excavation in trenches for foundations in hard soil (Note: Henceforth walls are considered in the same serial order in which they appear in item no. 2)						

(continued)

MEASUREMENT SHEET   (Continued)

Item No.	Description	No.	Length in m	Breadth in m	Depth or ht in m	Quantity	Total
	LONG WALLS						
		1	9.80	1.00	0.30	2.94	
		1	11.55	1.00	0.30	3.47	
		1	5.90	1.00	0.30	1.77	
		1	5.40	1.00	0.30	1.62	
		1	4.35	1.00	0.30	1.31	
		1	8.20	1.00	0.30	2.46	
	SHORT WALLS						
		6	2.40	1.00	0.30	4.32	
		2	4.50	1.00	0.30	2.70	
		2	2.90	1.00	0.30	1.74	
		1	0.60	1.00	0.30	0.18	
						22.51 cu. m	22.51 cu. m
4.	Cement concrete M 75 in foundation trenches.						
	LONG WALLS						
		1	9.80	1.00	0.15	1.47	
		1	11.65	1.00	0.15	1.73	
		1	5.90	1.00	0.15	0.89	
		1	5.40	1.00	0.15	0.81	
		1	4.35	1.00	0.15	0.65	
		1	1.00	0.70	0.15	0.11	
		1	8.20	1.00	0.15	1.23	
	SHORT WALLS						
		6	2.40	1.00	0.15	2.16	
		2	4.50	1.00	0.15	1.35	
		2	2.90	1.00	0.15	0.87	
		1	0.60	1.00	0.15	0.09	
	Steps:	1	1.80	0.50	0.15	0.14	
		1	1.35	0.50	0.15	0.10	
						11.60 cu. m	11.60 cu. m
5.	UCR masonry in cm (1 : 6 prop.) in foundation and plinth						

(*continued*)

MEASUREMENT SHEET   (Continued)

Item No.	Description	No.	Length in m	Breadth in m	Depth or ht in m	Quantity	Total
	FIRST LAYER						
	LONG WALLS						
	Length = C to C length + one width of course except wall below first step of stair						
	$L = 8.80 + 0.70 = 9.50$ m	1	9.50	0.70	0.40	2.66	
	$L = 10.55 + 0.70 = 11.25$ m	1	11.25	0.70	0.40	3.15	
	$L = 4.90 + 0.70 = 5.60$ m	1	5.60	0.70	0.40	1.57	
	$L = 4.40 + 0.70 = 5.10$ m	1	5.10	0.70	0.40	1.43	
	$L = 3.35 + 0.70 = 4.05$ m	1	4.05	0.70	0.40	1.13	
	$L = 1.35 - 0.35 = 1.00$ m	1	1.00	0.40	0.40	0.16	
	$L = 7.20 + 0.70 = 7.90$ m	1	7.90	0.70	0.40	2.21	
	SHORT WALLS						
	Length = (C to C length) − (One width of course)						
	$L = 3.40 - 0.70 = 2.70$ m	6	2.70	0.70	0.40	4.54	
	$L = 5.50 - 0.70 = 4.80$ m	2	4.80	0.70	0.40	2.69	
	$L = 3.90 - 0.70 = 3.20$ m	2	3.20	0.70	0.40	1.79	
	$L = 1.60 - 0.70 = 0.90$ m	1	0.90	0.70	0.40	0.25	
	SECOND LAYER						
	LONG WALLS						
	$L = 8.80 + 0.60 = 9.40$ m	1	9.40	0.60	0.30	1.69	
	$L = 10.55 + 0.60 = 11.15$ m	1	11.15	0.60	0.30	2.00	
	$L = 4.90 + 0.60 = 5.50$ m	1	5.50	0.60	0.30	0.99	
	$L = 4.40 + 0.60 = 5.00$ m	1	5.00	0.60	0.30	0.90	
	$L = 3.35 + 0.60 = 3.95$ m	1	3.95	0.60	0.30	0.71	
	$L = 1.35 - 0.30 = 1.05$ m	1	1.05	0.40	0.30	0.13	
	$L = 7.20 + 0.60 = 7.80$ m	1	7.80	0.60	0.30	1.40	
	SHORT WALLS						
	$L = 3.40 - 0.60 = 2.80$ m	6	2.80	0.60	0.30	3.02	
	$L = 5.50 - 0.60 = 4.90$ m	2	4.90	0.60	0.30	1.76	
	$L = 3.90 - 0.60 = 3.30$ m	2	3.30	0.60	0.30	1.19	
	$L = 1.60 - 0.60 = 1.00$ m	1	1.00	0.60	0.30	0.18	

(*continued*)

MEASUREMENT SHEET   (Continued)

Item No.	Description	No.	Length in m	Breadth in m	Depth or ht in m	Quantity	Total
	PLINTH						
	LONG WALLS						
	$L = 8.80 + 0.50 = 9.30$ m	1	9.30	0.50	0.75	3.49	
	$L = 10.55 + 0.50 = 11.05$ m	1	11.05	0.50	0.75	4.14	
	$L = 4.90 + 0.50 = 5.40$ m	1	5.40	0.50	0.75	2.03	
	$L = 4.40 + 0.50 = 4.90$ m	1	4.90	0.50	0.75	1.84	
	$L = 3.35 + 0.50 = 3.85$ m	1	3.85	0.50	0.75	1.44	
	$L = 1.35 - 0.25 = 1.10$ m	1	1.10	0.30	0.50	0.17	
	$L = 7.20 + 0.50 = 7.70$ m	1	7.70	0.50	0.75	2.89	
	SHORT WALLS						
	$L = 3.40 - 0.50 = 2.90$ m	6	2.90	0.50	0.75	6.53	
	$L = 5.50 - 0.50 = 5.00$ m	2	5.00	0.50	0.75	3.75	
	$L = 3.90 - 0.50 = 3.40$ m	2	3.40	0.50	0.75	2.55	
	$L = 1.60 - 0.50 = 1.10$ m	2	1.10	0.50	0.75	0.41	
	Steps: Front	1	1.50	0.90	0.15	0.20	
	(Allowing for 4 cm	1	1.50	0.60	0.19 m	0.17	
	Shahabad stone paving in						
	cm over UCR masonry)	1	1.50	0.30	0.10	0.09	
	Side:	1	1.00	0.90	0.15	0.14	
		1	1.00	0.60	0.19	0.11	
		1	1.00	0.30	0.19	0.05	
						65.55 cu. m	65.55 cu. m
6.	External angles in facework Corner stone size 30 m × 15 cm × 15 cm Height = 0.40 + 0.30 + 0.75 = 1.45 m	9			1.45	13.05	13.05
7.	Damp proof course 12 cm thick in c.c M150 (Note: The width of d.p.c on external walls is 45 cm and on internal walls it is 40 cm The walls are regrouped for this item only and the lengths calculated with the help of foundation plan						

(continued)

MEASUREMENT SHEET   (Continued)

Item No.	Description	No.	Length in m	Breadth in m	Depth or ht in m	Quantity	Total
	LONG WALLS						
	$L = 8.80 + 0.50 = 9.30$ m	1	9.30	0.45		4.19	
		1	9.30	0.40		3.72	
		1	1.75	0.45		0.79	
	$L = 4.90 + 0.40 = 5.30$ m	1	5.30	0.40		2.12	
	$L = 4.40 + 0.25 + 0.20$ $= 4.85$ m	1	4.85	0.40		1.94	
	$L = 3.35 + 0.20 + 0.25$ $= 3.80$ m	1	3.80	0.45		1.71	
	$L = 7.20 + 0.50 = 7.70$ m	1	7.70	0.45		3.47	
	SHORT WALLS						
	$L = 3.40 - 0.40 = 3.00$ m	4	3.00	0.45		5.40	
		2	3.00	0.40		2.40	
	$L = 5.50 - 0.40 = 5.10$ m	2	5.10	0.45		4.59	
	$L = 3.90 - 0.40 = 3.50$ m	2	3.50	0.40		2.80	
						33.13 sq. m	33.13 sq. m
8.	Filling in plinth with selected excavated material Bed, kitchen and study $L = 4.40 - 0.50 = 3.90$ m $B = 3.40 - 0.50 = 2.90$ m	3	3.90	2.90	0.75	25.45	
	Front verandah $L = 2.30 - 0.50 = 1.80$ m $B = 5.50 - 0.50 = 5.00$ m	1	1.80	5.00	0.75	6.75	
	Living room $L = 4.90 - 0.50 = 4.40$ m $B = 3.90 - 0.50 = 3.40$ m	1	4.40	3.40	0.75	11.22	
	WC bath, verandah $L = 5.50 - 0.50 = 5.00$ m $B = 3.35 - 0.50 = 2.85$ m	1	5.00	2.85	0.75	10.69	
	Passage $L = 4.40 - 2.30 = 2.10$ m $B = 1.60 - 0.50 = 1.10$ m	1	0.50	1.10	0.75	0.41	
	Staircase $L = 2.80 - 0.50 = 2.30$ m						

(*continued*)

MEASUREMENT SHEET   (Continued)

Item No.	Description	No.	Length in m	Breadth in m	Depth or ht in m	Quantity	Total
	$B = (1.60 + 3.40) - 0.50$						
	$= 4.50$ m	1	2.30	4.50	0.75	7.76	
						62.28 cu. m	
	Deduct for : stair footings	1	1.00	0.40	0.75	0.30	
						61.98 cu. m	61.98 cu. m
9.	Brickwork II sort in cm (1:6) in superstructure						
	LONG WALLS						
	$L = 8.80 + 0.40 = 9.20$ m	1	9.20	0.40	3.45	12.70	
	$L = 10.55 + 0.40 = 10.95$ m	1	10.95	0.40	3.45	15.10	
	$L = 4.90 + 0.40 = 5.30$ m	1	5.30	0.40	3.45	7.31	
	$L = 4.40 + 0.40 = 4.80$ m	1	4.80	0.40	3.45	6.62	
	$L = 3.35 + 0.40 = 3.75$ m	1	3.75	0.40	3.45	5.18	
	$L = 7.20 + 0.40 = 7.60$ m	1	7.60	0.40	3.45	10.49	
	SHORT WALLS						
	$L = 3.40 - 0.40 = 3.00$ m	1	3.00	0.40	3.45	4.14	
	$= 5.50 - 0.40 = 5.10$ m	2	5.10	0.40	3.45	14.08	
	$= 3.90 - 0.40 = 3.50$ m	2	3.50	0.40	3.45	9.66	
	$= 1.60 - 0.40 = 1.20$ m	1	1.20	0.40	3.45	1.67	
						86.95 cu. m	
	Deduct for:						
	Door openings including						
	lintels: Door D	2	0.20	0.40	2.65	2.54	
	Door D_1	7	1.00	0.40	2.15	6.02	
	Window openings including lintels and sills						
	Window W_1	2	2.10	0.40	1.80	3.02	
	W_2	9	1.10	0.40	1.80	7.13	
	W_3	2	1.00	0.40	0.90	0.72	
	W_4	1	0.70	0.40	1.50	0.42	
	Grill work in verandah including sills and lintels	4	1.50	0.40	1.70	4.20	
	Cupboards	5	1.00	0.30	1.75	2.63	
						26.68	
			Net	quantity		60.27 cu. m	60.27 cu. m

(*continued*)

MEASUREMENT SHEET   (Continued)

Item No.	Description	No.	Length in m	Breadth in m	Depth or ht in m	Quantity	Total
10.	Half brick thick walls in cm (1 : 4) WC and bath side walls $L = 1.75 - (0.10 + 0.05)$ $= 1.60$ m	2	1.60		3.45	11.04	
	Front walls $L = 1.50 - (0.20 + 0.05)$ $= 1.25$ m	1	1.25		3.45	4.31	
	$L = 1.10 - (0.05 + 0.05)$ $= 1.00$	1	1.00		3.45	3.45	
						18.80 sq. m	
	Deduct for: Door openings including lintels D_2		0.80		2.00	3.20	
			Net	quantity		15.60 sq. m	15.60 sq. m
11.	Brickwork II sort in cm (1:6) in superstructure in *first floor* **LONG WALLS** Bed and kitchen $L = 8.80 + 0.40 = 9.20$ m	2	9.20	0.30	3.43	18.93	
	Bath and WC side wall $L = 10.95 \times 9.20 = 1.75$	1	1.75	0.30	3.00	1.58	
	Living $4.90 + 0.40 = 5.30$ m	1	5.30	0.30	2.88	4.58	
	Stair case $L = 2.80 + 0.30 = 3.10$		3.10	0.30	2.58	4.80	
	Parapet wall $L = 4.40 + 0.20 - 0.10$ $= 4.50$ m	1	4.50	0.20	1.00	0.90	
	$L = 3.5 + 0.20 - 0.10$ $= 3.45$ m	1	3.45	0.20	1.00	0.69	
	SHORT WALLS Bed and kitchen $L = 3.40 - (0.10 + 0.15)$ $= 3.15$ m						

(*continued*)

MEASUREMENT SHEET   (Continued)

Item No.	Description	No.	Length in m	Breadth in m	Depth or ht in m	Quantity	Total
	Front and rear	2	3.15	0.30	3.43	6.48	
	Middle	1	3.15	0.30	3.50	3.31	
	Gable portion Mean height of triangular portion $1.57/2 = 0.78$ m	1	3.15	0.30	0.78	0.74	
	Living room $L = 3.90 - (0.15 + 0.10)$ $= 3.65$ m	2	3.65	0.30	3.00	6.59	
	WC and bath: rear wall $L = (1.50 + 1.10) - 0.10 + 0.05 = 2.55$ m	1	2.55	0.30	3.00	2.30	
	Staircase $L = 3.40 - 0.40 = 3.00$ m	2	3.00	0.30	2.58	4.64	
	Parapet walls: Front $L = (5.50 + 3.40) - 0.10 - 0.15 = 8.65$	1	8.65	0.20	1.00	1.73	
	Rear $L = (5.50 - 2.60) - 0.05$ $= 2.85$ m	1	2.85	0.20	1.00	0.57	
						57.84 cu. m	
	Deduct for: Door openings including lintels Door D_1	6	1.00	0.30	2.10	3.87	
	Window openings including sill and lintels Window W_1	1	2.10	0.30	1.80	1.14	
	W_2	9	1.10	0.30	1.80	5.35	
	W_3		1.00	0.30	0.90	0.54	
	W_4	1	0.70	0.30	1.50	0.32	
	Cupboards	3	1.00	0.20	1.75	1.05	
	Slab bearing $L = 4.90 + 0.40 + 0.40$ $= 5.70$	1	5.70	0.30	0.30	0.21	
						12.48	
						45.36 cu. m	45.36 cu. m

(continued)

MEASUREMENT SHEET   (Continued)

Item No.	Description	No.	Length in m	Breadth in m	Depth or ht in m	Quantity	Total
12.	Half brick thick wall in cm (1 : 4) WC and bath Side walls $L = 1.75 + 0.05 - 0.10$ $\quad = 1.70$ m	2	1.70		3.00	10.20	
	Front walls $L = 1.50 - (0.10 + 0.05)$ $\quad = 1.35$	1	1.35		3.00	4.05	
	$L = 1.10 - 0.10 = 1.00$	1	1.00		3.00	3.00	
						17.25 sq. m	
	Deduct for: Door openings including lintel Door D_2	2	0.80		2.00	3.20	
						14.05 sq. m	14.05 sq. m
13.	Providing and fixing country teak wood frames of 70 mm $\times$ 95 mm size scantlings of doors, windows and cupboards. *Doors* D type 2 nos						
	Vertical post	4	2.60	0.07	0.095	0.069	
	Head and sill	4	1.48	0.07	0.095	0.039	
	Transom	2	1.18	0.07	0.095	0.014	
	D_1 type: Ground and first floor $7 + 6 = 13$ nos						
	Vertical posts	26	2.10	0.07	0.095	0.363	
	Head	13	1.28	0.07	0.095	0.111	
	D_2 type: Ground and first floor $2 + 2 = 4$ nos						
	Vertical posts	8	2.00	0.07	0.095	0.106	
	Head	4	1.08	0.07	0.095	0.029	
	Windows W_1 type: Ground and first floor $= 2 + 1 = 3$ nos						

(continued)

MEASUREMENT SHEET   (Continued)

Item No.	Description	No.	Length in m	Breadth in m	Depth or ht in m	Quantity	Total
	Vertical posts $3 \times 3 =$ 9 nos	9	1.58	0.07	0.095	0.095	
	Sill and head	6	2.38	0.07	0.095	0.095	
	Transom	3	2.08	0.07	0.095	0.042	
	W_2 type: Ground and first floor = 9 + 9 = 18 nos						
	Vertical posts	36	1.58	0.07	0.095	0.380	
	Sill and head	36	1.38	0.07	0.095	0.330	
	Transom	18	1.08	0.07	0.095	0.130	
	Window W_3 Type: Ground and first floor 2 + 2 = 4 nos						
	Vertical posts	8	0.98	0.07	0.095	0.052	
	Sill and head	8	0.98	0.07	0.095	0.052	
	Cupboards Ground and first floor = 5 + 3 = 8 nos						
	Verticals	16	1.98	0.07	0.095	0.211	
	Sill and head	16	1.28	0.07	0.095	0.136	
						2.25 cu. m	2.25 cu. m
14.	Form work to lintels and beams For openings of:						
	Door D: Soffit	2	1.10	0.40		0.88	
	Sides	24	1.40		0.15	0.84	
	Edges	4	0.40		0.15	0.24	
	Door D_1: Soffit (Gr. floor)	7	1.00	0.40		2.80	
	Soffit (First floor)	6	1.00	0.30		1.80	
	Sides (Gr. Floor + F F)	26	1.30		0.15	5.07	
	Edges Gr. Floor	14	0.40		0.15	0.84	
	Edges First floor	12	0.30		0.15	0.54	
	Door D_2: Soffit (Gr.floor + F F)	4	0.80	0.10		0.32	
	Sides ("+")	8	1.10		0.10	0.88	
	Edges ("+")	8	0.10		0.10	0.08	

(continued)

MEASUREMENT SHEET (Continued)

Item No.	Description	No.	Length in m	Breadth in m	Depth or ht in m	Quantity	Total
	Windows W_1: Soffits	2	2.10	0.40		1.68	
	Sides	4	2.40		0.20	1.92	
	Edges	4	0.40		0.20	0.32	
	Window W_2: Soffits G F	9	1.10	0.40		3.96	
	F F	9	1.10	0.30		2.97	
	Sides	36	1.40		0.15	7.56	
	(G F + F F)						
	Edges (G F)	9	0.40		0.15	0.54	
	(F F)	9	0.30		0.15	0.41	
	Window W_3: Soffits						
	(G F)	2	0.70	0.40		0.56	
	(F F)	2	0.70	0.30		0.42	
	Sides						
	(G F + F F)	8	1.00		0.15	1.20	
	Edges G F	4	0.40		0.15	0.24	
	F F	4	0.30		0.15	0.18	
	Window W_4: Soffits						
	F F	1	0.70	0.30		0.21	
	Sides	2	1.00		0.15	0.30	
	Edges	2	0.30		0.15	0.09	
	Grill work in verandah						
	Soffits	4	1.50	0.40		2.40	
	Sides	8	1.80		0.20	2.88	
	Edges	8	0.40		0.20	0.64	
	Cupboards						
	Soffits (G F)	4	1.00	0.40		1.60	
	Soffits (F F)	3	1.00	0.30		0.90	
	Sides (G F + F F)	14	1.30		0.15	2.73	
	Edges (G F)	8	0.40		0.15	0.48	
	Edges (F F)	6	0.30		0.15	0.27	
	Beam in staircase Soffit	1	2.40	0.30		0.72	
	Sides	2	2.80		0.30	1.68	
	Edges	2	0.30		0.30	0.18	
	Beam in back verandah						
	for WC wall: Soffits	1	2.95	0.20		0.59	
	Sides	2	3.35		0.20	1.34	
	Edges	2	0.20		0.20	0.08	

(continued)

MEASUREMENT SHEET   (Continued)

Item No.	Description	No.	Length in m	Breadth in m	Depth or ht in m	Quantity	Total
	For front wall: Soffit	1	2.35	0.20		0.47	
	Sides	2	2.70		0.20	1.08	
	Edges	2	0.15		0.20	0.06	MC
						54.95 sq. m	54.95 sq. m
15.	Formwork to chajja						
	Door D: Bottom	2	1.49	0.60		1.68	
	Sides	2	1.40		0.10	0.28	
		4	0.60		0.10	0.24	
	D_1 Bottom (F F)	3	1.30	0.60		2.34	
	Sides (F F)	3	1.30		0.10	0.39	
		6	0.60		0.10	0.36	
	Window W_1: Bottom						
	$2 + 1 = 3$	3	2.40	0.60		4.32	
	(Ground + F F)						
	Sides $2 + 1 = 3$	3	2.40		0.10	0.72	
	$4 + 2 = 6$	6	0.60		0.10	0.36	
	Window W_2: Bottom						
	(Gr. Floor + F F)						
	$7 + 9 = 16$	16	1.40	0.60		13.44	
	Sides	16	1.40		0.10	2.24	
		32	0.60		0.10	1.92	
	Window W_3: Bottom						
	(G F + F F)	4	1.00	0.60		2.40	
	Sides	4	1.00		0.10	0.40	
		8	0.60		0.10	0.48	
	Verandah grill work						
	Bottom	4	1.80	0.60		4.32	
		4	1.80		0.10	0.72	
		8	0.60		0.10	0.48	
						37.09 sq. m	37.09 sq. m
16.	Providing and laying cement concrete M150 in lintels and beams lintel over:						
	Door D	2	1.40	0.40	0.15	0.17	
	" D_1 (Ground floor)	7	1.30	0.40	0.15	0.55	
	" D_1 (First floor)	6	1.30	0.30	0.15	0.35	

(continued)

MEASUREMENT SHEET   (Continued)

Item No.	Description	No.	Length in m	Breadth in m	Depth or ht in m	Quantity	Total
	" D_2 (Ground floor) and first floor)	4	1.10	0.10	0.10	0.04	
	Window W_1 (Ground floor)	2	2.40	0.40	0.20	0.38	
	(First floor)	1	2.40	0.30	0.20	0.14	
	Window W_2 (Ground floor)	9	1.40	0.40	0.15	0.77	
	(First floor)	9	1.40	0.30	0.15	0.57	
	Window W_3 (Ground floor)	2	1.00	0.40	0.15	0.12	
	(First floor)	2	1.00	0.30	0.15	0.09	
	Window W_4 (First floor)	1	1.00	0.30	0.15	0.05	
	Verandah grillwork	4	1.80	0.40	0.20	0.58	
	Cupboards (Ground floor)	4	1.30	0.40	0.15	0.31	
	(First floor)	3	1.30	0.30	0.15	0.18	
	Beam in staircase	1	2.80	0.30	0.30	0.25	
	Beam in WC and bath	1	3.35	0.20	0.20	0.13	
	Beam in WC and bath	1	2.70	0.20	0.20	0.11	
						4.79 cu. m	4.79 cu. m
17.	Providing and laying PCC M 150 in *chajja*						
	Over door D	2	1.40	0.60	0.10	0.17	
	Over door D_1	3	1.30	0.60	0.10	0.23	
	Over window W_1	3	2.40	0.60	0.10	0.43	
	Over window W_2	16	1.40	0.60	0.10	1.34	
	Over window W_3	4	1.00	0.60	0.10	0.24	
	Over verandah grillwork	4	1.80	0.60	0.10	0.43	
						2.84 cu. m	2.84 cu. m
18.	Formwork to floor and roof slab						
	Floor slab over Bed, kitchen and study	3	4.00	3.00		36.00	
	Living room	1	4.50	3.50		15.75	
	Front verandah	1	5.10	1.90		9.70	
	Passage						
	$4.50 + 0.40 + 0.40 = 5.30$ m	1	5.30	1.20		6.36	

(*continued*)

MEASUREMENT SHEET   (Continued)

Item No.	Description	No.	Length in m	Breadth in m	Depth or ht in m	Quantity	Total
	Back ver., passage, WC and bath	1	5.10	2.95		15.04	
	Formwork to sides:						
	$L = 8.80 + 0.40 = 9.20$ m	1	9.20		0.12	1.10	
		1	3.40		0.12	0.41	
		1	1.75		0.12	0.21	
	$L = 5.50 + 0.40 = 5.90$ m	1	5.90		0.12	0.71	
	$L = 3.35 + 0.40 = 3.75$ m	1	3.75		0.12	0.45	
	$L = 4.40 + 0.40 = 4.80$ m	1	4.80		0.12	0.58	
	$L = 2 \times 3.40 + 5.50 + 0.40$ $= 12.70$ m	1	12.70		0.12	1.52	
	Roof slab						
	Over living room	1	4.70	3.65		17.16	
	Over WC and bath	1	2.40	1.75		4.20	
	Projection						
	$L = 4.70 + 0.30 + 0.30$ $= 5.30$ m	1	5.30	0.20		1.06	
	$L = 3.65 + 0.30 + 0.20$ $= 4.15$	2	4.15	0.20		1.66	
	WC and bath						
	$L = 2.60 + 0.20 + 0.10$ $+ 0.40 = 3.30$ m	2	3.30	0.20		1.32	
	$L = 1.75 + 0.40 + 0.10$ $= 2.25$ m	2	2.25	0.20		0.90	
	Sides of slab	1	5.30		0.12	0.64	
		2	4.15		0.12	1.00	
		2	3.30		0.12	0.79	
		2	2.25		0.12	0.54	
	Landing						
	$L = 2.80 - 0.40 = 2.40$ m	1	2.40	0.75		1.80	
	Sides	1	2.80		0.15	0.42	
	$L = 0.75 + 0.20 = 0.95$ m	2	0.95		0.15	0.29	
						119.61 sq. m	119.61 sq. m
19.	Providing and laying cement concrete M 150 in slab						
	Bed and kitchen	1	9.20	3.40	0.12	3.75	

(continued)

MEASUREMENT SHEET   (Continued)

Item No.	Description	No.	Length in m	Breadth in m	Depth or ht in m	Quantity	Total
	Front verandah, living and back verandah	1	10.95	5.90	0.12	7.75	
	Over study	1	4.80	3.40	0.12	1.96	
	Roof slab						
	Living room	1	5.30	4.15	0.1	2.64	
	WC and bath	1	3.30	2.25	0.12	0.89	
	Landing	1	2.80	0.95	0.12	0.32	
						17.31	
	Deduct for T-beams	1	2.80	0.30	0.12	0.10	
		1	3.35	0.20	0.12	0.08	
		1	2.70	0.20	0.12	0.07	
						0.25	
			Net	quantity		17.06 cu. m	17.06 cu. m
20.	Providing formwork to risers, treads and waist slab of stair						
	Waist slab bottom:						
	$L = \sqrt{(2.20)^2 + (1.65)^2}$						
	$= 2.75$ m						
	Width $= 1.10 + 0.10$						
	$= 1.10$ m	2	2.75	1.10		6.05	
	Sides	2	2.75		0.27	1.49	
	Risers	20	1.10		0.18	3.96	
	CC block below: first step						
	Sides	2	1.10		0.25	0.55	
	Ends	1	0.25		0.25	0.06	
						12.11	
	Deduct for triangular portion at sides						
	$2 \times 9 = 18$ nos	18	0.25		0.18	0.41	
					2	11.70 sq. m	11.70 sq. m
21.	Providing formwork to RCC parapet						
	Sides						
	$L = \sqrt{(2.25)^2 + (1.62)^2}$						
	$= 2.77$ m						

(*continued*)

MEASUREMENT SHEET   (Continued)

Item No.	Description	No.	Length in m	Breadth in m	Depth or ht in m	Quantity	Total
	Two flights $2 \times 2 = 4$ nos	4	2.77		1.00	11.10	
	Ends	2	0.10		1.00	0.20	
	At landing level	2	0.40		1.00	0.80	
						12.10 sq. m	12.10 sq. m
22.	Providing and laying CCM 150 in steps						
	Waist slab	2	2.75	1.10	0.12	0.73	
	Steps	20	1.10	$\dfrac{(0.25 \times 0.18)}{2}$		0.50	
	Block below first step	1	1.10	0.25	0.25	0.07	
						1.30 cu. m	1.30 cu. m
23.	Providing and laying C C M 150 in parapet Length as per item No. 21	2	2.77	0.10	1.00	0.55	
		1	0.40	0.10	1.00	0.04	
						0.59 cu. m	0.59 cu. m
24.	Providing m.s reinforcement as per design. Reinforcement in half brick thick walls 6 mm dia. 2 bars in every 30 cm						

Ground floor

Total length

$2 \times 1.60 =$ 3.20

$1 \times 1.25 =$ 1.25

$1 \times 1.00 =$ $\underline{1.00}$

            5.45 m

Deduct $2 \times 0.80 = \underline{1.60\ m}$

(Door portion)   3.85 m

In first 2 m

no. of bars $= 200/30$

        $= 7$ nos   7   3.85

In next 1.45 m

no. of bars $= 145/30$

        $= 5$ nos   5   5.45 m   0.129 q

(*continued*)

MEASUREMENT SHEET   (Continued)

Item No.	Description	No.	Length in m	Breadth in m	Depth or ht in m	Quantity	Total
	Total length 54.20 @ 0.22 kg/m						
	First floor						
	Total length =						
	$2 \times 1.70 =$ 3.40						
	$1 \times 1.35 =$ 1.35						
	$1 \times 1.00 =$ 1.00						
	5.75 m						
	Deduct $2 \times 0.80 =$ 1.60 m						
	(Door portion) 4.15 m						
	In first 2 m						
	No. of bars = 200/30 = 7	7	4.15				
	In top 1 m = 100/30 = 4	4	5.75				
			52.00 m				
	Total length						
	52.00 @ 0.22 kg/m					0.11 q	
						0.23 q	
	Add 5% for waste					0.12 q	
	(*Note*: In the absence of detailed design of reinforcement in RCC members the quantity of steel may be approximately calculated from the cubic contents of PCC as below:						
	Item Concrete cu. m						
	10 – 4.69						
	17 – 2.84						
	19 – 17.07						
	22 – 1.29						
	23 – 0.60						
	26.49						
	Total quantity of PCC 26.49 cu. m						
	Approximate quantity of steel @ 85 q cu. m					22.52 q	
						22.76 q	22.76 q

(*continued*)

MEASUREMENT SHEET   (Continued)

Item No.	Description	No.	Length in m	Breadth in m	Depth or ht in m	Quantity	Total
25.	Providing 5 cm thick Shahabad stone sill in cm (1:4)						
	Window: W_1 (G.F)	2	2.20	0.40		1.76	
	" W_2 (G.F)	9	1.20	0.40		4.32	
	" W_2 (F.F)	9	1.20	0.30		3.24	
	" W_3 (G.F)	2	0.80	0.40		0.64	
	" W_3 (F.F)	2	0.80	0.30		0.48	
	Verandah grill work	4	1.60	0.40		2.56	
						13.00 sq. m	13.00 sq. m
26.	Providing and fixing in position country teakwoodwork in roof Ridge						
	Over bed and kitchen						
	$L = 9.20 + 1.00 - 4.75$ $= 5.45$ m	1	5.45	0.08	0.15	0.07	
	Over stair						
	$L = 4.65 - 4.00 = 0.65$ m	1	0.65	0.08	0.15	0.01	
	Wall plate						
	Bed and kitchen	2	8.80	0.08	0.13	0.18	
		2	3.40	0.08	0.13	0.07	
	Staircase	2	3.40	0.08	0.13	0.07	
		2	2.80	0.08	0.13	0.06	
	Purlins						
	Bed and kitchen						
	$L = 10.20 - 2.40 = 7.80$ m	2	7.80	0.08	0.13	0.16	
	Staircase						
	$L = 4.65 - 2.00 = 2.65$ m	2	2.65	0.08	0.13	0.06	
	Common rafters						
	Over bed and kitchen						
	Total roof length $= 10.20$ m						
	No. of rafters $= \dfrac{10.20}{0.45} + 1$						
	Say $= 24$ nos						
	Two slopes $= 2 \times 24 = 48$						

(continued)

MEASUREMENT SHEET (Continued)

Item No.	Description	No.	Length in m	Breadth in m	Depth or ht in m	Quantity	Total
	Add 2 for central rafters $\dfrac{2}{50\ nos}$ on hipped end						
	Length $(4.75/2) \times 1.155$ $= 2.74$ m	50	2.74	0.05	0.08	0.55	
	Over staircase						
	Total roof length $= 4.65$ m						
	No. of rafters						
	$= (4.65/0.45) + 1 = 11$ nos						
	Two slopes $= 2 \times 11$ $= 22$ nos						
	Add 2 as above $\dfrac{2}{24\ nos}$						
	$L = (4.00/2) \times 1.155$ $= 2.31$ m	24	2.31	0.05	0.08	0.22	
	Hip rafters						
	Over bed and kitchen						
	L in hor. plane						
	$= \sqrt{(2.37)^2 + (2.37)^2}$						
	$= 2.37 \times 1.414 = 3.35$ m						
	True length						
	$= \sqrt{(3.35)^2 + (2.37^2 \times \tan 30°)^2}$						
	$= \sqrt{(3.35)^2 + (1.35)^2}$ $= 4.62$ m	4	4.62	0.13	0.18	0.43	
	Over staircase						
	Length in hor. plane						
	$= \sqrt{(2)^2 + (2)^2}$ $= 2.83$ m						
	True length						
	$= \sqrt{(2.83)^2 + (1.16)^2}$ $= 3.28$ m	4	3.28	0.13	0.18	0.31	
	Eaves boards						
	Bed and kitchen	2	10.20	0.03	0.15	0.09	

(continued)

MEASUREMENT SHEET   (Continued)

Item No.	Description	No.	Length in m	Breadth in m	Depth or ht in m	Quantity	Total
		2	4.75	0.03	0.15	0.04	
	Staircase	2	4.65	0.03	0.15	0.04	
		2	4.00	0.03	0.15	0.04	
						2.40 cu. m	2.40 cu. m
	(Note: Cubic contents are worked out correct to 0.01 cu. m by rounding 0.005 or more to 0.01 and neglecting the lower values)						
27.	Providing and applying 3 coats of oil painting to woodwork in roof; Ridge						
	Girth $= 2 \times 0.08 + 2 \times 0.15$ $= 0.46$ m	1	5.45	0.46		2.51	
		1	0.65	0.46		0.30	
	Wall plates and purlins Girth $= 2 \times 0.08 + 2 \times 0.15$ $= 0.42$ m	2	8.80	0.42		7.39	
		2	3.40	0.42		2.86	
		2	3.40	0.42		2.86	
		2	2.80	0.42		2.35	
		2	7.80	0.42		6.55	
		2	2.65	0.42		2.23	
	Common rafters Girth $= 2 \times 0.05 + 2 \times 0.08$ $= 0.26$ m	50	2.74	0.26		35.62	
		24	2.31	0.26		14.41	
	Hip rafters Girth $= 2 \times 0.13 + 2 \times 0.18$ $= 0.62$ m	4	4.62	0.62		11.46	
		4	3.28	0.62		8.13	
	Eaves board Girth $= 2 \times 0.05 + 2 \times 0.15$ $= 0.25$ m	2	10.20	0.25		5.10	
		2	4.75	0.25		2.37	

(continued)

MEASUREMENT SHEET   (Continued)

Item No.	Description	No.	Length in m	Breadth in m	Depth or ht in m	Quantity	Total
		2	4.65	0.25		2.33	
		2	4.00	0.25		2.00	
						108.47 sq. m	108.47 sq. m
28.	Providing Mangalore tiled roof with battens 4.5 cm × 2.5 cm @35 cm c/c Slope width = length of rafter = 10 cm eaves projection (Over bed and kitchen) = 2.74 + 0.10 = 2.84 m	2	10.20	2.84		57.94	
	Over staircase slope width = 2.31 + 0.10 = 2.41 m	2	4.65	2.41		22.41	
						80.35 sq. m	80.35 sq. m
29.	Providing and laying in position Mangalore ridge tiles						
	Over bed and kitchen	1	5.45			5.45	
	″ stair	1	0.65			0.65	
	″ hips of bed and kitchen	4	4.62			18.48	
	″ hips of staircase	4	3.28			13.12	
						37.70 m	37.70 m
30.	Screwing eaves tiles Bed and kitchen	2	10.20			20.40	
		2	4.75			9.50	
	Staircase	2	4.65			9.30	
		2	4.00			8.00	
						47.20 m	47.20 m
31.	Providing 10 cm dia. AC downtake pipes Height = 0.75 + 3.60 + 0.20 = 4.55	5	4.55			22.75 m	22.75 m
32.	10 cm dia. AC pipe bends	5				5 no.	5 no.
33.	10 cm dia. AC pipe drop ends	5				5 no.	5 no.

(continued)

MEASUREMENT SHEET   (Continued)

Item No.	Description	No.	Length in m	Breadth in m	Depth or ht in m	Quantity	Total
34.	Providing sand faced cement plaster 25 mm thick in two coats to stone masonry						
	Exterior plastering						
	From 15 cm below ground up to plinth level (Measured from front left hand corner in clockwise direction)	1	9.30		0.95	8.83	
		1	3.40		0.95	3.23	
		1	1.75		0.95	1.66	
		1	6.00		0.95	5.70	
		1	3.35		0.95	3.18	
		1	7.70		0.95	7.32	
		1	12.80		0.95	12.16	
	Sides of steps: Front	2	0.90		0.72	1.30	
						43.38	
	Deduct Redundant area	2	0.30		0.18	0.11	
		2	0.60		0.18	0.22	
						0.33 sq. m	
						43.05 sq. m	43.05 sq. m
35.	Providing sand faced cement plaster 20 mm thick in two coats to external faces of brickwork up to 10 m ht (Measured as mentioned above)						
	Height = 3.60 + 3.50 −0.15 = 6.95 m	1	9.20		6.95	63.94	
		1	3.40		6.95	23.63	
	Side wall of WC and bath	1	1.75		6.60	11.55	
	Rear wall of WC and bath	1	2.85		6.60	18.81	
	Back verandah						
	Height = 3.60 + 1.00 = 4.60 m	1	3.05		4.60	14.03	

(continued)

MEASUREMENT SHEET   (Continued)

Item No.	Description	No.	Length in m	Breadth in m	Depth or ht in m	Quantity	Total
		1	3.45		4.60	15.87	
	Staircase (Height = 3.60 + 2.50 = 6.10 m)	1	3.40	6.10	20.74		
		1	3.10		6.10	18.91	
	Front walls	1	4.50		4.60	20.70	
		1	8.95		4.60	41.17	
		1	3.80		6.95	26.41	
	First floor						
	Side wall of bed (part)	1	2.30		3.35	7.70	
	Living room, long wall	1	5.30		3.00	15.90	
	Living room, short wall	2	3.95		3.00	23.70	
	Kitchen, part	1	1.35		3.35	4.52	
	WC and bath front wall	1	2.35		3.00	7.65	
	WC and side wall	1	1.70		3.00	5.10	
	Stair	1	3.10		2.50	7.75	
		1	3.70		2.50	9.25	
	Inside and top face of parapet wall Height = 1.00 + 0.20 = 1.20 m Rear side						
		1	3.05		1.20	3.66	
		1	3.45		1.20	4.14	
		1	4.50		1.20	5.40	
		1	8.95		1.20	10.74	
						381.27 sq. m	
	Deductions For openings of:						
	Doors–D	2	1.10		2.50	5.50	
	D_1(First floor)	4	1.00		2.00	8.00	
	D_2(First floor)	2	0.80		1.90	3.04	
	Windows:						
	W_1(2 + 1 = 3 nos)	3	2.10		1.60	10.08	
	W_2(7 + 9 = 16)	16	1.10		1.60	28.16	
	W_3(2 + 2 = 4)	4	0.70		1.00	2.80	
	W_4	1	0.70		3.00	2.10	

(*continued*)

MEASUREMENT SHEET   (Continued)

Item No.	Description	No.	Length in m	Breadth in m	Depth or ht in m	Quantity	Total
	Grillwork in verandah	4	1.50		1.50	9.00	
						68.68	
						312.59 sq. m	312.59 sq. m
36.	Providing cement plaster 12 mm thick, single coat with *neeru* finish	6	4.00		3.45	82.00	
	Ground floor						
	Bed, kitchen and study	6	3.00		3.45	62.10	
	Living room	2	4.50		3.45	31.05	
		2	3.50		3.45	21.15	
	Front verandah	3	5.10		3.45	52.79	
	Back verandah	1	1.90		3.45	6.55	
	Passage	1	4.00		3.45	13.80	
		1	2.10		3.45	7.25	
	Passage	1	2.80		3.45	9.66	
	Staircase	1	2.40		3.45	8.28	
		2	3.40		3.45	23.46	
	Back verandah	1	3.35		3.45	11.56	
		1	1.20		3.45	4.14	
	WC and bath front wall	1	2.45		3.45	8.45	
	WC and bath outside wall	1	1.60		3.45	5.52	
	WC and bath inside	4	1.50		3.45	20.70	
	Bath inside	2	1.25		3.45	8.64	
	WC inside	2	1.00		3.45	6.90	
	FIRST FLOOR						
	Bed and kitchen	4	4.15		3.50	58.10	
		4	3.15		3.50	44.10	
	Gable portion	2	$3.15 \times$	$\left(\dfrac{1.57}{2}\right)$		4.95	
	Living room	2	4.70		2.88	27.07	
		2	3.65		2.88	21.02	
	Staircase	2	3.00		2.65	15.90	
		2	2.40		2.65	12.72	
	WC and bath long walls	4	1.60		3.00	19.20	
		2	1.35		3.00	8.10	
		2	1.00		3.00	6.00	
						601.16 sq. m	

(*continued*)

MEASUREMENT SHEET (Continued)

Item No.	Description	No.	Length in m	Breadth in m	Depth or ht in m	Quantity	Total
	Deductions						
	For openings of:						
	Door: D_1(G F + F F)						
	(7 + 2) = 9	9	1.00		2.00	18.00	
	D_2 (Gr. floor)	2	0.80		1.90	3.04	
	Window: W_2	2	1.10		1.60	3.52	
	Ends of waist slab in stair	2	2.77		0.15	0.83	
	Ends of steps	18	(0.25 × 0.18/2)			0.41	
	Landing	1	2.40		0.15	0.36	
		2	1.00		0.15	0.30	
						26.46	
						574.70 sq. m	574.70 sq. m
37.	Providing and applying colour wash in two coats					952.20 sq. m	952.20 sq. m
	(43.03 + 312.67 + 596.50)						
	(Quantity of item and						
	No. 34 + 35 + 36)						
	= 952.20 sq. m						
38.	Providing and laying 12 cm thick c c M.60 in flooring						
	Bed, kitchen and study	3	4.00	3.00		36.00	
	Living room	1	4.50	3.50		15.75	
	Stair	1	2.40	3.40		8.16	
	Front verandah	1	5.10	1.90		9.69	
	Passage	1	2.10	1.20		2.52	
		1	2.80	1.20		3.36	
	WC, bath and rear Verandah	1	5.10	2.95		15.05	
						90.53 sq. m	90.53 sq. m
39.	Providing and laying 25 mm thick polished Shababad stone flooring in cm (1:4) prop.						
	Ground floor						
	Bed, kitchen and study	3	4.00	3.00		36.00	
	Living room	1	4.50	3.50		15.75	

(*continued*)

MEASUREMENT SHEET   (Continued)

Item No.	Description	No.	Length in m	Breadth in m	Depth or ht in m	Quantity	Total
	Stair	1	2.40	3.40		8.16	
	Front verandah	1	5.10	1.90		9.69	
	Passage	1	2.10	1.20		2.52	
	Passage	1	2.80	1.20		3.36	
	WC, bath and rear Veranda	1	5.10	2.95		15.05	
	Add for door sills						
	Door: D	2	1.20	0.40		0.96	
	D_1	7	1.00	0.40		2.80	
	FIRST FLOOR						
	Bed and kitchen	2	4.15	3.15		26.15	
	Living	1	4.70	3.65		17.16	
	Bath	1	1.65	1.35		2.23	
	WC	1	1.65	1.00		1.65	
	Add for door sills						
	Door: D_1	5	1.00	0.30		1.50	
	D_2	2	0.80	0.10		0.16	
						143.14	
	Deduct:						
	For portion below 10 cm partition walls on ground floor	2	1.60	0.10		0.32	
		1	1.25	0.10		0.13	
		1	1.00	0.10		0.10	
						0.55	
						142.59 sq. m	142.59 sq. m
40.	Providing 25 mm thick rough Shahabad stone flooring in cm (1:4 proportion)						
	Over steps: Front side	3	1.50	0.30		1.35	
	Side	3	1.00	0.30		0.90	
	ON TERRACE						
	From front to rear measured in rectangles						
	$L = (5.50 + 3.40) - 0.15$ $= 8.75$ m	1	8.75	7.40		64.75	

(*continued*)

MEASUREMENT SHEET (Continued)

Item No.	Description	No.	Length in m	Breadth in m	Depth or ht in m	Quantity	Total
	$W = (4.40 + 2.80) + 0.20$ $= 7.40$ m *Rear* $L = 5.50 - 0.15 = 5.35$ m $B = 3.35 - 0.20 = 3.15$ m	1	5.35	3.15		16.85	
						83.85	
	Deduct redundant area						
	Living room	1	3.95	5.30		20.94	
	Stair portion		3.30	3.10		10.23	
	W C and bath $L = 1.35 + 0.10 + 1.00 + 0.10 = 2.55$ m $W = 1.00 + 0.10 = 1.10$ m	1	2.55	1.10		2.81	
						33.98	
						49.87 sq. m	49.87 sq. m
41.	Providing and fixing 5 cm thick precast R C C slabs in cupboards 3 nos in each cupboards						
	Ground floor $4 \times 3 = 12$ nos	12	1.30	0.30		4.68	
	First floor $3 \times 3 = 9$ nos	9	1.30	0.20		2.34	
						7.02 sq. m	7.02 sq. m
42.	Providing country teakwood 35 mm thick fully panelled double shutters Door D type	2	1.06		1.88	3.99	
	Window W_1 $3 \times 2 = 6$ nos	6	0.96		1.46	8.41	
	Window W_2	18	0.96		1.46	25.23	
	Cupboards	7	0.86		1.86	11.20	
						48.83 sq. m	48.83 sq. m
43.	Providing 35 mm thick solid core flush door in single leaf						
	Ground floor + first floor $7 + 6 = 13$	13	0.86		1.91	21.35 sq. m	21.35 sq. m

(continued)

MEASUREMENT SHEET   (Continued)

Item No.	Description	No.	Length in m	Breadth in m	Depth or ht in m	Quantity	Total
44.	Providing ledged, braced and framed door shutters						
	Ground and first $2 + 2 = 4$	4	0.66		1.81	4.78 sq. m	4.78 sq. m
45.	Providing 6 mm thick glass louvred window shutter						
	Window W_3 $2 + 2 = 4$ nos	4	0.56		0.86	1.93 sq. m	1.93 sq. m
46.	Providing fully glazed shutters for ventilators						
	Over door D	2	1.06		0.35	0.24	
	Over windows						
	$W_1 - 3 \times 2 = 6$	6	0.96		0.45	2.59	
	Over windows W_2	18	0.96		0.45	7.67	
						10.5 sq. m	10.5 sq. m
47.	Hanging the shutters of doors, windows and ventilators with hinges						
	Door D	2	1.06		1.88	3.99	
	Door D_1	13	0.86		1.91	2.35	
	Door D_2	4	0.66		1.81	4.78	
	Window W_1 $3 \times 2 = 6$ nos	6	0.96		1.46	8.41	
	Window W_2	18	0.96		1.46	25.23	
	Ventilators over door D	2	1.06		0.35	0.74	
	Ventilator over						
	Window W_1	6	0.96		0.45	2.59	
	Window W_2	18	0.96		0.15	7.67	
	Cupboards	7	0.86		1.86	11.20	
						66.96 sq. m	66.96 sq. m
48.	Fixing the shutters to frame louvred window	4	0.56		0.86	1.93 sq. m	1.93 sq. m
49.	Providing and fixing in position CI grillwork in openings						
	In verandah	4	1.50		1.50	252 kg	
	9.00 sq. m @ 28 kg/sq. m					2.52 q	

(*continued*)

MEASUREMENT SHEET   (Continued)

Item No.	Description	No.	Length in m	Breadth in m	Depth or ht in m	Quantity	Total
50.	Providing and fixing in position CI grillwork in windows						
	Window W_1 $3 \times 2 = 6$ no.	6	0.96		1.46	8.41	
	Window W_2	18	0.96		1.46	25.23	
						33.64 sq. m	
			33.64 sq. m @ 28kg/ sq. m			94.92 kg	
						9.42 q q	9.42
51.	Providing and fixing in position precast c c jaali in staircase						
	Window W_4	1	0.70		3.00	2.10 sq. m	2.10 sq. m
52.	Providing and fixing with screws m.s holdfasts 20 cm $\times$ 4 cm $\times$ 4 cm to frames						
	Door D $2 \times 6 = 2$ nos	12				12	
	Door D_1 $13 \times 6 = 78$ nos	78				78	
	Door D_2 $4 \times 6 = 24$ nos	24				24	
	Window W_1 3×4 = 12 nos	12				12	
	Window W_2 18×4 = 72 nos	72				72	
	Window W_3 4×4 = 16 nos	16				16	
						214 no.	214 no.
53.	Providing and fixing with screw 10 cm size brass butt hinges						
	Door D $2 \times 2 \times 3 = 12$	12				12	
	Door D_1 $13 \times 3 = 39$	39				39	
	Door D_2 $4 \times 3 = 12$	12				12	
	Window W_1 $3 \times 2 \times 2 \times 2$ = 24	24				24	
	Window W_2 $18 \times 2 \times 2 = 72$	72				72	
	Cupboards $7 \times 2 \times 2 = 28$	28				28	
						187 no.	187 no.

(continued)

MEASUREMENT SHEET   (Continued)

Item No.	Description	No.	Length in m	Breadth in m	Depth or ht in m	Quantity	Total
54.	Providing and fixing with screws brass tower bolts 20 cm size						
	Door D $2 \times 2 = 4$	4				4	
	Door D_1 $13 \times 1 = 13$	13				13	
	D_2 $4 \times 1 = 4$	4				4	
	Windows W_1 $3 \times 4 = 12$	12				12	
	W_2 $18 \times 4 = 72$	72				72	
						105 no.	105 no.
55.	Providing and fixing brass handle						
	Door D $2 \times 2 = 4$	4				4	
	Door D_1 $13 \times 2 = 26$	26				26	
	Door D_2 $4 \times 2 = 8$	8				8	
	Window W_1 $3 \times 2 = 6$	6				6	
	Window W_2 $18 \times 2 = 36$	36				36	
	Cupboards $7 \times 1 = 7$	7				7	
						87 no.	87 no.
56.	Providing and fixing brass aldrop						
	D	2				2	
	D_1	13				13	
	D_2	4				4	
						19 no.	19 no.
57.	Providing and fixing stoppers						
	D	4				4	
	D_1	13				13	
	D_2	4				4	
						21 no.	21 no.
58.	Providing and applying two coats of oil paint to woodwork in doors and windows						
	Door $D = 2$ nos $1\frac{1}{8} \times 2$ face $= 4.5$	4.5	1.18		2.00	10.62	

(*continued*)

MEASUREMENT SHEET   (Continued)

Item No.	Description	No.	Length in m	Breadth in m	Depth or ht in m	Quantity	Total
	Door ventilator $2 \times \dfrac{1}{2} \times 2 = 2$ no.	2	1.18		0.48	1.13	
	Door D_1 Both faces	13	0.98		1.98	25.23	
	Door $D_2 = 4$ no. $\times 1\dfrac{1}{8} \times 2$ face $= 9$ no.	9	0.78		1.88	13.00	
	Windows: W_1 $= 3$ no. $\times 1\dfrac{1}{8} \times 2$ face $= 6.75$ m	6.75	2.08		1.10	15.44	
	Ventilator over window W_1 $3 \times 1\dfrac{1}{2} \times 2$ faces $= 3$ no	3	2.08		0.48	3.00	
	Window W_2: $18 \times 1\dfrac{1}{8} \times 2$ faces $= 40.50$	40.50	1.08		1.10	48.11	
	Ventilator over W_2: $= 18 \times \dfrac{1}{2} \times 2$ faces	18	1.08		0.48	9.33	
	Window W_3: $4 \times 1\dfrac{1}{2} \times 2$ faces $= 4$ no	4	0.68		0.98	2.67	
	Cupboards: $7 \times 1\dfrac{1}{8} \times 2$ faces $= 15.75$ no	15.75	0.98		1.98	30.56	
						159.09 sq. m	159.09 sq. m
59.	Water supply and sanitary fittings. Allow a provisional sum of rupees 5500		Provisional sum				
60.	Electrical work. Allow a provisional sum of rupees 4000		Provisional sum				

ABSTRACT SHEET

Item No.	Quantity	Description	Rate ₹	Per	Amount ₹
1.	137.87 sq. m	Excavate over site 15 cm deep to remove vegetable soil and remove and deposit excavated material with a lead of 50 m	40	sq. m	5,514.80
2.	45.65 cu. m	Excavate in trenches for foundation in ordinary soil and remove and stack the excavated material as directed within a lead of 50 m and lift of 1.5 m	207	cu. m	9,449.55
3.	22.51 cu. m	Excavate in trenches for foundation in hard soil and remove and stack the excavated material within a lead of 50 m and lift of 1.5 m including return, fill in and ram selected excavated material	365	cu. m	8,216.15
4.	11.60 cu. m	Providing and laying in foundation cement concrete M 75 including levelling, ramming and curing	6,359	cu. m	73,764.40
5.	65.55 cu. m	Providing UCR masonry in cement mortar (1:6) in foundation and plinth including curing	4,734	cu. m	3,10,313.70
6.	13.05 cu. m	Providing external angles in facework with corner stones of 30 cm × 15 cm × 15 cm size dressed on beds and joints and laid in cm (1:6) including curing	500	m	6,525.00
7.	33.13 sq. m	Providing and laying damp-proof course consisting of 12 cm thick cement concrete M 20 and two coats of hot bitumen including formwork, ramming and curing	1,000	sq. m	33,130.00
8.	61.98 cu. m	Filling in plinth selected excavated material in layers, including watering and compacting	750	cu. m	46,485.00
9.	60.27 cu. m	Providing brickwork II sort in cm (1:6) in superstructure of ground floor including scaffolding and curing	7,755	cu. m	4,67,393.85
10.	15.60 sq. m	Providing 10 cm thick brickwork in cm (1:4) in superstructure of ground floor including scaffolding and curing	802	sq. m	12,511.20
				Total of page, carried over	
				₹	9,73,303.65

(continued)

ABSTRACT SHEET   (Continued)

Item No.	Quantity	Description	Rate ₹	Per	Amount ₹
			Brought forward		9,73,303.65
11.	45.36 cu. m	Providing brickwork II sort in cm (1:6) in superstructure of first floor and parapet including scaffolding and curing	7,755	cu. m	3,51,766.80
12.	14.05 cu. m	Providing 10 cm thick brickwork in cm (1:4) in superstructure of first floor including scaffolding and curing	802	sq. m	11,268.10
13.	2.25 cu. m	Providing and fixing in position country wood frames (of 70 mm × 95 mm size scantling) of doors, windows and cupboards	1,18,981	cu. m	2,67,707.25
14.	54.95 sq. m	Providing formwork to lintels and beams	750	sq. m	41,212.50
15.	37.09 sq. m	Providing formwork to RCC chajja	750	sq. m	27,817.50
16.	4.79 cu. m	Providing and laying cement concrete M 150 in lintels and beams including ramming, curing and fair finish	6,359	cu. m	30,459.61
17.	2.84 cu. m	Providing and laying cement concrete M 150 in chajja including ramming, curing and fair finish	6721	cu. m	19,087.64
18.	119.61 sq. m	Providing formwork of approved type to roof slab and landing	750	sq. m	89,707.50
19.	17.06 cu. m	Providing and laying cement concrete M 20 in slab and landing including ramming, curing and fair finish	6,750	cu. m	1,15,155.00
20.	11.70 sq. m	Providing formwork to risers, treads and waist slab of stair	750	sq. m	8,775.00
21.	12.10 sq. m	Providing formwork to RCC parapet of stair	800	sq. m	9,680.00
22.	1.30 cu. m	Providing and laying cement concrete M 150 steps of stair including ramming, curing and fair finish	7,000	cu. m	9,100.00
23.	0.59 cu. m	Providing and laying cement concrete M 20 in parapet including ramming, curing and fair finish	6,750	cu. m	3,982.50
			Total of page, carried over		
			₹		19,59,023.05

(continued)

ABSTRACT SHEET   (Continued)

Item No.	Quantity	Description	Rate ₹	Per	Amount ₹
			Brought forward		19,59,023.05
24.	22.76	Providing, cutting, bending, hooking, conveying, placing in position and binding mild steel reinforcement as per design	6,100	q	1,38,836.00
25.	13.00 sq. m	Providing and laying 5 cm thick Shahabad stones in sills in cm (1:4)	650	sq. m	8,450.00
26.	2.40 cu. m	Providing and fixing in position country teak wood work in roof	1,18,981	cu. m	2,85,554.40
27.	108.47 sq. m	Providing and applying three coats of oil paint of approved colour to wood work in roof	120	sq. m	13,016.40
28.	80.35 sq. m	Providing and fixing Mangalore tiled roof on country teakwood battens of 4.5 cm × 2.5 cm size fixed 35 cm centre to centre apart	1,280	sq.m	1,02,848.00
29.	37.70 m	Providing and laying Mangalore ridge tiles in cement mortar (1:2)	200	m	7,540.00
30.	47.20 m	Screwing eaves tiles as directed.	50	m	2,360.00
31.	22.75 m	Providing and fixing in position 10 cm dia. AC down take pipes	300	m	6,825.00
32.	5.00 no.	Providing and fixing 10 cm diameter AC pipe bends	250	no.	1,250.00
33.	5.00 no.	Providing and fixing 10 cm diameter AC pipe drop ends	250	no.	1,250.00
34.	43.05 sq. m	Providing sand faced cement plaster (1:4) 25 mm thick in two coats to stone masonry including curing	750	sq. m	32,287.50
35.	312.59 sq. m	Providing sand faced cement plaster (1:4) 20 mm thick in two coats to brickwork up to 10 m high including scaffolding and curing	700	sq. m	2,18,813.00
36.	574.70 sq. m	Providing cement plaster (1:6) 12 mm thick in single coat to brickwork with neeru finish including scaffolding and curing	500	sq. m	2,87,350.00
		Total of page, carried over			
			₹		30,65,403.35

(continued)

ABSTRACT SHEET (Continued)

Item No.	Quantity	Description	Rate ₹	Per	Amount ₹
			Brought forward		30,65,403.35
37.	952.20 sq. m	Providing and applying colour wash of approved shade in two coats including scaffolding	50	cu. m	47,610.00
38.	90.53 sq. m	Providing and laying plain cement concrete M60 in flooring including ramming and curing	12,000	cu. m	10,86,360.00
39.	142.59 sq. m	Providing and laying 25 mm thick polished Shahabad stone flooring in cement mortar (1:4) including pointing and polishing	1,000	sq. m	1,42,590.00
40.	49.87 sq. m	Providing and laying 25 mm thick rough Shahabad stone flooring in cement mortar (1:4) including pointing	900	sq. m	44,883.00
41.	7.02 sq. m	Providing and fixing in position 5 cm thick precast RCC slabs in cupboards	937	sq. m	6,577.74
42.	48.83 sq. m	Providing 35 mm thick country teak wood fully panelled shutters for doors, windows and cupboards	8,750	sq.m	4,27,262.50
43.	21.35 sq. m	Providing 35 mm thick solid core flush door in single leaf	6,282	sq. m	1,34,120.70
44.	4.78 sq. m	Providing 25 mm thick country teak wood ledged, braced and framed door shutters	7,200	sq.m	34,416.00
45.	1.93 sq. m	Providing 6 mm thick glass louvred window shutters	7500	sq. m	14,475.00
46.	10.5 sq. m	Providing fully glazed shutters for ventilators with 35 mm thick country teakwood framework	4,804		50,442.00
47.	66.96 sq. m	Hanging the shutters of doors, window, ventilators and cupboards with hinges including fittings and fastenings	800	sq. m	53,568.00
48.	1.93 sq. m	Fixing the shutters to frames (labour only)	750	sq. m	1,447.50
49.	9 sq. m	Providing and fixing in position CI grillwork	3,258	sq.m	29,322.00
50.	33.64 sq. m	Providing and fixing CI grillwork to timber frame of windows	3,258	q	1,09,599.12
			Total of page, carried over		
				₹	52,48,074.91

(continued)

ABSTRACT SHEET   (Continued)

Item No.	Quantity	Description	Rate ₹	Per	Amount ₹
			Brought forward		52,48,074.91
51.	2.10 sq. m	Providing and fixing in position precast cement concrete jaali in stair including wire reinforcement	800	sq. m	1,680.00
52.	159.09 no.	Providing and applying two coats of oil paint to woodwork in doors and windows	120	sq. m	19,090.80
		Total			52,68,845.71
53.	p.s	Water supply and sanitary fittings. Allow a provisional sum (about 10% to 15% of the cost of all items above)			7,90,326.85
54.	p.s	Electrical work. Allow a provisional sum (about 8% to 12% of the total cost)			6,32,261.48
		Total cost			66,91,434.05
		Add contingencies at 5%, say,			3,34,571.70
				Total	70,26,005.75
				Total say	70,26,005.00

A Two Span Arch Bridge

MEASUREMENT SHEET

Item No.	Description	No.	Length	Breadth	Depth or height	Quantity	Total
1.	Excavation in trenches for foundations in hard soil Lead 50 m, Lift 1.5 m *Abutment* (left bank) Avg. GL $\dfrac{600.40 + 600.50}{2} = 600.45$ Stratum level = 598:50 Depth of excavation = 1:95 m	1	7.80	3.00	1.50	35.10	
	Pier Avg. GL $\dfrac{599.15 + 598.85}{2} = 599.00$ Stratum level = 598.50 Depth of excavation = 0.50	1	9.64	2.20	0.50	10.60	
	Abutment (right bank) Avg. GL $\dfrac{600.85 + 600.95}{2} = 600.90$ Stratum level = 598.50 Depth of excavation = 2.40 m	1	7.80	3.00	1.50	35.10	
	Wing walls *Length*: measured at right angles to the centre line of road $= 5.60 - 0.15 = 5.45$ m *Width*: measured parallel to the centre line of road at midway At the junction with abutment $= 2.50 \times \sqrt{2} = 3.535$						

(continued)

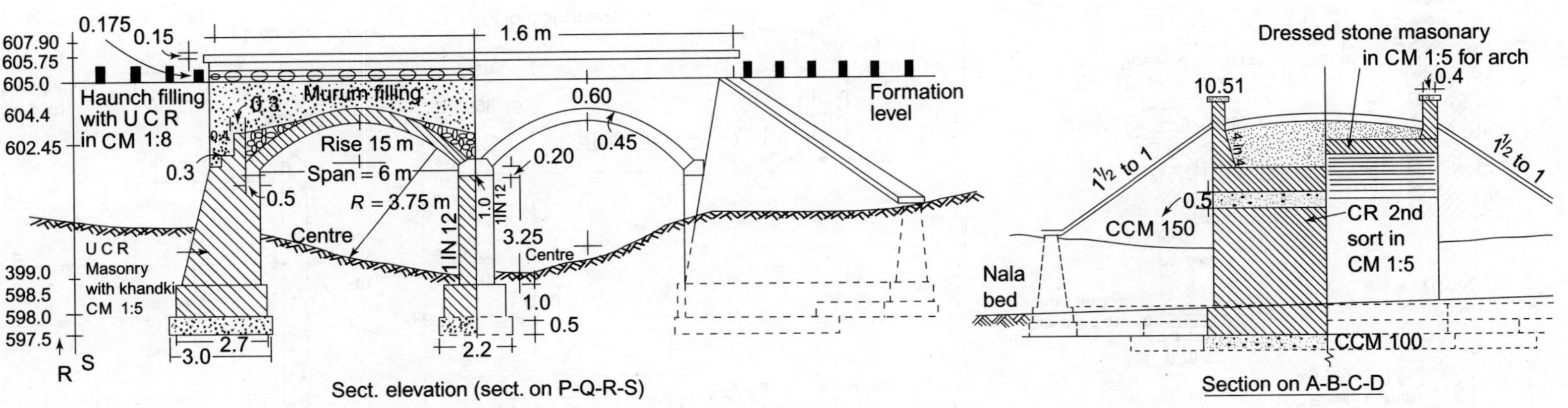

Estimate 2   A two span arch bridge

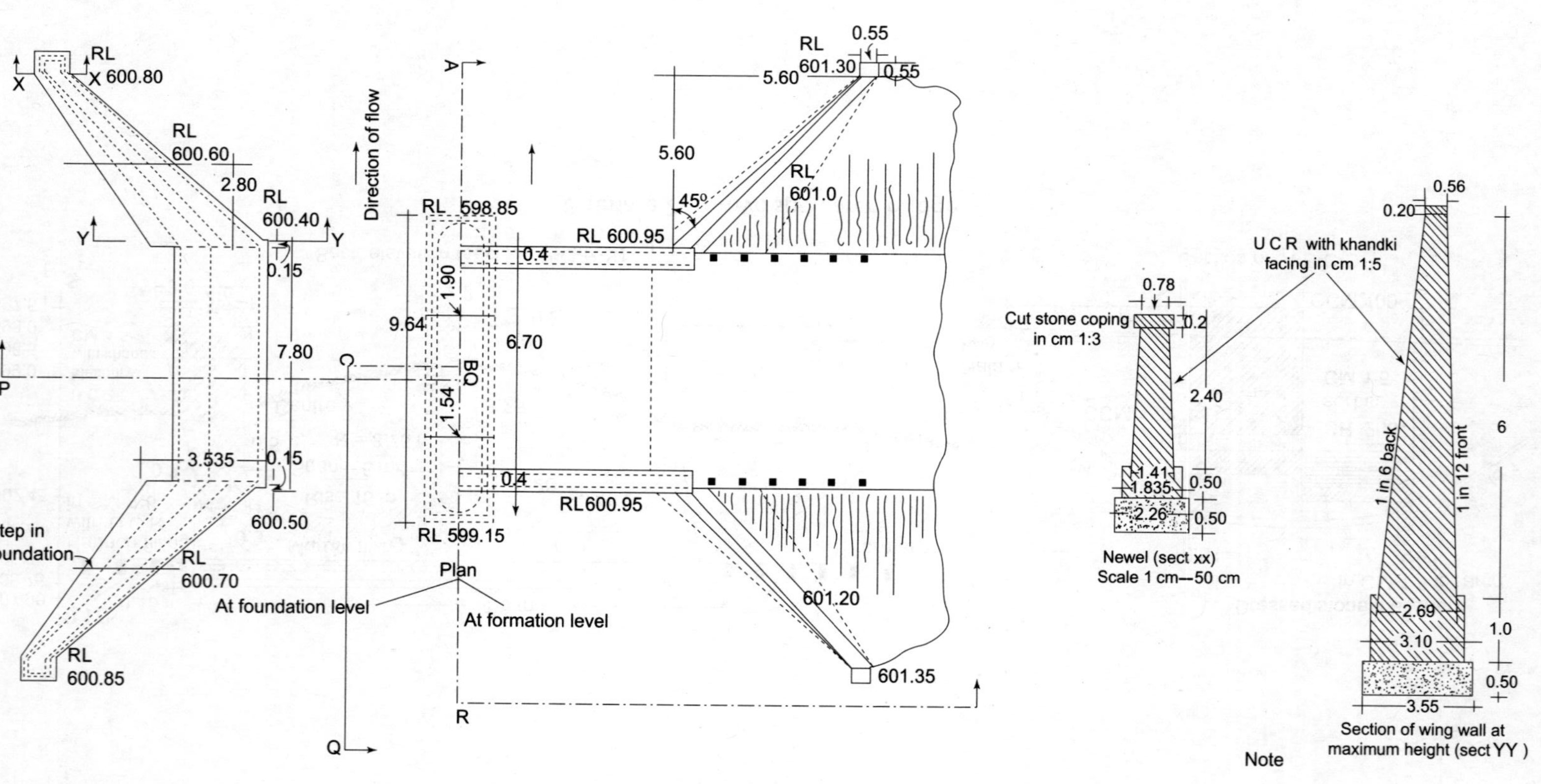

Estimate 2   Continued

MEASUREMENT SHEET   (Continued)

Item No.	Description	No.	Length	Breadth	Depth or height	Quantity	Total
	At the junction with newel $= 1.60 \times \sqrt{2} = 2.262$ Mean width $= 2.90$ m Wing wall–U/S left bank Avg. GL $$\frac{600.50 + 600.70 + 600.80}{3} = 600.66$$ Stratum level $\underline{= 598.50}$ Depth of excavation $= 2.16$ m	1	5.45	2.90	1.50	23.71	
	Wing wall: U/S right bank Avg. GL $$\frac{600.95 + 601.20 + 601.35}{3} = 601.16$$ Stratum level $\underline{= 598.50}$ Depth of excavation $= 2.66$ m	1	5.45	2.90	1.50	23.71	
	Wing wall: D/S left bank Avg. GL $$\frac{600.40 + 600.60 + 600.80}{3} = 600.60$$ Stratum level $\underline{= 598.50}$ Depth of excavation $= 2.10$ m	1	5.45	2.90	1.50	23.71	
	Wing wall: D/S right bank Avg. GL $$\frac{600.85 + 601.00 + 601.39}{3} = 601.05$$ Stratum level $\underline{= 598.50}$ Depth of excavation $= 2.55$ m	1	5.45	2.90	1.50	23.71	
	Newels U/S–left bank GL 600.85 Stratum level $\underline{598.50}$ Depth of excavation 2.35 m Length $= 1.60 \times \sqrt{2} = 2{:}26$ m	1	2.26	0.85	1.50	2.88	
	U/S–right bank GL 601.35 Stratum level $\underline{598.50}$ Depth of excavation 2.85 m	1	2.26	0.85	1.50	2.88	

(continued)

MEASUREMENT SHEET   (Continued)

Item No.	Description		No.	Length	Breadth	Depth or height	Quantity	Total
	D/S–left bank							
	GL	600.80						
	Stratum level	598.50						
	Depth of excavation:	2.30 m	1	2.26	0.85	1.50	2.88	
	D/S–right bank							
	GL	601.30						
	Stratum level	598.50						
	Depth of excavation:	2.80 m	1	2.26	0.85	1.30	2.88	
							187.16 cu. m	187.16 cu. m
2.	Excavation in trenches for foundation in hard soil Lead 50 m, Lift 3 m Depth of exc. in hard soil: (Total 1.50 m measured above) *Abutment*: left bank							
	Total depth	1.95 m						
	Less	1.50						
		0.45 m	1	7.80	3.00	0.45	10.53	
	Abutment: right bank							
	Total depth	2.40 m						
	Less	1.50						
		0.90 m	1	7.80	3.00	0.90	21.06	
	Wing walls: U/S–left bank							
	Total depth	2.16 m						
	Less	1.50						
		0.66	1	5.45	2.90	0.66	10.43	
	U/S–right bank							
	Total depth	2.66						
	Less	1.50						
		1.16 m	1	5.45	2.90	1.16	18.43	
	D/S–left bank							
	Total depth	2.10						
	Less	1.50						
		0.60 m	1	5.45	2.90	0.60	9.48	

(continued)

MEASUREMENT SHEET   (Continued)

Item No.	Description	No.	Length	Breadth	Depth or height	Quantity	Total
	D/S–right bank						
	Total depth 2.55						
	Less 1.50						
	1.05 m	1	5.45	2.90	1.05	16.60	
	Newels						
	U/S–left bank						
	Total depth 2.35						
	Less 1.50						
	0.85 m	1	2.26	0.85	0.85	1.63	
	U/S–right bank						
	Total depth 2.85						
	Less 1.50						
	1.35 m	1	2.26	0.85	1.35	2.59	
	D/S–left bank						
	Total depth 2.30						
	Less 1.50						
	0.80 m	1	2.26	0.85	0.80	1.54	
	D/S–right bank						
	Total depth 2.80						
	Less 1.50						
	1.30 m	1	2.26	0.85	1.30	2.50	
						94.79 cu. m	94.79 cu. m
3.	Excavation in trenches for foundation in soft rock, Lead 50 m, Lift 4.5 m						
	Depth of excavation = stratum level–Foundation level						
	= 598.50 – 597.50 = 1.00 m						
	For part of wing walls and newels						
	= 598.50 – 598.00 = 0.50 m						
	Abutments	2	7.80	3.00	1.00	46.80	
	Pier	1	9.64	2.20	1.00	21.21	
	Wing walls:						
	length of low foundation level						
	= 2.80 m						

(*continued*)

MEASUREMENT SHEET   (Continued)

Item No.	Description	No.	Length	Breadth	Depth or height	Quantity	Total
	Width at the junction of abutment $= 3.54$ m Width at the step $= 2.90$ m Mean width $= \dfrac{6.44}{2}$ $= 3.22$ m	4	2.65	3.22	1.00	34.13	
	Mean width of the remaining length $= \dfrac{2.90 + 2.20}{2} = 2.58$	4	2.80	2.58	0.50	14.45	
	Newels	4	2.26	0.85	0.50	3.82	
						120.41 cu. m	120.41 cu. m
4.	Providing and laying cement concrete M100 in foundation Abutments	2	7.80	3.00	0.50	23.40	
	Pier	1	9.64	2.20	0.50	10.60	
	Wing walls	4	2.65	3.22	0.50	17.06	
		4	2.80	2.58	0.50	14.45	
	Extra at step	4	0.50	2.90	0.50	2.90	
	Newels	4	2.26	0.85	0.50	3.84	
						72.25 cu. m	72.25 cu. m
5.	Providing UCR masonry in cm (1:5) Abutment From RL 598.00 to RL 599.00	2	7.50	2.70	1.00	40.50	
	Pier From RL 598.00 to 599.00	1	9.34	1.90	1.00	17.75	
	Wing walls From RL 598.00 to 599.00 Length $= 5.60$ Mean width $= 2.90 - 0.30 \times \sqrt{2}) = 2.48$ m	4	5.75	2.48	1.00	57.04	
	Newels From RL 598.50 to RL 599.00	4	1.96	0.70	0.50	2.74	
	Abutment From RL 599.00 To RL 602.45 Height $=$ 3.45 m						

(continued)

MEASUREMENT SHEET   (Continued)

Item No.	Description	No.	Length	Breadth	Depth or height	Quantity	Total
	Width at RL 599.00 = 2.36 m Width at RL 602.45 = 1.50 3.86 m Mean width $= \dfrac{3.86}{2} = 1.93$	2	7.50	1.93	3.45	99.88	
	Pier From RL 599.00 To RL 602.25 Height = 3.25 m Width at RL 599.00 = 1.54 Width at RL 602.25 = 1.00 Mean width $= \dfrac{2.54}{2} = 1.27$ m	1	7.50	1.27	3.25	30.36	
	Area at RL 599.00 = 1.54 Area at RL 602.25 = 1.00 Mean width $= \dfrac{2.54}{2} = 1.27$ m	1	7.50	1.27	3.25	30.96	
	Area at RL 599.00 $= \dfrac{1}{2}\left\{\dfrac{\pi}{4} \times (1.54)^2\right\}$ $= 0.927$ sq. m Area at RL 602.25 $= \dfrac{1}{2}\left\{\dfrac{\pi}{4} \times (1.00)^2\right\}$ $= (0.493$ sq. m$)$ Mean area $= \dfrac{0.927 + 0.493}{2}$ $= 0.71$ sq. m	2	0.71		3.25	4.62	
	Wing walls Area at the junction with abutment $= \dfrac{1.90 \times \sqrt{2} + 0.40 \times \sqrt{2 \times 6}}{2}$ $= 9.72$ sq. m $= A_1$						

(*continued*)

MEASUREMENT SHEET (Continued)

Item No.	Description	No.	Length	Breadth	Depth or height	Quantity	Total
	Area at the junction with newel $$= \dfrac{1.00 \times \sqrt{2} + 0.40 \times \sqrt{2} \times 2.40}{2}$$ $= 2.36$ sq. m $= A_2$ $$\text{Volume} = \frac{L}{3}\left(A_1 + A_2 + \sqrt{A_1 + A_2}\right)$$	4	$\dfrac{5.75}{3}$	$(9.72 + 2.36$ $+ \sqrt{9.72 + 2.36})$		119.26	
	Newels Width at RL 599.00 = 1.41 Width at RL 601.40 = 0.56 Mean length = $\dfrac{1.97}{2}$ Say = 0.99 m Width = 0.55 m	4	0.99	055	2.40	5.22	
	Head walls Portion over arches Cross-sectional area at max. height 2.25 m (RL 605.00 − 602.75) Top width = 0.40 m Bottom width = 0.96 m $$A_1 = \frac{0.40 + 0.96 \times 2.25}{2} = 1.53 \text{ sq. m}$$ Cross-sectional area at min. height that is, over the crown 0.60 m (RL 605 − RL 604.40): Top width: 0.40 m Bottom width: 0.55 m $$\text{Area} = \frac{0.40 + 0.55}{2} \times 0.60$$ $= 0.285$ sq. m $$\text{Volume} = \frac{L}{3}\left(A_1 + A_2 + \sqrt{A_1 + A_2}\right)$$ where L is half span at RL 602.75 $$= \frac{6.60}{2} = 3.30 \text{ m}$$	4×2	$\left[\dfrac{3.30}{3}\right.$ $\times \left(1.53 + 0.29 + \sqrt{1.53 + 0.29}\right)\Big]$			27.89	

(*continued*)

MEASUREMENT SHEET   (Continued)

Item No.	Description	No.	Length	Breadth	Depth or height	Quantity	Total
	Volume of head walls over skewbacks and pier						
	Cross-sectional area = 1.53 sq. m						
	Length = (Total length at parapet) $-4 \times$ Length (considered above)						
	$= 16.00 - 4 \times 3.30 = 2.80$ m	2	2.80	$\times$ (1.53)		8.57	
						444.79 cu. m	
	Deductions						
	From foundation masonry						
	Redundant volume at the stepped foundation	4	2.80	2.48	1.00	27.78	
	Extra concrete in steps	4	0.50	2.48	0.50	2.48	
	From abutment PCC skewbacks	2	7.50	0.50	0.20	1.50	
						31.76	
						413.03 cu. m	413.03 cu. m
6.	*Extra over* for khandki facing						
	Abutments	2	7.50		3.45	51.75	
	Pier						
	Height 3.25 m						
	Sloping height						
	$= \sqrt{(3.25)^2 + (0.27)^2} = 3.26$ m	2	7.50		3.26	48.90	
	Cut and ease water = 4.00						
	Mean girth						
	$= \frac{1}{2}\pi - 1.54 + \frac{1}{2}\pi.1.00 = 4.00$	2	4.00		3.26	26.08	
	Wing wall = U/S left bank						
	From RL 600.51						
	To RL 605.00						
	Max height: 4.49 m						
	Face height						
	$= \sqrt{(4.49)^2 + (\frac{4.49}{12})^2} = 4.51$ m						
	From RL 600.51						
	To RL 601.40						
	Min. height = 0.89 m						

(continued)

MEASUREMENT SHEET   (Continued)

Item No.	Description	No.	Length	Breadth	Depth or height	Quantity	Total
	Face height						
	$= \sqrt{(0.89)^2 + \left(\frac{0.89}{12}\right)^2} = 0.892$ m						
	Mean face height $\quad = 4.51$						
	$\qquad\qquad\qquad\quad + 0.892$						
	$\qquad\qquad\qquad\quad \overline{}$						
	$\qquad\qquad\qquad\qquad\quad 2$						
	$\qquad\qquad\qquad = 2.701$ m						
	Length $= 5.60 \times \sqrt{2} = 7.92$ m	1	7.92		2.70	21.38	
	$Newel \left(\dfrac{0.56 + 0.62}{2}\right)$	1	0.56		0.89	0.50	
	Mean length $= 0.59$	1	0.59		0.89	0.53	
	U/S-Right bank						
	From RL $\qquad$ 601.00						
	To RL $\qquad\quad$ 605.00						
	Max. height $= \qquad$ 4.00 m						
	Face height						
	$= \sqrt{(4)^2 + \left(\frac{4}{12}\right)^2} = 4.01$ m						
	From RL $\qquad$ 601.00						
	To RL $\qquad\quad$ 601.40						
	Min. height $= \qquad$ 0.40						
	Face height						
	$= \sqrt{(0.40)^2 + \left(\frac{0.40}{12}\right)^2} = 0.40$ m						
	Mean face height						
	$= \dfrac{4.01 + 0.40}{2} = 2.21$ m	1	7.92		2.21	17.50	
	Newel	1	0.56		0.40	0.22	
	Mean length						
	$= \dfrac{0.55 + 0.59}{2} = 0.57$ m	1	0.57		0.40	0.23	
	D/S left bank.						
	From RL $\qquad$ 600.45						
	To RL $\qquad\quad$ 605.00						
	Max. height $\qquad$ 4.55 m						

(continued)

MEASUREMENT SHEET   (Continued)

Item No.	Description	No.	Length	Breadth	Depth or height	Quantity	Total
	Face height $$=\sqrt{(4.55)^2 + \left(\frac{4.55}{12}\right)^2} = 4.57\ \text{m}$$						
	From RL 600.45						
	To RL 601.40						
	Min. height 0.95 m						
	Face height $$=\sqrt{(0.95)^2 + \left(\frac{0.95}{12}\right)^2} = 0.954\ \text{m}$$						
	Mean face height $$=\frac{4.570 + 0.954}{2} = 2.76\ \text{m}$$	1	7.92		2.76	21.86	
	Newel faces	1	0.56		0.95	0.52	
	Mean length $$=\frac{0.56 + 0.64}{2} = 0.60\ \text{m}$$	1	0.60		0.95	0.57	
	D/S-Right bank						
	From RL 600.90						
	To RL 605.00						
	Max. height = 4.10						
	Face height $$=\sqrt{(4.10)^2 + \left(\frac{4.10}{12}\right)^2} = 4.114\ \text{m}$$						
	From RL 600.90						
	To RL 601.40						
	Min. height = 0.50 m						
	Face height $$=\sqrt{(0.50)^2 + \left(\frac{0.50}{12}\right)^2} = 0.502\ \text{m}$$						
	Mean face height $$=\frac{4.114 + 0.50}{2} = 2.31\ \text{m}$$	1	7.92		2.31	18.30	
	Newel						
	Mean length	1	0.56		0.50	0.28	
	$$=\frac{0.56 + 0.60}{2}$$	1	0.58		0.50	0.29	

(*continued*)

MEASUREMENT SHEET   (Continued)

Item No.	Description	No.	Length	Breadth	Depth or height	Quantity	Total
	Head walls	2	16.00		2.25	72.00	
						280.91	
	Deduct. For redundant area in segmental arch portion	$2 \times \frac{2}{3}$	6.60		1.65	14.52	
						266.39 sq. m	266.39 sq. m
7.	Providing formwork to PCC bed blocks						
	Over abutments	2	7.50		0.30	4.50	
	Vertical face	2	7.50		0.20	3.00	
	Inclined face	2	7.50		0.45	6.75	
	Ends	4	0.50		0.20	0.40	
	Ends	4	$\dfrac{(0.50 + 0.20)}{2}$		0.30	0.42	
	Vertical faces	2	7.50		0.20	3.00	
	Inclined faces	2	7.50		0.45	6.75	
	Ends	2	1.00	0.20	0.40		
	Ends	2	$\dfrac{(1.00 + 0.40)}{2}$		0.30	0.42	
						25.64 sq. m	25.64 sq. m
8.	Providing and laying PCC M150 in bed blocks on abutment and piers						
	Over abutments						
	Rectangular portion	2	7.50	0.50	0.20	1.50	
	Trapezoidal portion	2	7.50	$\frac{0.50 + 0.20}{2}$	0.30	1.58	
	Over pier						
	Rectangular portion	1	7.50	1.00	0.20	1.50	
	Trapezoidal portion	1	7.50	$\frac{1.00 + 0.40}{2}$	0.30	1.58	
						6.16 cu. m	6.16 cu. m
9.	Providing and constructing dressed stone Masonry in cm (1:5) in arches including centering. Quantity of arch work						

(continued)

MEASUREMENT SHEET   (Continued)

Item No.	Description	No.	Length	Breadth	Depth or height	Quantity	Total
	Volume of arch masonry $= L \times l_m \times t$ where $L =$ length of the arch; $t =$ thickness of arch; $l_m =$ mean length of arch $$l_m = 1 \times \frac{R_m}{R}$$ where $R =$ radians of introdos of arch; $R_m = R + \frac{1}{2}$ and $l =$ length of introdos $= \dfrac{8b - 2a}{3}$ where $b = \sqrt{(a)^2 + (H)^2}$ Here, $a = \dfrac{S}{2} = \dfrac{6.00}{2} = 3.00$ m $$H = \frac{R}{4} = \frac{6.00}{4} = 1.5 \text{ m}$$ $\therefore b = \sqrt{(3)^2 + 1.5^2} = 3.35$ m $$l = \frac{8b - 2a}{3}$$ $$= \frac{8 \times 3.35 - 2 \times 3}{3} = 6.93 \text{ m}$$ $$l_m = 6.93 \times \frac{\left(3.75 + \dfrac{0.45}{2}\right)}{3.75}$$ $= 7.345$ m say 7.35 m $L = 7.50$ m	2	7.50	7.35	0.45	49.61 cu. m	49.61 cu. m
10.	'Extra-over' for dressing face stones in arches. 2	7.35			0.45	6.62 sq. m	6.62 sq. m

(continued)

MEASUREMENT SHEET　(Continued)

Item No.	Description	No.	Length	Breadth	Depth or height	Quantity	Total
11.	Providing UCR masonry in cm (1:8) for haunch filling. Walls retaining haunch filling Length = Total length – 2 × width of head walls At RL 602.45 $L = 7.50 - 2 \times 1:04 = 5.42$ m At RL 602.75 $L = 7.5 - 2 \times 0.96 = 5:58$ m Mean length $= \dfrac{5.42 + 5.58}{2} = 5.50$ m	2	5.50	0.40	0.30	1.32	
	From RL 602.75 to RL 603.45 Length at 602.75 = 5.58 m At 602.45 $L = 7.50 - 2 \times 0.79 = 5:92$ m Mean length $= \dfrac{5.58 + 5.92}{2} = 5.75$ m	2	5.75	0.30	0.70	2.42	
	Haunch filling Considering rectangular portion from RL 602.45 to RL 604.40 $L = 2 \times 0.50 + 2 \times 6.00 + 1.00$ $\quad = 14$ m Width between face walls At RL 602.75 = 5.58 m At RL 604.40 = 7.50 $-2 \times 0.55 = 6.40$ m Mean width $= \left(\dfrac{5.58 + 6.40}{2} \right) = 5.99$ m						
	Height = 604.40 − 602.45 = 1.95 m	1	14.00	5.99	1.95	163.53	
						167.27	
	Deduct Segmental portion: $\frac{2}{3} \times$ Span × Rise	2		$\frac{2}{3} \times 6.00 \times 1.50$		12.00	

(continued)

MEASUREMENT SHEET   (Continued)

Item No.	Description	No.	Length	Breadth	Depth or height	Quantity	Total
	Arch masonry Portion	2	5.99	7.35	0.45	39.62	
	cc bed block	2	5.99	$\dfrac{0.50+0.20}{2}$		4.19	
	Over pier	1	5.99	$\dfrac{1.00+0.40}{2}$		4.19	
	Triangular portions	4	5.99	$\dfrac{3.5+0.95}{2}$		39.83	
						99.83	
						67.44 cu. m	67.44 cu. m
12.	Providing and laying hard murum below roadway.						
	Triangular portions	4	5.99	$\dfrac{3.50+0.95}{2}$		39.83	
	Over abutments Up to RL 604.00	2	5.99	1.00	1.95	23.36	

Rectangular portion

From RL 604.40 to RL 604.75

Height   0.35 m

Length at RL 604.75

$= 7.50 - 2 \times 0.40 = 6.70$ m

At RL 604.40 = 6.40 m

13.10

Mean length $= \dfrac{13.10}{2} = 6.55$

Width = length of parapet

$= 2 \times 1.5 + 2 \times 6.0 + 1.00$

$= 16.00$ m

Item No.	Description	No.	Length	Breadth	Depth or height	Quantity	Total
		1	6.55	16.00	0.35	36.68	
						99.87 cu. m	
	Deduct Walls to retain haunch	2	5.50	0.40	0.30	1.32	
		2	5.75	0.30	0.70	2.41	
						3.73	
						96.14 cu. m	96.14 cu. m

(*continued*)

MEASUREMENT SHEET   (Continued)

Item No.	Description	No.	Length	Breadth	Depth or height	Quantity	Total
13.	Providing CR masonry II sort with Khandki facing on both sides in cm (1:6)						
	Parapet walls	2	16.00	0.40	0.75	9.60 cu. m	9.60 cu. m
14.	'Extra-over' walling for dressing stones on beds and joints						
	Parapet walls	4	16.00		0.75	48.00	
	Parapet ends	4	0.50		0.75	1.50	
						49.50 sq. m	49.50 sq. m
15.	Providing and laying cut stone coping in cm 1:3						
	Over parapet walls	2	16.00	0.50	0.15	2.40	
	Over wing walls:						
	Length in horizontal plane $= \sqrt{(5.60)^2 + (5.60)^2} = 7.92$ m						
	Inclined length $= \sqrt{(7.92)^2 + (3.60)^2} = 8.70$ m	4	8.70	0.40	0.15	2.08	
	Over newels:	4	0.65	0.65	0.15	0.25	
						4.73 cu. m	4.73 cu. m
16.	Providing cement pointing (1:3) to all exposed faces of masonry (Refer to item no. 6 for details)						
	Abutment left bank	1	7.50		2.15	16.13	
	Abutment right bank	1	7.50		1.70	12.75	
	Pier	2	7.50		3.26	48.90	
	Cut and ease water	2	4.00		3.26	26.08	
	Wing wall–U/S left bank	1	7.92		2.70	21.38	
	Newel U/S–Left bank	1	0.56		0.89	0.50	
		1	0.59		0.89	0.530	
	Wing wall–U/S right bank	1	7.92		2.21	17.50	
	Newel U/S–right bank	1	0.56		0.40	0.22	
		1	0.57		0.40	0.23	
	Wing wall D/S–left bank	1	7.92		2.76	21.86	
	Newel D/S–left bank	1	0.56		0.95	0.53	
		1	0.60		0.95	0.57	

(continued)

MEASUREMENT SHEET   (Continued)

Item No.	Description	No.	Length	Breadth	Depth or height	Quantity	Total
	Wing wall D/S–right bank	1	7.92		2.31	18.30	
	Newel D/S–right bank	1	0.56		0.50	0.28	
		1	0.58		0.50	0.29	
	Soffits of arches						
	(For details refer to item no. 9)	2	7.50		6.93	103.95	
	Faces (front and rear) of abutment (left bank)						
	Exposed width at RL 600.30 = 0.39 m and at RL 602.25 = 0:61 m						
	Height = 1.95 m	2	$\frac{(0.39 + 0.61)}{2}$		1.95	1.95	
	Abutment right bank						
	Exposed width at RL 600.75 = 0.44 m RL 602.25 = 0.61 m						
	Height = 1.50 m	2	$\frac{(0.44 + 0.61)}{2}$		1.50	1.58	
	Face walls						
	Length exposed at RL 602.45						
	Over abutments 2 × 0.46						
	Clear spans 2 × 6.00						
	Pier top width 1 × 1.00						
	Total: 13.92 m						
	Exposed length at						
	RL 605.00 =						
	Length of parapet = 16.00 m						
	Less 2 × 0.56 = 1.12 = 14.88 m						
	Height 605.00 − 602.45 = 2.55 m	2	$\frac{(13.92 + 14.92)}{2}$		2.55	73.44	
	Parapet walls						
	Outside and inside	4	16.00		0.90	57.60	
	Top	2	16.00		0.50	16.00	
	Ends	4	0.40		0.75	1.20	
		4	0.50		0.15	0.30	
	Top of wing walls	4	8.70		0.50	17.40	
	Newel—top	4	0.65		0.65	1.69	
						461.16	

(continued)

MEASUREMENT SHEET   (Continued)

Item No.	Description	No.	Length	Breadth	Depth or height	Quantity	Total
	Deduct						
	Segmental portion	2	$\left(\frac{2}{3}\times6.00\times1.50\right)$			12.00	
	Portion of skew back	4	$\dfrac{(0.50+0.20)}{2}$			1.40	
			$\dfrac{(1.00+0.40)}{2}$			1.40	
						446.36 sq. m	446.36 sq. m
	Bed block over pier	2					
17.	Providing metalled road surface over bridge	1	16.00	6.70		107.20 sq. m	107.20 sq. m

ABSTRACT SHEET

Item No.	Quantity	Description	Rate ₹	Per	Amount ₹
1.	187.16 cu. m	Excavate in trenches for foundation in hard soil and remove and deposit excavated material within a lead of 50 m and lift of 1.5 m including return, fill in and ram selected excavated material	265.10	cu. m	49,616.12
2.	94.79 cu. m	Excavate in trenches for foundation in hard soil and remove and deposit excavated material within a lead of 50 m and lift of 3m including return, fill in and ram selected excavated material	263.00	cu. m	24,929.77
3.	120.41 cu. m	Excavate in trenches for foundation in soft rock and remove and deposit excavated material within a lead of 100 m and lift 4.5 m including return, fill in and ram selected excavated material	612.70	cu. m	73,775.21
4.	72.25 cu. m	Providing and laying in layers in foundation cement concrete M100 including levelling, ramming and curing	5,360.00	cu. m	3,87,260.00
5.	413.03 cu. m	Providing uncoursed rubble masonry in cm (1:5) including scaffolding and curing	5,022.00	cu. m	20,74,236.66

(continued)

ABSTRACT SHEET   (Continued)

Item No.	Quantity	Description	Rate ₹	Per	Amount ₹
6.	266.39 cu. m	'Extra-over' uncoursed rubble masonry for khandki facing including dressing of khandki in beds and joints	5,022.00	sq. m	13,37,810.58
7.	25.64 cu. m	Providing formwork to plain cement concrete bed blocks below skew-backs	750.00	sq. m	19,230.00
8.	6.16 cu. m	Providing and laying plain cement concrete M150 in bed blocks on abutments and piers	6,975.00	cu. m	42,966.00
9.	49.61 cu. m	Providing dressed stone masonry in arches in cm (1:5) including centring	8,000.00	cu. m	3,96,880.00
10.	6.62 sq. m	'Extra-over' for dressing of face stones in arches	1,400.00	sq. m	9,268.00
11.	67.44 cu. m	Providing uncoursed rubble masonry in cm (1:8) in haunch filling including curing	5,022.00	cu. m	3,38,683.68
12.	96.14 cu. m	Providing and laying in layers hard murum below roadway including watering and ramming	316.00	cu. m	30,380.24
13.	9.60 cu. m	Providing coursed rubble masonry II sort in cement mortar (1:6) with khandki facing on both sides including curing	5,423.00	cu. m	52,060.80
14.	44.50 sq. m	'Extra-over' coursed rubble masonry for dressing stones on beds and joints	500.00	sq. m	22,250.00
15.	4.73 cu. m	Providing and laying cutstone coping in cement mortar (1:3) including scaffolding and curing	7,500.00	cu. m	35,475.00
16.	446.36 sq. m	Providing cement pointing (1:3) to all exposed faces of masonry including scaffolding and curing	173.00	sq. m	77,220.28
17.	107.20 sq. m	Providing metalled road surface over area between parapet walls as per detailed specifications	1476.00	sq. m	1,58,227.20
		Total cost			51,30,269.54
		Add contingencies at 5%			2,56,513.48
		Total			53,86,783.02
		Add 'overheads' and profit @ 20%			10,77,356.60
		Grand total			64,64,039.62

The estimated cost of the bridge is rupees sixty four lakhs, sixty four thousand only.

Earthwork and Roads

ESTIMATE OF A ROAD

DATA

Name of work: Construction of major district road from ……… to ………
Length: 3.00 km to 4.00 km
Metalled width: 3.36 m
Width of soling: 4.00 m
Formation width: 7.30 m
Side slopes in embankment: 2:1
Side slopes in cutting: 1½ : 1
Land width between road boundary: 25 m

For longitudinal section and typical cross section of road, see next page. Average area is calculated as explained in Chapter 16.

EARTHWORK CALCULATIONS

Formation width (B) =7.30 m

Side slope (s) = 2 for embankment and (s) = 1.5 for cutting

Change	Height of embankment in m	Depth of cutting in m	Area (b + sh)h in sq. m	Mean area in sq. m	Length in m	Volume in cu. m		Remarks
						Banking	Cutting	
3000	—	—	0.000	0.630	30	18.90		
3030	0.165	—	1.259	1.986	30	59.58		
3060	0.340	—	2.713	1.356	30	40.68		
3090	—	—	0.000	1.933	30	—	57.99	
3120	—	0.482	3.867	5.353	30	—	160.59	
3150	—	0.804	6.839	3.420	15	—	51.30	
3165	—	—	—	3.581	15	53.72		
3180	0.804	—	7.162	7.687	30	230.61		

(continued)

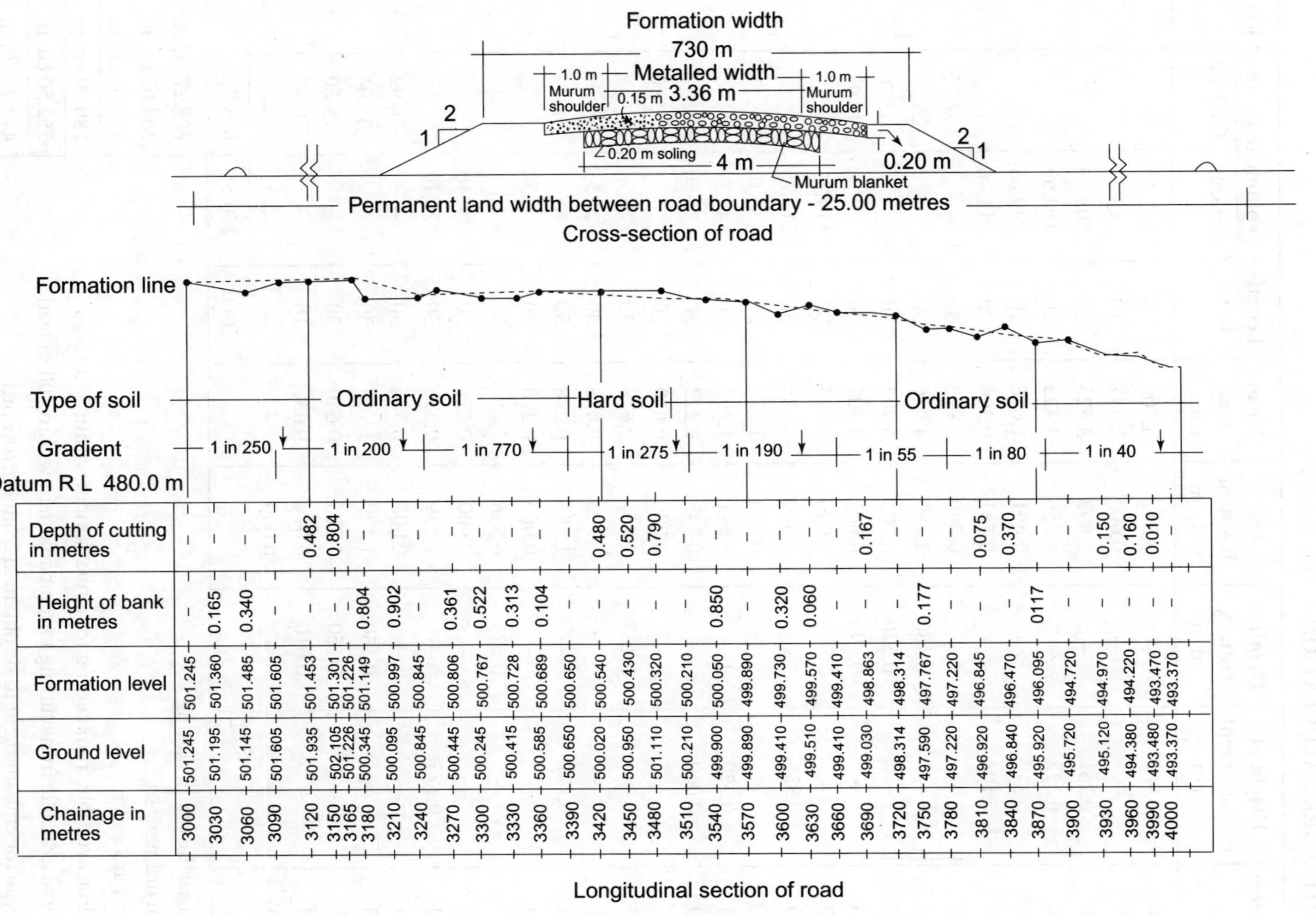

Chainage in metres	Ground level	Formation level	Height of bank in metres	Depth of cutting in metres
3000	501.245	501.245	—	—
3030	501.195	501.360	0.165	—
3060	501.145	501.485	0.340	—
3090	501.605	501.605	—	—
3120	501.935	501.453	—	0.482
3150	502.105	501.301	—	0.804
3165	501.226	501.226	—	—
3180	500.345	501.149	0.804	—
3210	500.095	500.997	0.902	—
3240	500.845	500.845	—	—
3270	500.445	500.806	0.361	—
3300	500.245	500.767	0.522	—
3330	500.415	500.728	0.313	—
3360	500.585	500.689	0.104	—
3390	500.650	500.650	—	—
3420	500.020	500.540	—	0.480
3450	500.950	500.430	—	0.520
3480	501.110	500.320	—	0.790
3510	500.210	500.210	—	—
3540	499.900	500.050	0.850	—
3570	499.890	499.890	—	—
3600	499.410	499.730	0.320	—
3630	499.510	499.570	0.060	—
3660	499.410	499.410	—	—
3690	499.030	498.863	—	0.167
3720	498.314	498.314	—	—
3750	497.590	497.767	0.177	—
3780	497.220	497.220	—	—
3810	496.920	496.845	—	0.075
3840	496.840	496.470	—	0.370
3870	495.920	496.095	0117	—
3900	495.720	494.720	—	—
3930	495.120	494.970	—	0.150
3960	494.380	494.220	—	0.160
3990	493.480	493.470	—	0.010
4000	493.370	493.370	—	—

EARTHWORK CALCULATIONS   (Continued)

Change	Height of embankment in m	Depth of cutting in m	Area (b + sh)h in sq. m	Mean area in sq. m	Length in m	Volume in cu. m		Remarks
						Banking	Cutting	
3210	0.902	—	8.212	4.106	30	123.18		
3240	—	—	0.000	1.447	30	43.41		
3270	0.361	—	2.895	3.625	30	108.75		
3300	0.522	—	4.356	3.423	30	102.69		
3330	0.313	—	2.491	1.636	30	49.08		
3360	0.104	—	0.781	0.390	30	11.70		
3390	—	—	0.000	1.925	30	—	57.75	
3420	—	0.480	3.850	4.026	30		120.78	
3450	—	0.520	4.202	5.402	30	—	162.06	Hard soil
3480	—	0.790	6.603	3.301	30		99.03	
3510	—	—	0.000	3.825	30	114.75		
3540	0.850	—	7.650	3.825	30	114.75		
3570	—	—	0.000	1.270	30	38.10		
3600	0.320	—	2.541	1.493	30	44.79		
3630	0.060	—	0.445	0.222	30	6.60		
3660	—	—	0.000	0.630	30		18.90	
3690	—	0.167	1.261	0.678	30	20.34		
3720	—	—	—	0.678	30	20.34		
3750	0.177	—	1.355	0.278	30	—	8.34	
3780	—	—	0.000	0.278	30	8.34		
3810	—	0.075	0.556	—	—	—	—	—
40	—	—	0.000	0.670	30	20.10		
70	0.175	—	1.339	0.670	30	20.10		
3900	—	—	0.000	0.564	30	—	16.92	
30	—	0.150	1.129	1.168	30		35.04	
60	—	0.160	1.206	0.640	30	—	19.20	
90	—	0.010	0.074	0.037	10	—	0.37	
4000	—	—	0.000	—	—	—	—	—
					Total	1250.51	808.27	

Total earthwork in cutting	808.27 cu. m
(a) In ordinary soil =	368.65 cu. m
(b) In hard soil =	439.62 cu. m
Material available from cutting for embankment within 50 m lead	280.00 cu. m
Material available from cutting for embankment within 100 m lead	552.50 cu. m
Material for embankment to be obtained from borrow pits	442.24 cu. m
say	442.00 cu. m

Name of work: Construction of major district road from .... to .... district ....
section .... From km 3 to km 4.

MEASUREMENT SHEET

Item No.	Description	No.	Length	Breadth	Depth or height	Quantity	Total
	PRELIMINARY						
1.	Surveying, layout and jungle clearance	1.00	1.00 km			1.00 km	1.00 km
2.	Land acquisition (Permanent)	1.00	1000	25.00		25,000 sq. m	2.50 ha
3.	Land acquisition (Temporary)	—	—	—		—	—
4.	Acquisition of stone quarries	1.00				1.00 no.	1 no.
5.	Providing and maintaining diversion	1.00 km				1.00 km	1.00 km
	EARTHWORK						
6.	Earthwork in cutting in ordinary soil. Lead 100 m and lift 1.50 m	As per separate sheet on earthwork calculations				368.65 cu. m	368.65 cu. m
7.	Earthwork in cutting hard soil including boulders up to 45 cm size. Lead 100 m, lift 1.5 m	As per separate sheet				439.62 cu. m	439.62 cu. m
8.	Spreading the excavated material for embankment in layers of 20 cm including consolidation. Lead 50 m	As per separate sheet				280.00 cu. m	280.00 cu. m
9.	Spreading the excavated material for embankment in layers of 20 cm including consolidation. Lead 100 m	As per separate sheet				552.50 cu. m	552.50 cu. m
10.	Earthwork from borrow pits for embankments in ordinary soil including spreading in layers of 20 cm and consolidation. Lead 50 m and lift 1.5 m	As per separate sheet				442.00 m	442.00 cu. m

(continued)

EARTHWORK CALCULATIONS   (Continued)

Item No.	Description	No.	Length	Breadth	Depth or height	Quantity	Total
	SURFACING						
11.	Preparation of subgrade (dressing to camber)	1	1000.00	4.00		4000.00 sq. m	4000.00 sq. m
12.	Collection of hard trap stone rubble 15 cm to 20 cm gauge for soling including quarrying and carting at site of work. Rubble of this size will be: 60%	0.60	1000.00	4.00	0.20	480.00 cu. m	480.00 cu. m
13.	Collection of hard trap stone rubble 8 cm to 10 cm gauge for soling including quarrying and carting at site of work. Rubble of this size will be: 40%	0.40	1000.00	4.00	0.20	320.00 cu. m	320.00 cu. m
14.	Collection of hard trap stone metal 3.75 cm gauge including quarrying, carting and stacking at site of work. Thickness before compaction 15 cm	1	1000.00	3.36	0.15	504.00 cu. m	504.00 cu. m
15.	Collection of hard murum including quarrying and stacking at site of work						
	(a) Over blanketing below soling	1	1000.00	4.00	0.08	320.00	
	(b) Over soling	1	1000.00	4.00	0.08	320.00	
	(c) Over metal	1	1000.00	3.36	0.05	168.00	
	(d) Over berms	2	1000.00	1.00	0.20	400.00	
						1208.00	1208.00 cu. m
16.	Collection of sand including quarrying, carting and stacking at site of work. A thin layer of 6 mm spread before final rolling of blindage. Width inclusive of side berms	1	1000.00	5.66	0.006	33.96 cu. m	say 34.00 cu. m

(continued)

EARTHWORK CALCULATIONS   (Continued)

Item No.	Description	No.	Length	Breadth	Depth or height	Quantity	Total
17.	Spreading soling 15 cm to 20 cm size and hand packing with 8 cm to 10 cm size rubble	As per item No. 12				480.00	
		As per item No. 13				320.00	
						800.00 cu. m	800.00 cu. m
18.	Consolidation of soling with 10 to 12 t weight roller	As per item No. 17				800.00 cu. m	800.00 cu. m
19.	Spreading and hand packing of 3.75 cm gauge metal	As per item No. 14				504.00 cu. m	504.00 cu. m
20.	Consolidation of metal with 10 t to 12 t weight roller	As per item No. 14				504.00 cu. m	504.00 cu. m
21.	Spreading murum and sand	As per item No. 15				1208.00	
		As per item No. 16				34.00	
						1242.00 cu. m	1242.00 cu. m
22.	Watering and tamping hard murum over berms and blacketing	As per item No. 15 (a) and (d)				320.00	
						400.00	
						720.00 cu. m	720.00 cu. m
23.	Kilometre stone of RCC including fixing	1.00				1.00 no.	1.00 no
24.	Hectometre stone including fixing	9.00				9.00 no.	9.00 no
25.	Boundary stones including fixing. 50 m c/c on each side of length	2	$\dfrac{1000}{50}+1$			42.00 nos	42.00 nos
26.	Guard stones including fixing Over high embankment @ 2.0 m c/c on both sides						
	Ch 3180.00 to 3210.00	2	$\dfrac{30}{2}$			30	
	On curves. 10.0 m c/c on both sides	2	$\dfrac{150}{10}$			30.00 no.	60.00 no.
						60 nos	

(continued)

EARTHWORK CALCULATIONS   (Continued)

Item No.	Description	No.	Length Breadth	Depth or height	Quantity	Total
27.	Supplying and fixing caution and sign boards	1			1.00 no.	1.00 no.
28.	Supplying and fixing village and river name boards	1			1.00 no.	1.00 no.
29.	Supplying and fixing junction boards	1			1.00 no.	1.00 no.

ABSTRACT SHEET

Name of work: Construction of major district road

From.........To .... District .... Sect .... From kilometre 3.00 to kilometre 4.00.

Item No.	Quantity	Description	Rate ₹	Per	Amount ₹
		PRELIMINARY			
1.	1.00 km	Surveying, layout and jungle clearance	10,000.00	km	10,000.00
2.	2.50 ha	Land acquisition permanent	7,50,000.00	ha	18,750,00.00
3.	—	Land acquisition temporary		—	
4.	1.00 no.	Acquisition of stone quarries	35,000.00	no.	35,000.00
5.	1.00 km	Providing and maintaining diversion* (200 m) *If necessary	30,000.00	km	30,000.00
					19,50,000.00
		EARTHWORK			
6.	368.65 cu. m	Earthwork in cutting in ordinary soil and depositing the excavated material either as spoil or using the same in bank work up to a lead of 100 m and lift 1.5 m	119.00	cu. m	43,869.35
7.	439.62 cu. m	Earthwork in cutting hard soil including boulders up to 45 cm size and depositing the excavated material either as a spoil or using the same in bankwork up to a lead of 100 m and lift 1.5 m	139.00	cu. m	61,107.18
8.	280.00 cu. m	Spreading the excavated material for embankment in layers of 20 cm including consolidation by hand roller with an average lead of 50 m	79.00	cu. m	22,120.00

(continued)

ABSTRACT SHEET   (Continued)

Item No.	Quantity	Description	Rate ₹	Per	Amount ₹
9.	552.50 cu. m	—do— Except that lead will be 100 m	90.00	cu. m	49,725.00
10.	442.00 cu. m	Earthwork from borrow pits for embankment in ordinary soil including spreading in layers of 20 cm and consolidation by hand roller up to a lead 50 m and a lift of 1.5 m	225.00	cu. m	99,450.00
					2,76,271.53
		SURFACING			
11.	4000.00 sq. m	Preparation of subgrade dressing to camber	10.00	sq. m	40,000.00
12.	480.00 cu. m	Collection of hard trap stone rubble 15 cm to 20 cm gauge for soling including quarrying and carting at site of work	800.00	cu. m	3,84,000.00
13.	320.00 cu. m	Collection of hard trap stone rubble 8 cm 10 cm gauge for soling including quarrying, carting and stacking at site of work	950.00	cu. m	3,04,000.00
14.	504.00 cu. m	Collection of hard trap stone metal 3.75 cm gauge including quarrying, carting and stacking at site of work	1,100.00	cu. m	5,54,400.00
15.	1208.00 cu. m	Collection of hard murum including quarrying and stacking at site of work	316.00	cu. m	3,81,728.00
16.	34.00 cu. m	Collection of sand including quarrying, carting and stacking at site of work	1,670.00	cu. m	56,780.00
17.	800.00 cu. m	Spreading 15 cm to 20 cm gauge soling stone and packing with 8 cm to 10 cm gauge soling on the road formation	80.00	cu. m	64,000.00
18.	800.00 cu. m	Consolidation of soling with steam (or diesel) road roller	50.00	cu. m	40,000.00
19.	504.00 cu. m	Spreading and hand packing 3.75 cm gauge metal on the road surface	90.00	cu. m	45,360.00
20.	504.00 cu. m	Consolidation of metal with steam road roller	65.00	cu. m	32,760.00
21.	1242.00 cu. m	Spreading murum and sand for blindage, for berms and for blanketing below soling	79.00	cu. m	98,118.00
22.	720.00 cu. m	Watering and tamping hard murum over berms and blanketing	21.00	cu. m	15,120.00
					20,16,266.00

(continued)

ABSTRACT SHEET   (Continued)

Item No.	Quantity	Description	Rate ₹	Per	Amount ₹
		MISCELLANEOUS ITEMS			
23.	1.00 no.	Supplying reinforced cement concrete (1:2:4 prop). Kilometre stone 27 cm × 37 cm × 111 cm size and fixing at site in a bed of PCC (1:4:8) including painting, lettering and finishing as per type design	3,249.00	no.	3,249.00
24.	9.00 nos	Supplying hard trap hectometre stone of 10 cm × 15 cm × 75 cm size and fixing in bed of lime concrete (1:2:4 prop.) including painting and finishing complete as per type design	950.00	no.	8,550.00
25.	42.00 nos	Supplying boundary stones 15 cm × 20 cm × 75 cm size, type, medium dressed to exposed faces and fixing in bed of lime concrete (1:2:4) including painting and finishing as per design	550.00	no.	23,100.00
26.	60.00 nos	Supplying guard stones 10 cm × 10 cm × 75 cm size of hard trap type, medium dressed to exposed faces and fixing in lime concrete (1:2:4) bedding including white washing complete	350.00	no.	21,000.00
27.	1.00 no.	Supplying m.s; caution and sign board of 45 cm × 38 cm size and 45 cm equilateral triangular piece on top fixed to country wood post 10 cm square and 3.6 m high including fixing in cement concrete (1:4:8) bed	2,500.00	no.	2,500.00
28.	1.00 no.	Supplying and fixing in position village and river name boards of m.s 60 cm × 60 cm size fixed to m.s frame 10 cm square and 3.6 m high including painting, figuring and fixing in cement concrete (1:4:8) bed	4,739.00	no.	4,739.00
29.	1.00 no.	Supplying and fixing in position junction boards 1.5 m × 1.00 m size and 2 cm thick, as per design	14,129.00	no.	14,129.00
					77,267.00

Name of work: Constructing major district road

GENERAL ABSTRACT

1.	Preliminary	₹	19,50,000.00
2.	Amount of earthwork	₹	25,96,873.00
3.	Amount of surfacing	₹	17,11,190.00
4.	Amount of cross-drainage works	₹	—
5.	Amount of miscellaneous Items	₹	69,250.00
	Total	₹	63,27,313.00
6.	Add 5% towards contingencies	₹	3,16,366.00
	Total	₹	66,43,679.00
7.	Add 20% towards overheads and profit	₹	13,28,735.00
8.	Add ₹15,000 per km towards arboriculture	₹	15,000.00
	Grand total	₹	79,87,414.00
		₹	80,00,000.00

Say Rupees eighty lakhs only.

ESTIMATES OF DIFFERENT TYPES OF ROAD SURFACES

Table No. III. 1, gives useful information regarding the different materials of construction, labour days and plant hours required for providing different types of road surfaces. An estimate for a given type of road surface can then be easily framed with the help of this table. As an illustration an estimate for one kilometre length of a two-coat surface dressing is prepared.

TABLE III. 1 Materials of construction, labour days and plant hours required per 100 sq. m of road surface

S. No.	Type of pavement	Materials — Mineral aggregate in cu. m								Coal kg	Asphalt binder kg	Labour days	Plant hours		
		5 cm or above	4 cm	2.5 to 3.0 cm	2 cm	12 mm	10 mm	6 mm	Sand				Boiler	Mixer	Roller
1.	*Water-bound macadam*														
	First layer—compacted thickness (10 cm)	15.00	—	—	—	—	1.60 to	—	—	—	—	—	—	—	—
	Second layer—compacted thickness (10 cm)	—	—	15.00	—	—	1.50	—	—	—	—	—	—	—	—
	Wearing coat compacted thickness 5 cm	—	—	—	7.50	—	—	—	1.20	—	—	—	—	—	—
	(Murum or prepared earth blindage required—3 cu. m/100 sq. m) *Surface treatment*														
2.	Primer coat	—	—	—	—	—	—	—	—	—	75 to 145	2.7 to 3	—	—	—
3.	Single coat surface dressing with asphalt 80 + 100 (hot process) or emulsion or cutback (cold process)	—	—	—	—	1.50 to 1.55	—	—	—	50 to 55	200 to 220	8 to 8.5	1.00	—	1.00
4.	Two-coat surface dressing. Hot or cold process as above														

S. No.	Type of pavement	Materials — Mineral aggregate in cu. m								Coal kg	Asphalt binder kg	Labour days	Plant hours		
		5 cm or above	4 cm	2.5 to 3.0 cm	2 cm	12 mm	10 mm	6 mm	Sand				Boiler	Mixer	Roller
	(a) First coat	—	—	—	—	1.50 to 1.55	—	—	—	42 to 50	170 to 200	8 to 8.5	1.00	—	1.0
	(b) Second coat	—	—	—	—	—	—	0.90 to 1.20	—	30 to 36	120 to 150	3.5 to 4	1.00	—	1.0
	Penetration macadamor grouting														
5	Full grout-hot or cold process														
	(a) 8 cm thick	8.00	1.00		1.80	—	1.20	—	—	230	925	20	—	—	1.25
	(b) 5 cm thick	3.60	1.80	0.60	1.50	—	1.2 to 1.5	—	—	144	675	16	—	—	1.25
	Semi-grout-hot or cold process 5 cm thick	—	6.00 to 7.00		—	—	1.20 to 1.50	—	—	110 to 115	450	20	—	—	1.25
	Premixes														
	Tack coat	—	—	—	—	—	—	—	—	—	50 to 70	2.5 to 3	—	—	—
	Premix carpets														
	(a) 2.5 cm thick (hot or cold process)	—	—	—	1.85	—	1.2	0.60	—	100	200	11	—	—	1.0
	(b) 2.0 cm thick	—	—	—	—	1.85	—	0.90	—	80	160	9	—	—	1.00
	Asphaltic concrete (Hot mix-cold laid) 6 cm thick	—	—	—	3.60	2.50	—	—	3.00	315.50	600 + 24.37 (solvent)		—	—	1.00
	5 cm thick	—	—	—	3.00	1.80	—	—	2.40	265.00	500 + 12.30 (solvent)		—	—	1

Note: (i)   *Coal will be required only when hot process is used.*

(ii)   *For materials and plant hours required for cement concrete pavement—refer to Table 20.13 Chapter 20.*

MEASUREMENT SHEET

Item No.	Description	No.	Length	Breadth	Depth or height	Quantity	Total
	Example: Estimate the cost for two-coat surface treatment (hot process) of a 4 m wide road, per kilometre length	—	—	—	—	—	—
	Name of road area		4.00×1000 $= 4000$ sq. m				
	Nature of existing surface		Water-bound macadam				
	Condition					good	
1.	Preparation of road surface	1	1000	4.00		4000 sq. m	4000 sq. m
	A. *Mineral aggregate*						
2.	Stone metal 12 mm size 1.53 cu. m per 100 sq. m $\dfrac{1.53 \times 4000}{100} = 61.20$ cu. m					61.20 cu. m	61.20 cu. m
3.	Stone metal 6 mm size at 1.2 cu. m per 100 sq. m $\dfrac{1.2 \times 4000}{100} = 48.00$ cu. m					48.00 cu. m	48.00 cu. m
	B. *Asphalt binder*						
4.	Paving asphalt 80–10 at 200 kg/100 sq. m for first coat, 150 kg/100 sq. m for second coat					8000	
						6000	
			Total			14000	
	Add 2% for waste					280	14.28
5.	Furnace oil @ 25 kg/100 sq. m					3.20 t	3.20 t
6.	Laying two coat surface treatment (hot process) as per specifications	1	As per item no. 2 and 3			1090 cu. m	1092.00 cu. m
		1	1000.00	4.00		40,000 sq. m	4000.00 sq. m
7.	Rolling	1.00				1.00 km	1.00 km

Name of work: Two-coat surface dressing to existing water-bound macadam

ABSTRACT SHEET

Item No.	Quantity	Description	Rate ₹	Per	Amount ₹
		Name of road. … Area 4 m × 1 kilometre (4000 sq. m) Nature of existing surface water bound macadam condition … good			
1.	4000.00 sq. m	Preparation of existing road surface for treatment including filling ruts, dressing to proper camber and grade and sweeeping clean	18.00	sq. m	72,000.00
2.	61.20 cu. m	Collection of hand trap stone metal 12 mm size including quarrying, carting and stacking at site of work	1,050.00	cu. m	64,260.00
3.	48.00 cu. m	Collection of hard trap stone metal 6 mm size including quarrying, carting and stacking at site of work	1,050.00	cu. m	50,400.00
4.	14.28 tonnes	Providing asphaltic binder of specified penetration including carting to the site of work	49,862.00	tonne	7,12,029.00
5.	3.20 tonnes	Providing adequate fuel of specified quality including carting and stacking at the site of work	23,000.00	tonne	73,600.00
6.	4000.00 sq. m	Laying two coat surface dressing (hot process) as per specifications	108.00	sq. m	4,32,000.00
7.	1.00	Consolidation of metal with power-driven road roller	20,000.00	km	20,000.00
		Total			14,24,289.36
		Add 5% for contingencies			71,214.47
					14,95,503.83
		Add 20% for overheads and profits			2,99,100.77
					17,94,604.59
		Grand total for 4000 sq. m		say ₹	17,94,605.00
		Rate per sq. m ₹275			

IV | A Canal

An estimate of a canal generally consists of the following heads.

1. *Preliminary work:*   For the preliminary work involved, a lump sum provision based on the past experience of similar work is made in the estimate.
2. *Land acquisition:*   Land required for the construction and maintenance of the canal is determined in hectares. Suitable rates per hectare of land are ascertained from the revenue authorities, who are guided by the sale and purchase deeds of similar land in the neighbourhood in recent years. Knowing the area of land to be acquired and the rate per unit area, the cost of land acquisition is worked out.
3. *Masonry works:*   Masonry works on a canal may include:
    - outlets,
    - drops or falls,
    - cross-drainage works,
    - road bridges, and
    - escapes.

    These are considered under the following heads.
4. *Regulation works:*   These include cross regulators, distributory head regulators, outlets, etc. Type plan and detailed estimate of a typical work of each type (for example, outlet) on the canal are prepared. Provision for the remaining works of the same type is made on the basis of the cost so estimated.
5. *Dropsor falls:*   As per above.
6. *Cross-drainage works:*   Type plan and detailed estimate of one CD work of each type (for example, an aqueduct) are prepared. The estimated cost is divided by the ventway provided and the probable cost of construction per square metre of the ventway is worked out. Provision for the remaining cross drainage works of the same category is made on the basis of this cost per square metre of ventway.
7. *Road bridges:*   Type plans and detailed estimates of bridges on different types of roads such as cart-tracks, village roads, district roads and so on, are prepared. Provision of the remaining types of road bridges is made on the basis of the detailed estimates of respective categories.
8. *Escapes:*   Either a detailed estimate is prepared or a lump sum provision based on the past experience of the work of similar type and magnitude is made in the estimate of the canal.

9. *Earthwork:*   Detailed estimate of earthwork for the entire length of the canal is prepared. A typical estimate of earthwork for a portion of a canal is prepared in the following example to illustrate the procedure.

An estimate of a branch canal or a distributory may be prepared in a similar way.

Example IV.1   Prepare a detailed estimate of a portion of a canal with the following data.

Chainage	Ground level	Proposed bed
500	726.80	725.50
600	726.20	—
700	726.00	—
800	725.80	—
900	725.50	—
1000	724.80	—

Trial pit results

Chainage	Ordinary soil	Hard soil	Soft rock
500	0.50	0.40	0.30
600	0.50	0.30	0.30
700	0.35	0.35	0.30
800	0.45	0.75	0.30
900	0.50	0.60	0.30

Bed width	1.5 m
Bed slope	1 in 2000
Full supply depth	0.75 m
Top width of banks	1.0 m
Side slopes in bank and cutting	1.5:1

Solution: Volume of earthwork is computed with the help of typical cross-sections. Cross-section at Ch 500.00

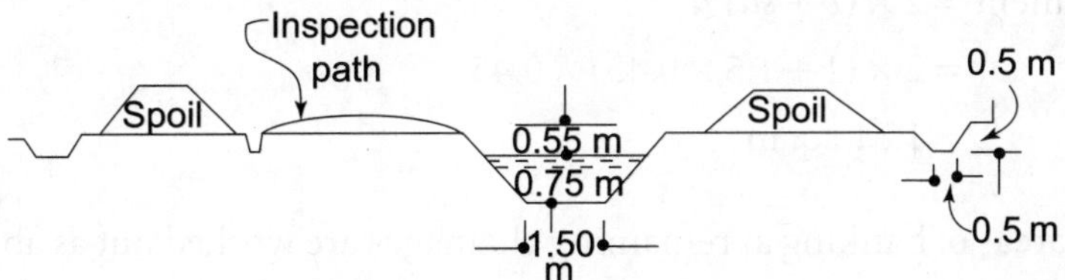

Fig. IV.1   Canal in full cutting

Area of excavation $= (b + sd)\,d$

$$= (1.5 + 1.5 \times 1.3) \times 1.3$$

$$= 4.48 \text{ sq. m}$$

Similarly, area of excavation at other cross-sections is worked out and tabulated in column 6 of the table. Mean area of two consecutive sections is entered in column 7 and the volume of cutting (Length × Mean area), Col. 3 × Col. 7 is entered in column no. 8.

Area of cross-section of the catchwater drain is $= (0.50 + 1 \times 0.5) \times 0.50 = 0.50$ sq. m at the minimum depth. As there is sufficient ground fall the depth will be minimum at all points. The volume of cutting for the two catch water drains between two cross-sections will be: $2 \times 0.50 \times 100 = 100$ cu. m. This is entered in column no. 9.

Embankment

CBL at Ch. 500.00	725.50
Full supply depth	0.75
Free board	0.50
Top of bank at Ch. 500.00	726.75

With a fall of 0.1 m in 200 m the top of bank levels at remaining chainages are worked out and entered in column 10.

Height of embankment = (Top of bank level) − (Ground level), that is, (col. 10 − col. 2). The results are entered in column 11.

Cross-section at Ch. 700.00

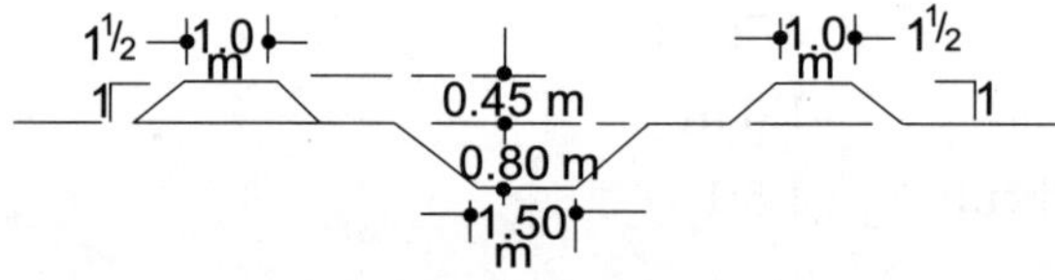

Fig. IV.2   Canal in partial cutting

Area of embankment $= 2 \times (b + sd)\,d$

$$= 2 \times (1 + 1.5 \times 0.45) \times 0.45$$

$$= 1.51 \text{ sq. m}$$

Cross-sectional areas of banking at remaining chainages are worked out as above and entered in column 12.

Cross-section at Ch 1,000.00

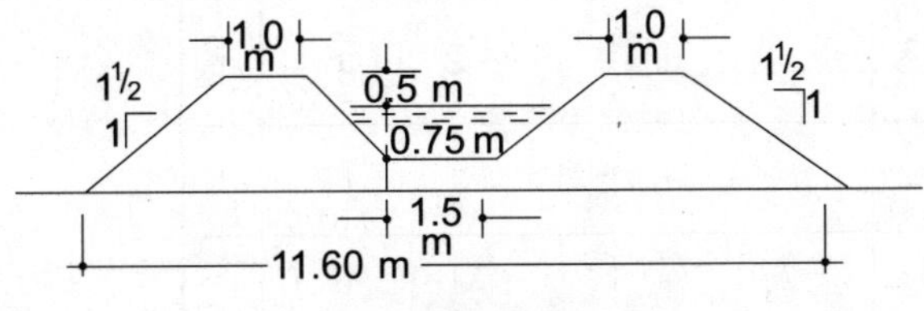

Fig. IV.3

$$\text{Area of banking} = \text{Gross area} - \text{Area of canal section}$$

$$\text{Gross area} = (7.25 + 1.5 \times 1.45) \times 1.45$$
$$= 13.66 \text{ sq. m}$$

$$\text{Area of canal section} = (1.5 + 1.5 \times 1.25) \times 1.25$$
$$= 4.22 \text{ sq. m}$$

$$\text{Area of banking} = 13.66 - 4.22 = 9.44 \text{ sq. m}$$

Mean areas are worked out and entered in column 13. Length and mean area are multiplied to determine the volume of banking. Thus, Col. 14 = Col. 3 × Col. 13.

The quantity available from cutting or required for banking is worked out and entered in columns 15 and 16 respectively.

$$\text{Col. 15} = (\text{Col. 8} + \text{Col. 9}) - \text{Col. 14}$$

$$\text{Col. 16} = \text{Col. 14} - (\text{Col. 8} + \text{Col. 9})$$

Initial lead is normally 50 m and if the excess from cutting is available within this lead no extra provision for lead is made. Excess available from cutting may be utilised up to an extra lead of 100 m. Beyond this lead, however, it may be economic to obtain the material from borrow pits. The completed table will be as given on the next page.

$$\text{Average depth} = \frac{3.80}{5} = 0.76 \text{ m}$$

$$\text{Average depth in ordinary soil} = \frac{2.20}{5} = 0.44 \text{ m}$$

$$\text{Average depth in hard soil} = \frac{1.20}{5} = 0.24 \text{ m}$$

$$\text{Average depth in soft rock} = \frac{0.40}{5} = 0.08 \text{ m}$$

Table IV.1 Calculation of quantities

Earthwork in cutting and in embankment

| CH m | GL | Length m | Bed level | Cutting | | | | Qty of catch-water cu. m | TBL | Height m | Area sq. m | Bank mean area sq. m | Qty cu. m | Net available from cutting cu. m | Net required for banking cu. m | Initial lead cu. m | 50 m extra lead cu. m | 100 m extra lead cu. m | Net from borrow pit cu. m | Remarks |
				Depth m	Area sq. m	Mean Area sq. m	Qty cu. m													
1	2	3	4	5	6	7	8	9	10	11	12	13	14	15	16	17	18	19	20	21
500	726.80	—	725.50	1.30	4.48	—	—	—	726.75	—	0.00	—	—	—	—	—	—	—	—	—
—	—	100	—	—	—	3.32	332	100	—	—	—	0.76	76	356	—	76	—	—	—	—
600	726.20	—	225.40	0.80	2.16	—	—	—	726.65	0.45	1.51	—	—	—	—	—	—	—	—	—
—	—	100	—	—	—	1.97	197	100	—	—	—	1.76	176	121	—	176	—	—	—	—
700	726.00	—	725.30	0.70	1.78	—	—	—	726.65	0.55	2.00	—	—	—	—	—	—	—	—	—
—	—	100	—	—	—	1.61	161	100	—	—	—	2.29	229	32	—	229	—	—	—	—
800	725.80	—	725.20	0.60	1.44	—	—	—	726.45	0.65	2.57	—	—	—	—	—	—	—	—	—
—	—	100	—	—	—	1.14	114	100	—	—	—	3.19	319	—	95	256	—	—	63	—
900	725.50	—	725.10	0.40	0.84	—	—	—	726.35	0.85	3.82	—	—	—	—	—	—	—	—	—
—	—	100	—	—	—	0.42	42	100	—	—	—	6.63	663	—	521	142	—	—	521	—
1000	724.80	—	725.00	0.00	0.00	—	—	—	726.25	1.45	9.44	—	—	—	—	—	—	—	—	—
						TOTAL	846	500					1463	509	616	879			584	

Table IV.2   Classification of excavated material

Chainage	Total depth of excavation	Ordinary soil	Hard soil	Soft rock
500	1.30	0.50	0.40	0.40
600	0.80	0.50	0.30	0.00
700	0.70	0.35	0.35	0.00
800	0.60	0.45	0.15	0.00
900	0.40	0.40	0.00	0.00
Total	3.80 m	2.20 m	1.20 m	0.40 m

$$\text{Average area} = (1.5 + 1.5 \times 0.76) \times 0.76 = 2.00 \text{ sq. m}$$

$$\text{Area in ordinary soil} = \left(\frac{3.78 + 2.46}{2}\right) \times 0.44 = 1.37 \text{ sq. m}$$

$$\text{Percentage of ordinary soil} = \frac{1.37}{2.00} \times 100 = 68.50\%$$

$$\text{Area in hard soil} = \left(\frac{2.46 + 1.74}{2}\right) \times 0.24 = 0.50 \text{ sq. m}$$

$$\text{Percentage of hard soil} = \frac{0.50}{2.00} \times 100 = 25\%$$

$$\text{Area in soft rock} = \left(\frac{1.74 + 1.50}{2}\right) \times 0.08 = 0.13 \text{ sq. m}$$

$$\text{Percentage of soft rock} = \frac{0.13}{2} \times 100 = 6.5\%$$

Total quantity of excavation = 846 cu. m

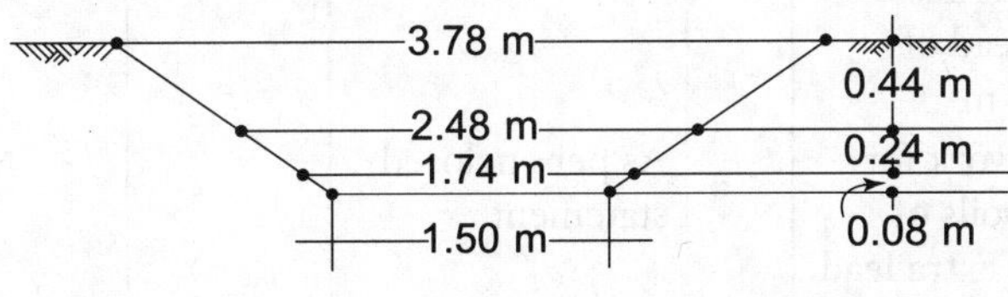

Fig. IV.4

1.   Excavation in ordinary soil $846 \times \dfrac{68.50}{100} = 579.50$ cu. m

    Add excavation of catch-water drains $= 500.00$ cu. m

                   Total $= 1{,}079.50 \ 25$

2.   Excavation in hard soil $= 846 \times \dfrac{25}{100}$

                  $= 211.5$ cu. m

3.   Excavation in soft rock $= 846 \times \dfrac{6.5}{100}$

                  $= 54.99$ cu. m

                  say 55.00 cu. m

The measurements are now recorded on measurement sheets and finally the abstract is prepared.

MEASUREMENT SHEET

Item No.	Description	No.	Length Breadth	Depth or height	Quantity	Total
1.	Excavation in ordinary soil including dressing the sides of trenches and stacking the materials within a lead of 50 m and lift 1.5 m		As per enclosed statement		1,079.00 cu. m	1,079.00 cu. m
2.	Excavation in hard soil including dressing the sides of trenches and stacking the materials neatly within a lead of 50 m and lift 1.5 m		As per enclosed statement		211.50 cu. m	211.50 cu. m
3.	Excavation in soft rock including dressing the sides of trenches and stacking the materials neatly within a lead of 50 m and lift 1.5 m		As per enclosed statement		55.00 cu. m	55.00 cu. m
4.	Providing earthwork for banking from spoils of excavation, with extra lead		As per enclosed statement		NIL	NIL

(continued)

MEASUREMENT SHEET   (Continued)

Item No.	Description	No.	Length Breadth	Depth or height	Quantity	Total
5.	Providing earthwork for embankment from borrow pits in soil and murum including spreading in layers of 15 cm to 20 cm thickness		As per enclosed statement		584.00 cu. m	584.00 cu. m
6.	Consolidating the earth work in layer of 10 cm to 20 cm with power-driven roller including labour required for spreading, sectioning, dressing, watering		As per enclosed statement		1,463.00 cu. m	1,463.00 cu. m
7.	Providing kilometre stones of dressed trap stone as per design including fixing in lime concrete, engraving number, painting complete	1			1.00 no.	1.00 no.
8.	Providing half kilometre stones of dressed trap stone 75 cm × 20 cm × 15 cm as per design including fixing in lime concrete engraving number and painting	1			1.00 no.	1.00 no.
9.	Providing and fixing canal boundary stones 75 cm × 20 cm × 5 cm as per design including engraving lettering and painting 100 m spacing on both sides (5 × 2)	10			10.00 no.	10.00 no.
10.	Providing canal bad stones 30 cm × 20 cm × 15 cm size including fixing in lime concrete @ 4 per kilometre	2			2.00 no.	2.00 no.
11.	Providing 50 m stones 75 cm × 20 cm × 15 cm size of dressed trap stone including fixing in lime concrete, engraving lettering, painting	10			10.00 no.	10.00 no.

Name of work : ESTIMATE OF A CANAL

ABSTRACT SHEET

Item No.	Quantity	Description	Rate ₹	Per	Amount ₹
1.	1079.00 cu. m	Excavation in ordinary soil including dressing the sides of trenches and stacking the materials within a lead of 50 m and lift 1.5 m	98.90	cu. m	1,06,713.10
2.	211.50 cu. m	Excavation in soft rock including dressing the sides of trenches and stacking the materials neatly within a lead of 50 m and lift 1.5 m	138.70	cu. m	29,335.05
3.	55.00 cu. m	Excavation in soft rock including dressing the sides of trenches and stacking the materials neatly within a lead of 50 m and lift 1.5 m	138.70	cu. m	7,628.50
4.	NIL	Providing earthwork for banking from spoils of excavation with 50 m extra lead including laying in layer of 15 cm to 20 cm, sectioning and dressing	NIL	cu. m	NIL
5.	584.00 cu. m	Providing earthwork for embankment from borrow pits in soil and murum including spreading in layer of 15 cm to 20 cm	227.35	cu. m	1,32,772.40
6.	1463.00	Consolidating the earthwork in layers of 15 cm to 20 cm with power-driven roller including labour required for spreading, sectioning, dressing and watering	40.10	cu. m	58,666.30
7.	1.00	Providing kilometre stones of dressed trap stone as per design including fixing in lime concrete, engraving the number and painting	953.80	no.	953.80
8.	1.00	Providing half kilometre stone of dressed trap stone 75 cm × 20 cm × 15 cm as per design including fixing in lime concrete, engraving number and painting	953.80	no.	953.80
9.	10.00 nos	Providing and fixing canal boundary stones 75 cm × 20 cm × 15 cm as per design including engraving, lettering and painting	804.65	no.	8,046.50
10.	2.00 nos	Providing canal bed stones 30 cm × 20 cm × 15 cm size including fixing in lime concrete	350.00	no.	700.00

(continued)

ABSTRACT SHEET   (Continued)

Item No.	Quantity	Description	Rate ₹	Per	Amount ₹
			Brought forward		3,45,769.45
11.	10.00 nos	Providing 50 m stones 75 cm × 20 cm × 15 cm size of dressed trap stone including fixing in lime concrete, engraving, lettering and painting	300.00	no.	3,000.00
					3,48,769.45
		Add 5% contingencies			17,438.47
					3,66,207.92
		Add 20% overheads and profit			73,241.58
		Grand Total			4,39,449.50
				Say	4,40,000.00

Say, ₹4,40,000 only.

ESTIMATE

V | A Railway Track

Estimate for one km length of Permanent Way Track

Table V.1 contains the 'Standard dimensions of a track'.
Table V.2 contains the 'List of track materials, their dimensions and weight'.
Table V.3 contains the details of 'materials required per kilometre of single track'.
These are followed by the measurement sheets of the Estimate and these are followed by the
format of Abstract with rates and cost to be filled in.

Table V.1 Standard dimensions of a track

Type of gauge	Gauge	Min. formation width B in m embankment		Min. formation width B in m cutting		Ballast		Centre to centre dist. of track
	m	Single track	Double track	Single track	Double track	Top width b in m	Depth d m	
Broad gauge	1.676	6.10	10.67	5.49	10.06	3.95	0.2032	4.270
Metre gauge	1.00	4.88	8.52	4.26	7.92	2.99	0.1524	3.600
Narrow gauge	0.762	3.66	7.32	3.35	7.00	1.82	0.1524	3.510

Table V.2 List of track materials, their dimensions and weight

S. No.	Material		Dimensions in m	Weight in kg
1.	Rails (BG) 90 R		Length–12.80	44.70 kg per m
	(MG) 60 R		Length–11.89	29.80 kg per m
2.	Sleepers			
	(i) Timber	(BG)	$2.74 \times 0.254 \times 0.127$	–
		(MG)	$1.83 \times 0.203 \times 0.114$	–
		(NG)	$1.52 \times 0.178 \times 0.114$	–

(continued)

Table V.2 (Continued)

S. No.	Material	Dimensions in m	Weight in kg
	(ii) Steel	$2.667 \times 0.1524 \times 0.106$	BG 80 approx.
			MG 36
	(iii) CI Sleepers	2.261	BG 100
			MG 58
	(iv) Concrete	$2.59 \times 0.245 \times 0.165$	154 to 250
3.	Fish plates	Length = 0.4572	9.695 per no.
	For 90 R		
4.	Fish bolts and nuts	For 90 R–25 mm dia.	4.31 per pair of
		0.1492 m long	four
5.	Bearing plates		
	BG	$0.254 \times 0.2286 \times 0.01905$	8.69 per number
	MG	$0.2032 \times 0.2286 \times 0.015875$	6.72 per number

Table V.3   Materials required per kilometre of single track

S. No.	Materials	Number or quantity	
		BG	**MG**
1.	Ballast[1]	1115 cu. m	768 cu. m
2.	Sleepers[2]	1328 no. to 1560	1347 no. to 1598
3.	Rails[3]	156·25 no.	168·21 no.
4.	Fish plates 2 × no. of rails	313 no.	337 no.
5.	Fish bolts 24 × no. of rails	626 no.	674 no.
6.	Bearing plates 1 per rail	157 no.	169 no.

Note: 1.   Quantity of ballast required, when timber sleepers are used, is slightly less.

2.   Quantity of ballast required on curve will be slightly more.

3.   Check rails for level crossings, if any, are not included.

Example V.1   Prepare a detailed estimate of laying 1 km of broad gauge single track with $(n + 6)$ timber sleepers.

Width of land to be acquired—30 m

Formation width—6·10 m

Average depth of banking—1·50 m

Side slopes banking—2 : 1

**Name of work: ESTIMATE OF ONE KILOMETRE OF
PERMANENT WAY (Broad Gauge)**

MEASUREMENT SHEET

Item No.	Description	No.	Length	Breadth	Depth or height	Quantity	Total
1.	Land acquisition	1	1000	30.00	—	3,000 sq. m 3 ha	
2.	Earthwork/in embankment: including dressing						
	Lead—50 m	1	1000		(9.10×1.5)	13,650 cu. m	13,650 cu. m
	Area = (b + nd)						
	$\qquad = 6.1 + 2 \times 1.5$						
	$\qquad = 9.10$ sq.m						
3.	Broken stone ballast graded between 2 cm and 5 cm						
	or	From	Table			1,100.00	
	Alternately,						
	Area = (h + nd) d						
	$\qquad = (3\text{–}33 + 1 \times 0.33)$	1	1000	1.214		1,214.00	
	$\qquad 0.33$						
	$\qquad = 1.214$ sq. m	1328	2.74	0.25	0.13	118.00	
	Deduct for sleepers					1,096.00	
						1,100.00	1,100.0 cu. m
4.	Wooden sleepers with anticreep bearing plates with keys and round spikes including cost of treatment and freight (*n* + 6 density)	1560				1,560	1,560 no.
5.	Adzing and dating of sleepers including stacking	1560				1,560	1,560 no.
6.	Loading, unloading and leading of sleepers	1560				1,560	1,560 no.

(continued)

MEASUREMENT SHEET   (Continued)

Item No.	Description	No.	Length	Breadth	Depth or height	Quantity	Total
7.	Laying and fastening the sleeper in position	1	1.00			1.00 km	1.00 km
8.	Providing 44.70 kg/m (90 lb new rails 12.8 m long)	156.25	@ 44.70 kg per m			89.40 t	89.40
9.	Providing fish plates for every fifth joint No. of joints $= \dfrac{2 \times 1000}{12.8 \times 5}$ say $= 32$ No. of fish plates 2×32	64	@ 9.695 kg/no			620.28 kg 0.62 t	0.62 t
10.	Fish bolts for the above 32 sets of 4 each	32	@ 4.31 kg/set			13.92 Say 0.14 t	0.14 t
11.	Loading, leading and unloading of materials Total weight in t Item 8 — 89.40 Item 9 — 0.62 Item 10 — 0.14						
	90.16 t	1				90.16 t	90.16 t
12.	Freight charges @ 5% on item no. 8, 9 and 10	—	5% of item 8, 9 and 10				
13.	Laying and fixing of rails	1	1.0 km			1.00 km	1.00 km
14.	Welding of rails No. of welds = $156 - \left(\dfrac{1}{5} \times 156\right) =$ 124 nos	124				124	124 nos
15.	Testing of weld joints. 2% of item no. 14					Lump sum	
16.	Consumable stores and permanent way tools					Lump sum	
17.	Maintenance charges					Lump sum	IS lump sum

Name of Work: ESTIMATE OF QUANTITIES FOR ONE KILOMETRE OF PERMANENT WAY
(Broad Gauge)
ABSTRACT SHEET

Item	Quantity	Description	Per ₹	Per	Amount ₹
1.	3.00 ha	(i) *Land* Land acquisition		ha	
2.	13660 cu. m	(ii) *Earthwork* Providing earthwork in embankment from borrow pits, including dressing Lead up to 50 m	236.48	cu. m	
3.	1100 cu. m	(iii) *Ballast* Providing broken stone ballast graded between 2 cm to 5 cm size including quarrying, carting, spreading and packing	1187.77	cu. m	
4.	1560 no.	(iv) *Sleeper and fastenings* Providing wooden sleepers with anticreep bearing plates with keys and round spikes including cost of treatment		no.	
5.	1560 nos	Adzing and dating sleepers including stacking		no.	
6.	1560 nos	Loading, unloading and leading of sleepers		no.	
7.	1.00 km	Laying and fastening the sleepers in position		km	
8.	89.40 t	(v) *Rails and rail joints* Providing 44.70 kg/m (90 lb/yd) new rails 12.8 m long		t	
9.	0.62 t	Providing fish plates fer every fifth joint		t	
10.	0.14 t	Providing fish belts		t	
11.	90.16t	Loading, leading and unloading, of materials		t	
12.	Provisional sum	Freight charges Item no. Amout 8 — 9 — 10 — Total — 5 per cent of ₹			
13.	1.00 km	Laying and fixing of rails		km	
14.	124 nos	Welding of rails		r.o	
15.		Testing of joints 2 per rent of ₹			
16.	Lump sum	Consumable stores and permanent way tools			
17.	Lump sum	Maintenance charges		Total	

Add 2 per cent work charged establishment charges—

Grand Total

VI | A Tunnel

INTRODUCTION

A tunnel is an artificial underground passage through a hill or under a road or river for a railway line or road or water to pass through. The shape and size is dependent upon the purpose, such as a railway line or road to pass through. For example, Fig. VI.1 shows a tunnel to carry water.

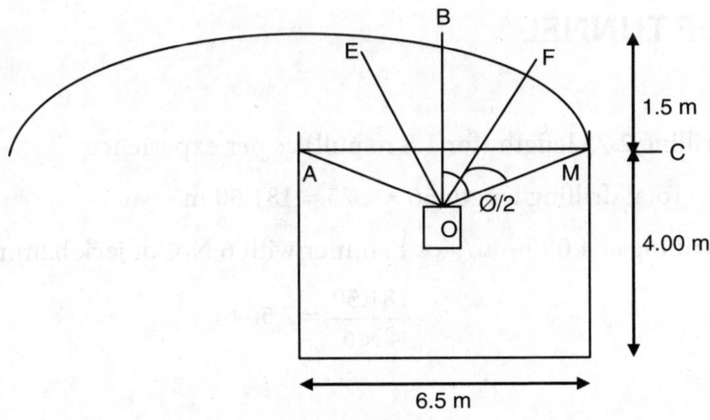

Fig. VI.1

The cost of this structure can be estimated either by *unit quantity method or total quantity method*. Examples of the use of unit quantity method are Estimates I to V above. For proper understanding of the total quantity method, the estimation of the cost of the tunnel has been used as an example.

The cost of any structure essentially includes the cost of five Ms:

 (i) Materials,
 (ii) Men (labour),
(iii) Machinery and equipment,
(iv) Management (overheads), and
 (v) Money (profit).

These factors are common to both the methods. The essential difference is that the cost of a unit quantity is worked out for every item of work and multiplied by the TOTAL UNITS under every item of work, whereas, as the name suggests, the total cost is ascertained for all the five elements by the use of blasting. The cross-section area is first ascertained by dividing the cross-section area into rectangles, segments, triangles and so on (Fig.VI.1). It works out to 33.73 sq.m. = ,

Drilling Section Area of tunnel	33.73 sq.m
No. of holes required for blasting	33.73/0.5625
(1 hole for 0.75 m × 0.75 m = 0.5625)	59.96 Nos
Add 10% hole for wastage	5.99
Total	65.95, say 66 holes

The illustration below is an attempt to indicate how the cost of tunnel is worked out by the total quantity method.

ESTIMATE OF TUNNEL

Excavation

Considering to drilling 2.75 length (for 1.8 m pull) as per experience

$$\text{Total drilling length } 66 \times 2.75 = 181.50 \text{ m}$$

No. of hours for drilling at 4.00/hour/jack hammer with 6 Nos of jack hammer

$$\frac{181.50}{4 \times 6} = 7.56 \text{ hr}$$

Mucking

As per observation result of Tunnel on nearby canal yield pe blast

$$= \text{Cross section area} \times \text{pull}$$

$$= 33.73 \times 1.80 = 60.71 \text{ cu.m/blast}$$

(Time required for mucking 60.71 cu.m)

$$\text{No. of trips} = \frac{60.71}{3.00} = 20.23 \text{ Nos, Say} = 21 \text{ trips}$$

$$\text{hrs required} = 21 \times 0.383 = 8.04 \text{ hrs}$$

Total working hours required per cycle

Bottom cleaning	1.00 hrs
Line out	1.00 hrs
Drilling	7.56 hrs
Charging and blasting	2.00 hrs
Degassing	1.00 hrs
Mucking	8.04 hrs
Scaialing	0.5 hrs
Total	**21.10 hrs**

No of blasting for pull section in one month

$$\text{No of blasting} = \frac{\text{Working days} \times \text{hrs}}{\text{Cycle time}}$$

$$= \frac{25 \times 24}{21.10} = 28.43 \text{ Nos}$$

Quantity of excavation per month = C/S Area × No of blasting × pull

$$= 33.73 \times 28.43 \times 1.80 = 1726.09 = \textbf{1726 cu.m}$$

Labour charges

Labour charges for line out in one month (as per observation):	
General Supervisor – 2 No × 165	₹ 330
A) Line out for	
a) One surveyor 1 @ 165/-	₹ 165
b) 2 Helpers @ 160/-	₹ 320
B) Drilling for 1 Month	
Foreman 2 @ 236/-	₹ 472
Driller 3 @ 160/-	₹ 480
Fitter 3 @ 200/-	₹ 600
C) Blasting	
a) Foreman 2 @ 236	₹ 472
b) Helper 2 @ 160	₹ 320

(*continued*)

Labour charges   (Continued)

D)	Scaialing	
	a) Tapkar 3 @ 236	₹ 708
	b) Helper 2 @ 160	₹ 320
E)	Bottom Cleaning	
	a) 8 Labour @ Rs.133/day	₹ 1,064
F)	Electrification	
	a) Electrification @ 236/day 1 No	₹ 236
	b) Wireman @ 236/day 1 No	₹ 236
G)	Ventilation	
	Fitters @ Rs.200/day 1 No	₹ 200
	Helper @ 160/day 2no	₹ 320
	Total from A to G	₹ 6,243/day
	Total for month 6,243× 25	₹ 1,56,075 month

Machinery Charges

A)	Compressor of 365 Cfm @ 708/hr	
	Hence total cost of drilling	
	7.56 × 4 × 708	₹ 21,410/Blast
B)	Jack hammer 6 @20/hr	
	Total one operationof drilling for 8 hrs is	
	7.56 × 6 × 20	₹ 907/Blast
C)	Tipper (Tata tipper) 7.50T capacity for all purpose	
	1 × 513 × 7.56 hrs	₹ 3,878/Blast
D)	Degassing compressor 105 Cfm one @ 260/hr	
	Degassing 1 hr per Blast	
	1 × 1 × 260	₹ 260/Blast
E)	Blower 21.10 hrs @ 260/hr	₹ 5,486/Blast
	Mucking poclain (100)	
	8.04 × 1793	₹ 14,416/Blast
	1. Dumper	
	1 × 888 × 8.04 × 4	₹ 28,558/Blast
	Generator 200 KVA 1 No	
	1 × 21.10 × 1440	₹ 30,384/Blast
	Total	**₹ 1,05,299/Blast**

Item No. 18.1 as per Mechanical DSR
(District Schedule of Rates)

Total 1,05,299 × 28.43   **₹ 29,93,651/Month**

Material Charges

A)	Line out (Tape, Paint, Staff, Level and so on)	₹ 2,000.00
B)	Drilling rod @ 9000/No rate as per market rate	
	0 Life of drilling rod 250m	
	Total drilling required = 181.50 × 28.43/250 = 20.64 No	₹ 1,85,760.00
	Say 20.64 × 9000	
	2 Pipe line with fitting and so onL.S	₹ 4,000.00
	3 Gum boots, helmets, cells	₹ 2,000.00
C)	Detonators	
	No of hole 66/Blast	
	Detonators required for 28.43 blasts will be	
	66 × 28.43	
	1,876.38 Nos	
	1,877.00 Nos	
	Rate of detonators with wire and so on 11/No as per DSR	
	1877 × 11	₹ 20,647.00
D)	Gelatin @ Rs 1.30 Kg/cu.m	
	Quantity required will be 1.30 × 1726 × 114.40	₹ 2,56,691.00
E)	Charge for galvanometer, battery and blasting wire	
	@ 50% cost of gelatin (C.W.C)	₹ 1,28,345.00
F)	Electric bulbs, holders, cables, switches and so on	
	Including charger of energy L.S.	₹ 4,000.00
Total		**₹ 6,03,443/Month**

ABSTRACT

1)	Labour charges	₹ 1,56,075.00/Month
2)	Machinery charges	₹ 29,93,651.00/Month
3)	Material	₹ 6,03,443.00/Month
	Total	**₹ 36,13,519.00/Month**
36,13,519.00/1726		₹ 2,094.00/cu.m
	Add for ventilation	₹ 61.08/cu.m
		₹ 2,155.08/cu.m
	Add for lead charges for 3.50 km	₹ 54.52/cu.m
	Total	**₹ 2,209.60/cu.m**

(*continued*)

ABSTRACT   (Continued)

Add 1% energy charges		₹ 2,210/cu.m
Add 10% O.H charges		₹ 220.96/Cu.m
Add 2% for sundries		₹ 44.20/cu.m
Add 3% insurance		₹ 66.30/cu.m
	Total	**₹ 2,563.19/cu.m**
Add wet condition 10%		₹ 256.32/cu.m
	Total	**₹ 2,819.51/cu.m**
	Say	*₹ 3,000.00 per cu m*

Estimated cost of tunnel per km = 33.72 cross-section area × 1000 m × 3000 = ₹10,11,60,000